W. MILLIGAN

the pictorial encyclopedia of insects

Consultant: GEORGE ASHBY

the pictorial encyclopedia of insects

V. J. Stanek

HAMLYN

LONDON · NEW YORK · SYDNEY · TORONTO

Designed and produced by Artia for
The Hamlyn Publishing Group Limited
London · New York · Sydney · Toronto

Hamlyn House, Feltham, Middlesex, England

Reprinted 1970, 1972

ISBN 0 600 03085 7

Printed in Czechoslovakia by Svoboda, Prague

FOREWORD

NOT SO many years ago entomologists might have been described as academic-looking gentlemen armed with butterfly nets and killing jars, who spent most of their time searching out and capturing vast numbers of insect specimens which they promptly killed, set out and carefully arranged in neat and tidy rows in special cabinets with glass-topped drawers. It is true that a certain amount of this sort of activity still occurs, for both reference and working collections of insects are just as important today as ever they were. However, it is true to say that the collection and determination of specimens is but a small facet of the work of present-day entomologists who are actively engaged in work on nearly every subject directly and indirectly connected with the insect world.

Despite the presence of extensive libraries of books and scientific papers full of accumulated knowledge about insects and other invertebrates, the whole field of entomology is continually becoming vastly more complex, and the acquisition of greater knowledge and experience merely serves to underline the fact that the 'insect man', both professional and amateur, is still indispensable.

The as yet uncontrollable army of insects still consumes tremendous quantities of vitally needed human food, while the war against insect-borne disease continues to be fought as actively as ever. Their persistent destruction of property, material and goods continues to cost the peoples of many countries large sums of money for replacements, and there is little likelihood of a short-term solution to this problem.

Even the discovery and invention of the more deadly and efficient chemical insecticides have not proved to be the complete and expected answer to insect control. In fact, often the widespread and inefficient use of these aids by unskilled and ignorant people has resulted in the development and multiplication of insect strains resistant to the poisons used, while most of the natural predators have been killed off. This has sometimes resulted in the pest itself continuing to flourish in even greater numbers than before.

Specialists in biological control have accomplished great things, as for example in California where the ladybird beetle, *Rodolia cardinalis,* was imported and introduced into the citrus groves to combat the menace of cottony cushion-scale which was causing extensive damage to orange and lemon trees. One hundred and twenty-nine ladybirds were released and in two years had increased to such numbers that the scale had disappeared. This is but one example of the many and successful uses of insects in biological control, encouraging extensive use of these methods. Large entomological research stations employ teams of technicians in whole-time breeding and production of vast numbers of specialized insects necessary for this kind of work. Insects are also bred and reared to serve as food supplements, both live and dead, for various animals and livestock; and as educational exhibits in schools, zoological departments and similar places. They are also bred to produce silk, edible dye-stuff and many other commercial materials. Many insects are still eaten in some countries as human food. It is possible in some of our modern cities to purchase tins of preserved insects prepared ready for table use. Such delicacies as fried bumblebees and chocolate-covered ants rather surprisingly found a ready sale recently in large stores in London and the provinces.

At this point, do not let us forget some of the other useful insects which are indispensable to our everyday life. It might be possible to do without the honey from the honeybees, but life without the benefit of all the pollination work effected by bees and other insects would be grim, for it is upon these activities that flowers and fruit crops depend.

Insecticides do not differentiate between useful and harmful insects and it will thus be realized that the control of insect pests without harming the beneficial insects such as bees, etc, is a far more complex subject than one might at first suppose.

Despite modern travel methods enabling easy travel to foreign countries to collect and examine insects in their own habitat, it is certainly not everybody who has the inclination to venture abroad to see the wonderful iridescent coloration and aerobatic displays of Morpho butterflies or the superb leaf-like camouflage of leaf insects and the columns of leaf-cutting ants marching back to their nests bearing parasol-like pieces of petals or leaves above themselves like canopies. This pictorial encyclopedia, therefore, brings all of these wonders of nature directly to your own home.

It is also a record of the insect world. With the spread of civilization and the resultant increase in land required for cultivation, together with the injudicious use of insecticides, many of the rarer and more curious species are becoming much harder to find, in some cases there is a real risk of them disappearing completely.

Dr. V. J. Stanek, former director of the famous Prague Zoological Gardens and an entomologist of note, besides producing many nature films and being a first-class photographer, has made it possible for everybody to have a look 'behind-the-scenes' at the fascinating world of insects.
During his many and extensive travels throughout the world, Dr. Stanek has made it his business to form a superb photographic record of all the insects and invertebrates that he has collected and studied.
The wonderful photographs in this encyclopedia, which together number over 1,000 black-and-white as well as coloured plates, have not been produced in a remote and academic style, but have been made in most cases direct from the living insects. Each insect has been described and Dr. Stanek has also provided a brief and systematic survey of the whole Insect Class.
In my opinion, as a practising entomologist, the publishers of this insect encyclopedia have rendered a service to us by making an English translation available in this country. I am sure that this book will be most useful to teachers, students and in fact anyone interested in insects who requires a photographic reference work of this nature.

GEORGE J. ASHBY

INTRODUCTION

Insects *Insecta* form by far the largest class of the phylum of arthropods *Arthropoda*. Moreover, they comprise the greatest number of all living creatures on earth and, although very ancient, the class has not yet reached the climax of its evolution. Most of them are terrestrial; only a very small proportion have comparatively recently adapted themselves to an aquatic existence. The present scientific view is that all the arthropods are descended from worm-like creatures which did live in water and were closely related to the present-day phylum earthworms or annelids *Annelida*, especially to the class of primitive earthworms *Archiannelida*, and perhaps also to one or two other classes of higher annelid worms. The construction of their bodies corresponds, in many details, to that of the annelids, though it has of course adapted to a terrestrial way of life as far as necessary. The immediate predecessors of the insects are thought to have been the class of millipedes *Diplopoda*, which can be classed together with the insects to form a subclass *Tracheata*.
The bodies of the insects are composed of 21 segments which, in the fully-grown state of most insects, tend to merge into one another and form three principle sections: head, thorax, and hind quarters (abdomen).
The head (caput) consists of six segments used together to form a rounded whole, and is usually protected by a hard outer skeleton.
The first, or front, section of the head (Segmentum oculare) consists of the forehead (Frons), upper lip (Labrum), a pair of compound eyes, and simple eyes (Ocelli) in pairs. The second section of the head (Segmentum antennale) bears the antennae and the unpaired simple middle eye. The third section (Segmentum postantennale) bears, in the case of the majority of forms of higher insects in the adult state, no appendages, but contains nerve ganglia, as do the two previous sections. Together with them it forms the front part of the head (Caput anterius – Procephalon), which contains the sense organs.
The rear part of the head (Caput posterius) serves for the preparation and digestion of the food, and is therefore called the Gnathocephalon. The first segment of this part—the fourth of the whole—bears the upper part of the mouthparts, the mandibles, and is called Segmentum mandibulare. The fifth segment of the head (Segmentum maxillare) forms the maxillae, the second or lower pair of mouthparts. On the sides of this there is usually a pair of feelers called maxillary palps.
The last, or sixth section of the head (Segmentum labiale), bears the lower lip or Labium. This, too, has a pair of feelers, the labial palps, on the sides.

The second part of the body of an insect consists of the thorax. The front section is called the Prothorax, the middle Mesothorax, and the back Metathorax.
Each of these sections has a pair of articulated legs attached to the lower part; in other words, the insects have six legs. In addition, in the case of winged forms, both the front and back sections of the thorax bear a pair of wings. The thorax of insects is generally fused together to form a strong, chitinous structure. Within it are contained the gullet, the main cord of the central nervous system, which is connected to every part of the body, and the tube-shaped heart. Here, too, are situated salivary glands, which have an outlet into the mouthparts and powerful muscles which operate the wings. These move in an up-and-down direction, effecting a contracting and expanding movement between the upper or back thoracic

plates (tergites) and the lower plates (sternites). Thin, weak pleurites on the sides of the thorax enable this movement to take place, while that of the wings, especially in the good fliers, is co-ordinated by the precise interaction of five special groups of muscles.

The legs have at their disposal their own muscular system, with the help of which they move in much the same way as do the legs of the vertebrates. The construction of individual parts of the legs is also similar, and is familiar to us from the anatomy of vertebrates. This is, however, only true of their order of succession and often of their method of functioning, and should not be interpreted as a homologous relationship—that is, as far as evolution is concerned, there is no connection with the parts of the limbs in the vertebrates which bear the same name. The leg of an insect begins with a basal articulation (coxa or hip), proceeds to a thigh-ring (trochanter), then into the thigh proper (femur) and the shin (tibia), and ends with the foot (tarsus), which, at the most, is composed of five components.

The wings, too, of the insects are not homologous with the wings of birds, since they cannot be viewed in the same way as proper members or limbs. They developed, instead, as two-layered outgrowths from the back part of the thoracic segments, and are interlaced with strong chitinous tracheate tubes, which form the so-called "venation", which is characteristic of some species and presents a convenient means of differentiation for purposes of classification.

Among the more primitive forms both pairs of wings are fully developed as organs of flight and are of equal size. Among the beetles the front pair of wings has taken on the function of covers (elytrae) for the rear pair, which do the actual flying. The same is true for most orthopterans. The rear wings are differently formed and are often folded together in a very complicated manner (for example beetles and earwigs). In the flies, on the other hand, only the front pair are used for flight and the rear wings have developed into a pair of club-like organs (halteres) used for balancing. There are some primitive insects—the *Apterygota*—which are wingless as a primary feature; the *Pterygota*, however, have either completely lost their wings as a secondary development or possess only atrophied remnants of them. The reasons for this may be the particular conditions of life adopted by the insect—for example, a parasitic existence (lice, fleas, etc.)—or immobility as a result of massive egg-production (for example, female termites and ants, the brachypterous females of butterflies, etc.).

The upper surface of the wings of the more primitive insects is bare, only occasionally a few hairs and often having a variegated pattern and colouring *Orthoptera, Neuroptera, Cicadina.* The butterflies, by contrast, are distinguished by the colourfulness of their wings and their shimmering patterns, which are brought about by the covering of overlapping scales with their extremely complicated inner structure. Due to the interference of the air in the thin layers of the scales, the widest and most subtle differences of shading in colouring and finish are seen.

The third section of an insect's body is the hind quarters (abdomen). Originally it was composed of twelve segments, but their number has been greatly reduced in recent forms: usually only ten or indeed as few as six segments remain. There are no real appendages to the hind parts, but in some forms so-called tail bristles (cerci) have developed on the last segment, with a terminal thread between them, which forms an extension of the last abdominal segment, the telson. The covering of the abdomen is usually less chitinous, and often quite soft, and therefore in several orders (the beetles and cockroaches) it is protected on top by the wing-covers. In such cases, these form a strong armour for the sensitive hind parts. In the abdomen, the important organs already mentioned are continued: the nervous system, the heart-tube and the tracheate breathing system. Above all, the digestive organs are situated here, as well as the anal passage and the sexual organs. In many females (several beetles, butterflies, flies and termites), the hind parts are filled to bursting with eggs and quite disproportionately enlarged. The females of many groups possess at the end of their bodies a tube for laying eggs (ovipositor or terebra) (examples: *Hymenoptera*—Sirex, *Saltatoria*—Acheta, *Panorpata*—Boreus, *Coleoptera*—Acanthocinus).

Insects breathe with the aid of the so-called tracheal system—namely, air tubes which pass through all parts of the body and begin on the surface in a number of openings (stigmata or spiracles). Moreover, the body of an insect is fitted with similar equipment to that which can be found in flying birds—that is, with so-called air-sacs, which lessen the weight and improve the aerostatic conditions. Thus, air-sacs of the most widely different forms may be found in the tracheal system of insects and are particularly advantageous to those insects which hover, glide, make use of air-currents or reach great heights.

The larvae of certain aquatic insects breathe with the help of particular leaf-like appendages,

called tracheal gills, which are permeated by a system of tiny, greatly ramified tracheoles.
The system of blood vessels is confined to the tubular heart near the back. Otherwise, the blood flows freely, making its own way through the empty spaces of the connective tissues. We speak, therefore, of an open circulatory system.
The intestinal system is always divided into fore intestine, middle intestine and hind intestine. In many species, it is equipped with special tubes or branched appendages, which contain symbionts—that is, micro-organisms (bacteria or fungi). These not only supply the insect with materials necessary for its survival (vitamins, materials to assist growth, etc.), but also have a share in the digestive process, such as in digestion of cellulose in the termites.
The insects are divided into two sexes, and reproduce sexually. After mating, the female lays anything from a small to a very large number of eggs; only in exceptional circumstances are they live-bearers). To this process, however, we must add what is called parthenogenesis, or virgin birth, by which the eggs of an unfertilised female as a rule bring forth only females. We do, however, know of cases (in some flies) in which parthenogenetic eggs are developed by immature larvae without fertilisation (paedogenesis), and these eggs, too, are capable of further development. Occasionally it happens that some insects—especially the little parasitic wasps *Encyrtidae*—are capable of reproduction by what is called polyembryony; that is, asexual reproduction of the larvae in the egg after it has been passed by the mother insect within the body of a host. From an egg of this type, hundreds of larvae can develop under favourable conditions.
With the exception of the *Apterygota,* almost all other insects undergo what is called metamorphosis—that is, a development which is only complete after it has gone through four stages, and is regarded as incomplete if one or more are omitted.
The four stages are: 1. the egg, in which the larva undergoes its initial development and awaits the appropriate time to emerge; 2. the larva, or stage of feeding and growth; 3. the pupa, or transitional stage; and 4. the mature insect, or imago, which is the reproductive stage. After reaching this fourth stage, the insect grows no further.
In the vast majority of different species of insects it is, of course, possible to observe a number of intermediate and additional stages in metamorphosis. However, in principle we distinguish insects with an incomplete metamorphosis *Hemimetabola* from those with a complete metamorphosis *Holometabola*.
In the hemimetabolans, the larva resembles the adult insect fairly closely throughout its whole growth period; only the wings are lacking. After several moults (ecdysis) these nymphs achieve the size of the adult insect, and during the later stages (instars) the first traces of wings appear, until finally, after the last ecdysis, the fully developed imago emerges (examples: bedbugs, cicadas).
Among the holometabolans, the larva bears no resemblance at all to the fully developed insect. It grows and passes through moults in much the same way, but eventually develops into the almost motionless pupa. After a certain time—usually very short, often only a few days—the original organs of the larva begin to be destroyed (histolysis) and the organs of the future imago developed (histogenesis). On completion of its development, it emerges from the cocoon by filling its tracheae (for example, in the wings), and so achieves its final form. Its chitinous armour hardens in the air, takes colour and the adult insect capable of reproduction is born. (Examples: ants, bees, and wasps—*Hymenoptera;* beetles—*Coleoptera;* stylopids—*Strepsiptera;* lacewings—*Neuroptera;* caddis flies—*Trichoptera;* butterflies—*Lepidoptera;* houseflies—*Diptera;* fleas—*Aphaniptera).*
The extraordinary fertility and reproductive powers of the Insect World has resulted in an almost infinite number both of individuals and of different species, many of which do a great deal of damage. Although much has been accomplished in the fight against these harmful insects, many of the problems still remain unsolved. New and more powerful insecticides have been produced but the use of these has still not brought about victory. The fight is often being waged in an inefficient, reckless and unprofessional manner so that beneficial insects as well as harmful ones are often being destroyed, thus upsetting the balance of nature.
There are also a great many attractive insects, such as bees, butterflies and some beetles, whose presence greatly enhances the pleasures of the countryside on a summer's day. Unfortunately more and more of the land is being cultivated and thus encroaching on what formerly consisted of undisturbed territory. This fact, together with the indiscriminate campaign against insects in general, has made many species extremely rare, and has necessitated many being placed under protection.
This book is intended as a contribution to the spread of more detailed information about them, so that they may also be better protected.

Dr. V. J. STANEK

1

People who know the world of insects only superficially will be surprised to learn that a large number of the most unpleasant carriers of disease do not belong to the Insects at all. In this illustrated encyclopedia of insects we could not, therefore, properly discuss the dangerous scorpions, poisonous spiders, parasitic ticks, biting centipedes and other creatures which molest and menace everyone who visits the warmer parts of the earth, especially the tropics. However, since anyone who spends any time in the tropics will be forced to have some dealings with them, as they infest not only the pockets and clothing of the visitor but even his bed, we will not deny the reader this "chamber of horrors", and begin therefore with those creatures which are not insects, but belong to the arthropods and are related to the insects, being among their ancestors.

During the palaeozoic era, or more precisely the Cambrian period—about 540 million years ago—when the sea was crawling with trilobites and the oozing mires of primeval forests were populated with creeping giants—centipedes, about 1.5 m long, resembling the arthropleura, then developed the first **pincered arthropods** *Subphylum Chelicerata,* most of which later

2

adopted a terrestrial way of life. Already during the Upper Silurian we find species of scorpions very similar to those which exist today. They mostly come from tropical and subtropical countries, there being about 600 species. Of all the representatives of the spider family *Arachnida,* they are the most feared.

The largest species of scorpions achieve a length of some 20 cm, and their sting can be fatal to human beings. **Palamnaeus fulvipes** [1] comes from the tropical Indies; while alive it is a brilliant black-green colour and measures up to 13 cm, including the pincers.

Euscorpius carpathicus [2] lives in Europe and is the most northerly distributed of all species of scorpion. It is brown and measures up to 4 cm. The remaining European species, which are smaller and less poisonous, are confined to the southern side of the Alps.

The tropical scorpions prefer damp and warmth. By day they hide under stones, in rotting tree stumps and under the bark of trees.

The Cuban species **Centruroides gracilis** [3] has a flat, slender body, long pincers and a tail (postabdomen) ending in a doubled sting. All scorpions give birth to live, fully-developed young, generally between 20 and 50. Until their first moult they are carried about on the back of their mother. Not until later does the young

3

4

scorpion separate from its mother. From that time, it lives the life of a recluse. The scorpion seizes its prey alive, but almost always kills it with its poisonous sting. The **Whip Scorpions** order *Pedipalpi* have a life pattern similar to that of the scorpions, except that they do not possess a poisonous sting. The flat body enables them to squeeze into extremely narrow cracks, where they prey on small arthropods and worms. Only the largest species attack

5

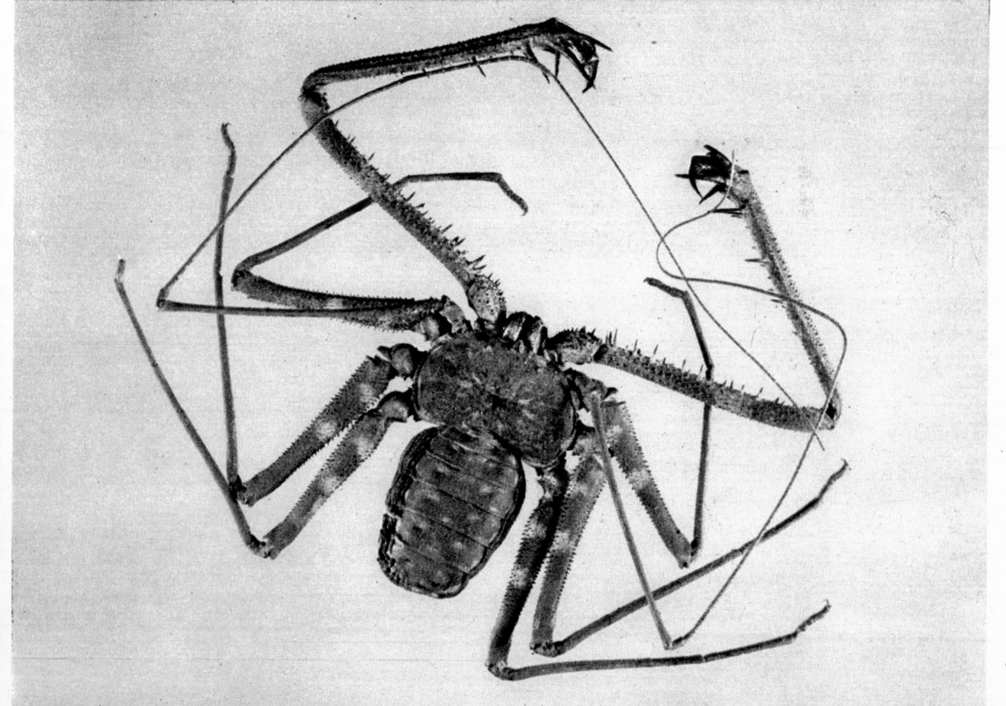

6

small vertebrates, especially frogs. One pair of its pedipalps are developed into strong pincers equipped with thorns and teeth. One member of the **Whip Scorpions** *Theliphonidae* is **Hypoctonus rangunensis,** shown in illustration [4], from southern Burma. Together with its tail appendage it measures about 6 cm. The thread-like, articulated appendage has only sensory functions; it is thin and equipped with sensitive hairs. The whip scorpions are not poisonous, but some species defend themselves by squirting an acrid liquid from a gland at the end of their bodies.

The suborder **Whip Spiders** *Amblypygi* consists of forms with a flat body and prominently developed grasping organs or fangs in the form of a first pair of legs extended greatly beyond the normal length, so that they resemble a pair of giant whips. There are no poison or other defensive glands. The species shown, the African **Phrynichus reniformis** [5], measures about 35 mm, excluding appendages. There are more than 20,000 species of spiders, and we can show here only a very few examples of the most important families. The order **Spiders** *Araneae* are, because of their biological make-up, among the most interesting creatures on earth. The four pairs of legs, the poisonous, claw-like fangs (chelicerae), the head- and breast-parts fused into one another, and the thin-skinned, limbless hind quarters (opisthosoma), which are equipped at the rear end with spinnerets, are the most important common characteristics.

The spiders are distributed almost everywhere, being absent only from the polar regions. Most species live in the warm and hot zones of the earth, though some species have been found high above the tree line on the crags of the Himalayas, up to a height of 7,500 m. Their size ranges from tiny creatures less than 1 mm long to the largest spiders (9 cm long excluding the legs). This is the **Guyanan Bird Spider,** *Theraphosa leblondi.* On the walls of heated cellars, where bananas imported from the tropics are ripened, one can find in large numbers a large, attractive grey-white spider, **Heteropoda venatoria** [6], one species of the large family *Sparassidae.* Including its legs, it reaches a length of about 8 cm. In the tropics

7

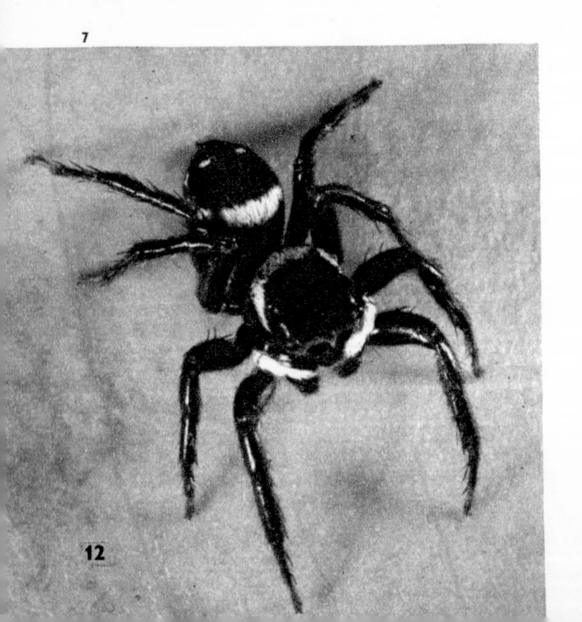

8

9

it is regarded, not unfavourably, almost as a domestic animal. By day it usually stays hidden behind furniture and pictures and preys at night on troublesome insects. The **Jumping Spider,** *Hasarius adansoni* [7] family *Salticidae,* is an important species, and known to us from the hot houses of botanical gardens. It is about 1 cm long and the males in particular have an attractive black and white colouring. The **Black Widow,** *Latrodectus mactans,* has a very bad reputation; it is certainly one of the most poisonous of all spiders. Related forms are found in various countries throughout the world. In Turkmenistan and neighbouring countries it goes under the name **Black Wolf** or **Karakurt** [8]; in the Mediterranean area there is a species called the **Malmignatte,** *Latrodectus tredecimguttatus.* These spiders also occur in Africa, Australia and New Zealand. By some authorities they are viewed as different species, by others as different forms of one and the same species. The spider illustrated is about 2 cm long and black with red spots. It comes from the southern steppes of the Soviet Union, being most common in Turkmenistan and Kirghizia. It is, however, also found in Arabia. The bite of the female, which is somewhat larger than the male, has more or less severe consequences for human beings. Only rarely does it actually result in death for men; the case is otherwise for cattle, however, which usually perish if they are bitten by the black widow.

The **Tarantula,** *Lycosa tarentula,* used to be greatly feared at one time. The body of the female [9] measures something over 25 mm. It inhabits the dry steppes of southern Europe and Asia Minor. At the beginning of the 17th-century in Italy and Spain people who had been bitten by the tarantula were supposedly cured by a dance which become more and more frenzied, ending in loss of consciousness. It was believed that this was the only way to avoid certain death. The dance has remained with us to this day, in the form of the "tarantella". Not until very much later was it realised that the bite of the tarantula was not always fatal.

In recent years, what is now the largest European spider, the **Tartar Wolf Spider,** *Lycosa sigoriensis* [10], has advanced across

10

11

12

Europe from its original habitat in south-east Europe and south-west Asia. The body of the female measures on occasions more than 3 cm and with outstretched limbs up to 7 cm. However, this spider too is not dangerous, and only bites man if it is seriously provoked. The effects are no worse than those of a bee-sting. It lives, like the tarantula, in holes in the earth.

In contrast to this, the **South American Wolf Spider,** *Lycosa raptatoria,* often attacks man. The result of a bite is that the skin around the bitten area undergoes necrotic disintegration. An effective anti-serum is at present being developed at the Serological Institute for Snake Poisons at Butantan near São Paolo.

There are in Brazil several species of aggressive spiders of the family **Combed Spiders** *Ctenidae,* which are dangerously poisonous. The bite of some of them has a haemolytic effect; of others, a neurotoxic. Occasionally, some of these poisonous spiders reach Europe with cargoes of bananas. If such a cargo of southern produce brings, for example, the **Bird Spider,** *Avicularia avicularia,* into our northern latitudes, there is usually a sensation, not only among those who are engaged in the banana-importing business but also among zoologists, collectors and breeders of exotic creatures. Although the bird spider very rarely bites and its poison has a very similar effect on human beings to that of a bee-sting, it must also be pointed out that the poison of some species of bird spiders is the principal ingredient of the notorious poison arrows of the South American Indians.

Another order of spider-like creatures is that of the **Barrel Spiders** or **Wind Scorpions,** *Solifuges.* About 600 different species of these fearsome-looking creatures live in the tropics and subtropics of almost all parts of the earth, except Australia. The length of their bodies varies around 5 cm. Their long legs are equipped with large numbers of sensitive hairs. The pedipalps are considerably extended, resemble legs, and are used as such, so that it could be said that the barrel spider runs on ten legs. This arachnid is pale yellow and covered with a film of hairs. It succeeds in crushing even the hard wing-covers of beetles in its sharp chelicerae. If it is driven into a corner, it prepares to defend itself, raises its head against the attacker, opens its powerful chelicerae, which look like a pair of pincers, and emits a spitting and hissing sound. With these sharp weapons it can deliver a very painful bite, but it possesses no poison glands. The barrel spiders are for the most part nocturnal creatures, living in steppes and deserts. By day they hide in holes and at night come out and hunt grasshoppers, beetles and larvae, as well as frogs and the smaller lizards. As with most arachnids, courting and mating among the solifuges is a very complex, interesting and dramatic procedure. As a rule, the male is eaten by the female after copulation. One of the 76 species which live in the southern lands of the Soviet Union is the **Common Barrel Spider,** *Galeodes araneoides* [11, 12].

The order of **mites** *Acari* is distributed throughout the whole earth, except for the polar regions. There are over 10,000 species so far recorded. Some of them do not achieve a length of $\frac{1}{10}$ mm. In their life-cycle they pass through an unusual number of forms. Mites are found almost everywhere in nature, on land and in water. They live on animal and plant debris, or are parasitic on animals and plants. Only a few species attack man. Among the most notorious is the **Itch Mite,** *Acarus siro* or *Sarcoptes scabiei* [13]. It measures about 0.25 mm, so that it cannot be detected with the naked eye. Today, there are effective and reliable ointments against the itch mite. However, in ancient times and in the Middle Ages it was far worse, and kings, popes and other notables (Herod, Antiochus, Philip II and Clement VIII) died of the affliction known as acariasis subcutanea. We can well understand

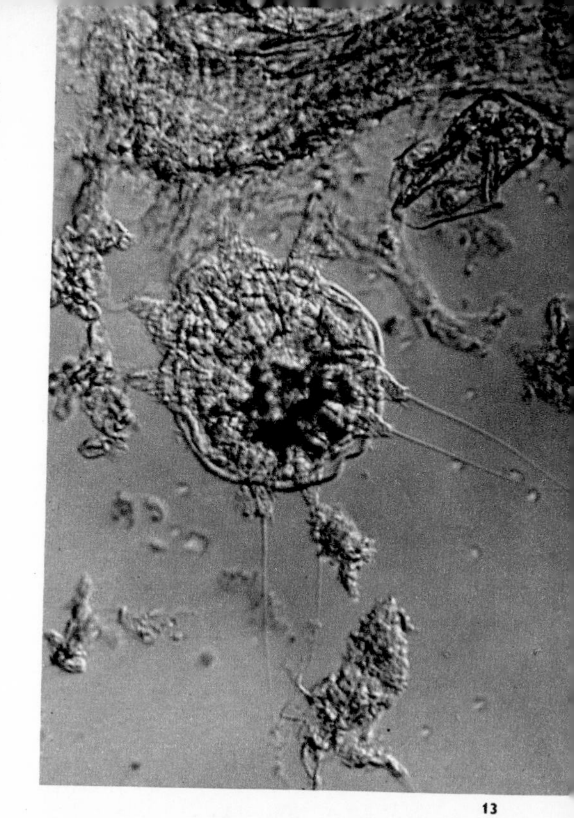

13

14

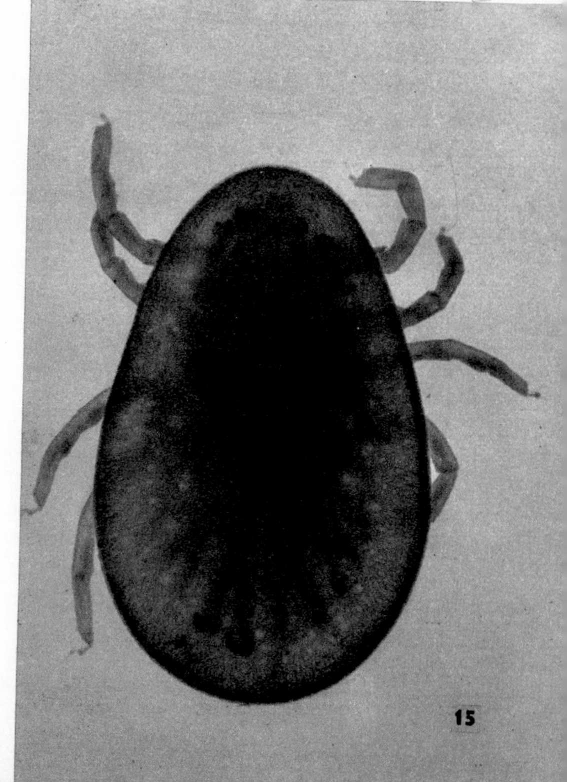

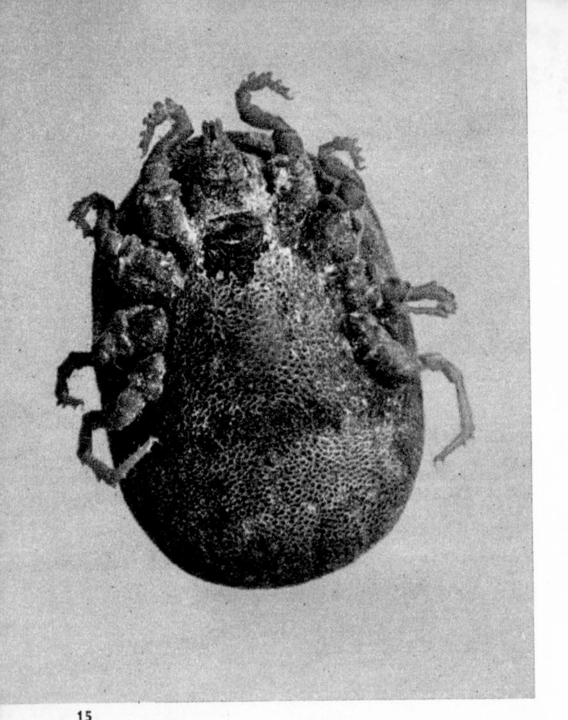

how terribly people used to suffer from this pestilence. On the body of the patient, boils up to the size of an egg would form under the skin, which were filled with living mites of the species **Harpyrhynchus tabescentium.** At a given moment in their development, great streams of larvae would pour out and infect the entire surroundings of the patient. After a hideous death, the corpse had to be sewn up in a leather sack before the ceremonial funeral could take place.

The largest mites are the **ticks** (superfamily *Ixodoidea*). When the female of the tropical tick, **Amblyomma clypeotatum,** has sucked herself full with blood, she can attain a length of up to 3 cm. The rest, however, are generally smaller forms, rarely exceeding 1 cm in length. A member of the **Leather Ticks** *Argasidae* is the **Dove Tick,** *Argas reflexus* [14]. It measures about 4.5 mm. The body is flattened and the head, thorax and hind quarters are all fused into one another. The mouthparts are adapted for biting, boring and sucking and the proboscis is situated on the underside. Ticks are able to go for months without food.

15

16

16

Ia Scorpion, *Centruroides gracilis*. Female with young. Cuba.

Ib Garden Spider, *Araneus diadematus*. Female. Central Europe.

II *Eurypelma spinicrus,* one of the Bird Spiders from Cuba.

Like bugs, they come out in the night and attack their prey: birds and mammals, including humans. Some species are world-wide in distribution: most of them, however, prefer tropical lands and the warmer parts of the temperate zone.

The *Argasidae* are dangerous carriers of many infectious diseases, with which they infect both animals and humans. One member of this family is the highly dangerous **Ornithodoros savignyi** [15]. The picture shows a female, stuffed tight with eggs. The female is greyish-brown, has yellowish legs and measures about 12 mm. This tick lives in North Africa and Arabia, also as far east as India. It is a carrier of the so-called relapsing fever *Febris recurrens*, also known as tick fever. The second family, much richer in genera and above all in species, is that of the **Shield Ticks,** *Ixodidae*. Included in this family are the **common tick** of Europe, the **Wood Ram** or **Wood Tick,** *Ixodes ricinus*. The males measure 2.5 mm and the young females 4 mm, though they reach a length of up to 11 mm when they have sucked themselves full with blood.

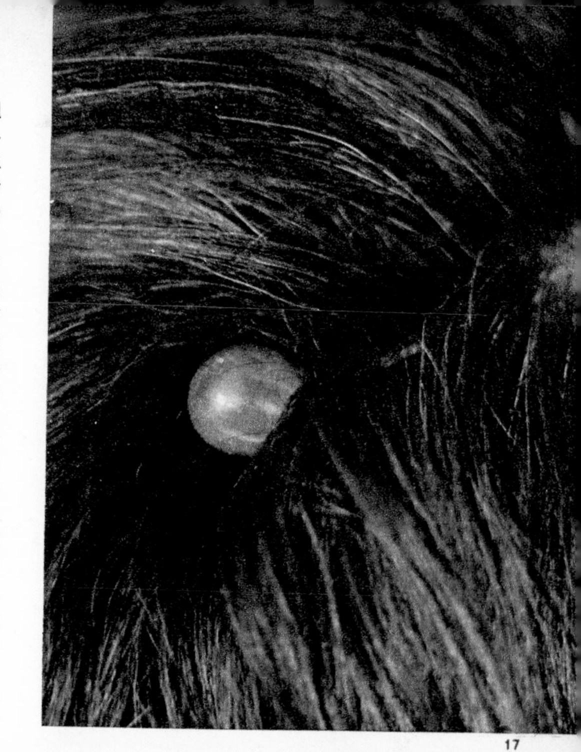

17

18

19

20

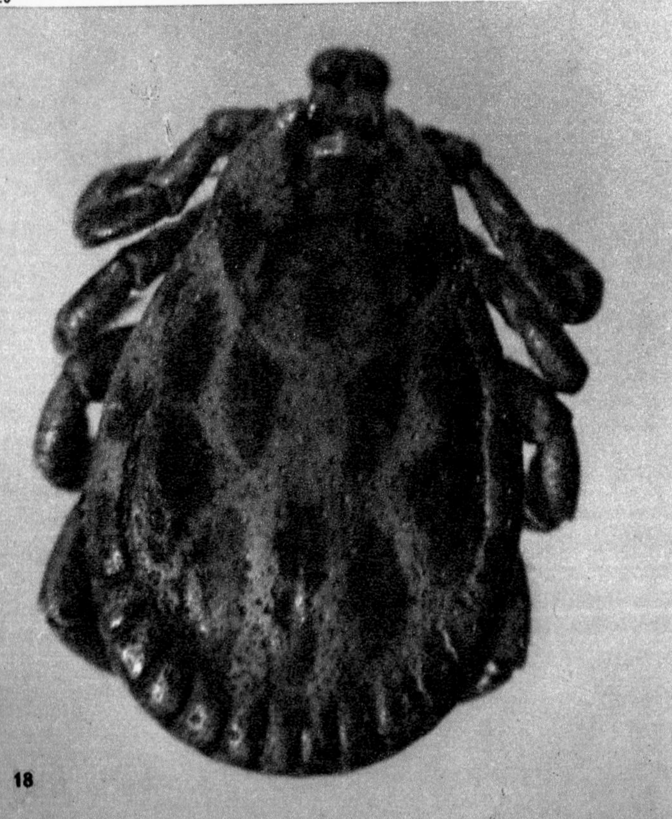

The upper side of the bladder-shaped, unsegmented body is dominated by the mouthparts. The tick finds its prey mainly by means of its sense of smell, which is situated on its front legs. It feels out with its segmented maxillary antennae a suitable place to make its bite. Between the antennae is the mouth, which ends in the so-called hypostome. On the underside of this are a number of tiny barbs. With this quill-like process the tick tears open the skin, sticks the whole hypostome into the wound, anchors itself with the barbs of the chelicera and pumps blood out of the bite with its oesophagus, which operates on a hydraulic principle. At the same time, the tick allows ixodin to flow into the wound mixed with its saliva, a liquid which prevents the blood from congealing, at the same time introducing pathogenic agents into the bloodstream of the host. Once it has bored its way into the skin, the parasite is best removed by first smearing it with some fatty substance such as Vaseline, butter or oil which obstructs its respiratory organs. After a short time the parasite suffocates and the deeply entrenched mouth comes loose of its own accord. If the tick has not finished boring into the wound, we can attempt to pull it out forcibly. However, in doing this we run the risk that the mouthparts will get torn off and remain in the wound, in which case they would have to be removed surgically.

21

The female of the common tick [17] lays up to about 3,000 round eggs, from which after a few weeks or months emerge six-legged larvae about 1 mm in length [16]. These little larvae attach themselves amidst low undergrowth to small mammals, sometimes to reptiles, where they remain for three to five days. Then they drop off and live in the earth, shed their skin and develop into eight-legged nymphs [18] which then, concealed in bushes and branches, await their future host. Once the host has been found, the tick sucks its fill over a period of several days, drops off again, sheds its skin a second time on the ground in among the roots of the undergrowth and becomes a fully-grown tick, ready to propagate its kind. The female in illustration [17] has bored its way into the skin of a dog. The wood tick is a carrier of dangerous viruses, which cause, for example, the so-called "tick encephalitis" and tularaemia.

In central Europe and the nearer parts of Asia lives **Haemaphysalis concinna** [19], a flat, eyeless tick, dark brown in colour and 3.5 mm long. In the larval stage it attacks small vertebrates, but as a nymph and when fully grown it attacks practically all the larger warm-blooded animals, including man. Among the hosts are found reptiles and various species of small birds. It is a carrier of haemosporidiosis among ungulates, spotted fever in East Asia, and among men the encephalitis virus.

In the warmer parts of central Europe live two species of shield-tick belonging to the genus **Dermacentor.** A further 12 species are found in central Asia, others in America and Africa. The two European species resemble each other very closely, displaying only slight anatomical differences. However, they do differ in their habitat.

The **Steppe Shield-tick,** *Dermacentor marginatus* [20], lives in the warmer parts of the steppes and the surrounding scrub lands, where it changes host in much the same way as the wood-tick. The male measures 4 to 5 mm and the female up to 14 mm when bloated after bloodsucking. They are very brightly coloured, with a red-brown effect. All species have eyes. At first they attack small rodents, insectivores and the smaller beasts of prey of the steppes. When they are fully grown, they suck the blood of larger animals, such as horses, donkeys, camels, deer, buffalo, cattle, dogs and cats, and in particular sheep. Man, too, is not infrequently attacked by them. It is an interesting fact that this species has been found as a fossil with the remains of a pleistocene rhinoceros. It carries animal diseases, especially those of horses.

In forested river valleys in the warmer parts of Europe and Asia lives the closely related *Shield-tick,* **Dermacentor pictus** [21]. In districts where forested areas are only found on high ground, they live at lower levels. In the southern USSR they carry piroplasmosis of horses, and tularaemia. The American species attack man and cause tick paralysis. **Dermacentor albipictus** and **D. venustus andersoni** of North America transmit the dreaded and often fatal Rocky Mountain spotted fever. The ticks of this family lay several thousand eggs and can go without food for up to 18 months.

In the whole southern part of central Asia, in Asia Minor, in Iraq, Iran, on the Red Sea

22

23

coasts and in north-east Africa lives the tick **Hyalomma anatolicum.** The male [22] is some 4 mm long, the female 5 mm when empty and hungry [24] and about 9 mm when it has sucked its fill [23]. The species of the genus *Hyalomma* have yellowish, remarkably long legs with a diagonal pattern. The body is yellow-brown, and the shield behind the head chestnut-brown; the male is reddish-brown in colour. The ticks shown in the illustrations were collected from camels in Cairo, Egypt.

Other poisonous and much feared arthropods are the **centipedes** *Chilopoda,* which form a separate class with some 1,700 species distinguished to date, having connections in the evolutionary scale with the most primitive forms of insects. Their narrow extended bodies

24

are made up of anything between 15 and 173 rings. On every ring is a pair of legs. On the head segment, which is already distinguishable, are a pair of long, sensitive antennae, two pairs of little jaws for breaking down food and a pair of strong claw-like organs for holding things, which close from both sides like tongs and are equipped with poison glands. The centipedes are carnivorous. They hunt, mostly at night, for small arthropods, larvae, molluscs and worms. The long tropical giant centipedes or scolopenders—up to 25 cm long—also attack small vertebrates. The **Brown Centipede,** *Lithobius forficatus* [25], grows to a length of up to 3 cm. It has 15 pairs of legs and is coloured glossy brown. On each side of its head it has about 40 single eye-points. The brown centipede lives in damp earth in gardens, deciduous forests and on the banks of streams and rivers. By day it can be found by digging in the earth, under a piece of wood on the ground, or behind the bark. Its bite is in most cases without effect.

In the Mediterranean area lives the **Waisted Scolopender,** *Scolopendra cingulata* = *morsitans,* which is about 10 cm long. It feeds chiefly on insects and spiders. Its bite is unpleasant, but has no real after-effects. In the tropics of the Old and New Worlds there are species of up to more than 25 cm in length, such as the **Giant Scolopender,** *Scolopendra gigantea,* of South and Central America. Its poisonous bite can also be dangerous to humans, especially children.

Inhabiting a similar area of distribution as the brown centipede is the European **Long-antennaed Centipede,** *Geophilus longicornis* [26], an interesting, eyeless creature which wriggles violently like a snake. It is ochre-coloured to red and measures about 4 cm. Usually, it has between 41 and 82 pairs of legs of equal length, of which only the rear pair, the so-called trailing legs, are of greater length. These centipedes prey on rain worms and larvae, but also nibble at plant roots. The species **Scolioplanis crassipes** from the Vosges in France turns a brilliant green when excited.

25

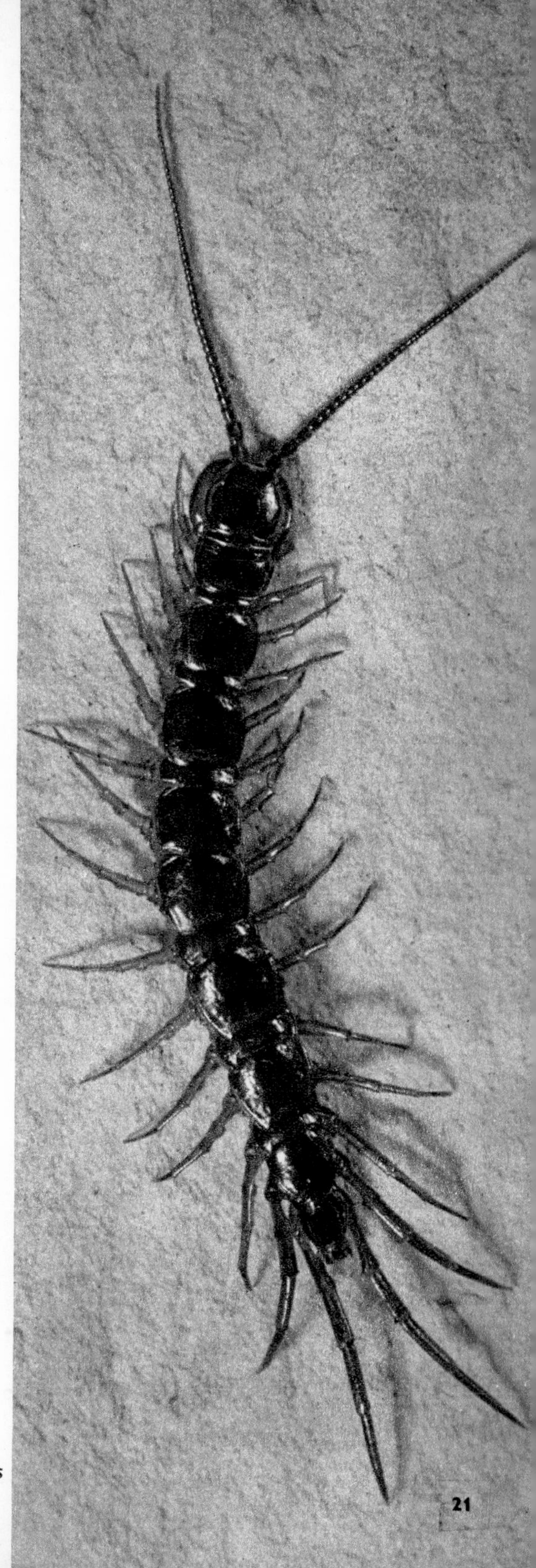

26

27

In contrast to the swift and aggressive centipedes, the **Millipedes** *Diplopoda* move very slowly. As their scientific name suggests, each segment of their body carries a double number of legs, that is, two pairs. The body is usually cylindrical and only rarely, in the case of a few species, flattened. The body segments are strongly calcined, in distinction to the chitinous bodies of the centipedes. Occasionally we find under a stone the "skeleton" of a dead millipede: a little heap of decayed body rings, which look like a row of tiny white bracelets. Most of the species so far classified are small and measure only a few centimetres. Only in the tropics are there found millipedes up to 30 cm long.

In the forests of central Europe we can find the rare little **Millipede,** *Polydesmus complanatus* [27]. It grows to a length of 2.5 cm. On the whole it is grey-brown, but sometimes reddish individuals are found. The back part of the body rings form 20 flattened, cornered shells. This polydesmial feeds in particular on rotting leaves in beech and alder woods. The female builds for her heap of yellow eggs a bell-shaped nest of earth in humus and under stones, in the form of a flat mound, round which she curls her body. Certain centipedes and scolopenders protect their eggs in the same way. However, they do not actually build nests.

Some of the large exotic **giant millipedes** occasionally arrive as sightless passengers, hidden in the foliage of bananas, in

28

countries of the temperate and cold zones. There, they are eagerly awaited by zoologists and collectors, who do their best to keep these unusual immigrants alive for as long as possible and to bestow every possible attention upon them. The 14 cm long **Spirobolus** [28] which I had came from West Africa, drank greedily fresh water, took apples, the poppy-seed filling from cakes and liver sausage. In their natural habitat, these giant millipedes apparently feed on various kinds of garbage. They do defend themselves if they get very excited, although there is no danger that they will bite like the poisonous centipedes; they only excrete a pungent smelling, corrosive fluid which does not damage the skin, but irritates the mucous membrane.

In central, northern, and eastern Europe, and also in Canada, we can find in later summer the little **Spotted Millipede,** *Blaniulus guttulatus.* It is yellowish to pale brown and studded with reddish spots on both sides. It often appears in large numbers in cultivated soils, among root vegetables and fallen fruits.

Since the geological Devonian period—that is, for about 350 million years—the surface of our planet has become populated by a countless army of insects. The **insects (Insecta = Hexapoda)** form a class of the family *Arthropoda* which is by far the most extensive of all living beings. The number of species in this class forms some 70 per cent. of all creatures and the percentage of individuals is even greater. In the literature of the subject, 850,000 classified species are often given, by others over 1¼ million, and in England the number of types of insect has been estimated at 20 million. It seems that this number is no exaggeration, for insects are found in practically every part

29

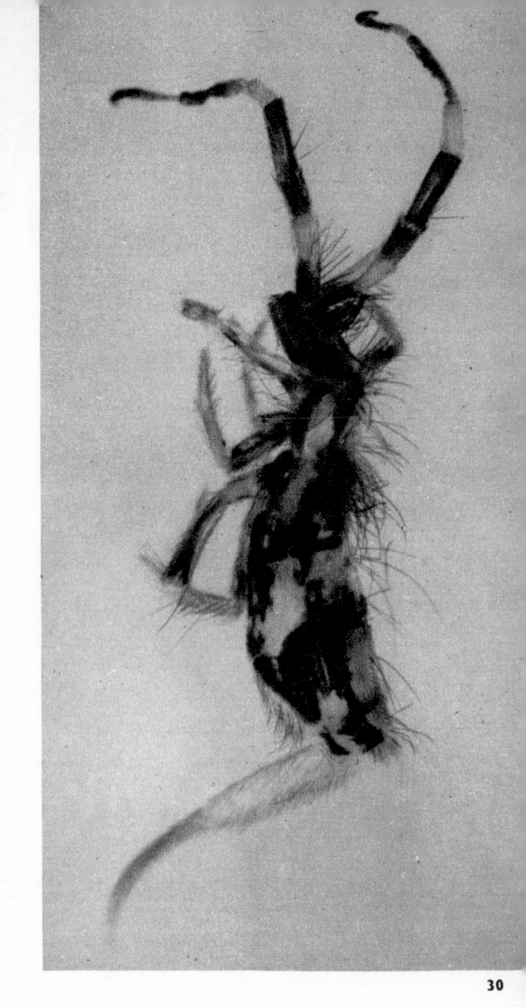
30

of the earth, except for some areas of polar regions. Of the extinct species, some 12,000 have been described and there are many new discoveries in store for entomologists.
During the long period of their development, insects have adapted themselves so completely to the most varied environments and conditions, that we are aware of the most improbable looking specimens. This bears witness to the enormous vitality of these creatures, having world-wide occurrence.
In the fully-grown state the body of the insect has six legs. It is divided into head, thorax and hind quarters. It is protected by an outer

armour of chitin, which is divided into moving segments. The head carries a pair of antennae. The mouthparts are designed either to bite or to suck, but can also be formed into a sting or adapted to suck semi-liquid food. Attached to the thorax are, besides the three pairs of legs, either one or two pairs of wings. Some groups of insects are primarily wingless, but in others the wings have in the course of time become vestigial. Among the extinct species, some are known which have three pairs of wings.

The most primitive and therefore the most ancient insects are the *Apterygota.* The first subclass, the **Springtails,** *Collembola,* comprises tiny, inconspicuous insects only a few millimetres long, most of which have a forked springing organ at the tail end of their body. They are distributed throughout the world, including the polar regions.

The **Green Glacier Springtail,** *Isotoma viridis* [29], inhabits central Europe and measures about 4.5 mm in length. In the early spring it can be observed on the snow. It is generally green, yellow, blue or violet and feeds on rotting foliage, wood and other vegetable remains which are available in whatever habitat it is in. Related species are sometimes present in such numbers on the snow that they seem to change the colour of the surface.

A large species is **Orchesella flavescens** [30], with a brightly coloured, hairy body. It measures up to 5 mm and lives in central Europe in moss, which also serves as its food. The largest of the non-jumping species—that is, those without the forked springing organ—is **Tetrodontophora bielanensis** [31]. It can measure as long as 9 mm, is blue-grey in colour and lives in central Europe in damp decaying vegetable matter.

31

One of the harmful species is the **Cucumber Springtail,** *Sminthurus/Bourletiella viridis,* which at present is distributed over the whole earth. It has a short, hunched body, is brightly coloured and measures about 1 to 2 mm. Since it preys principally on young plant growth, it does great damage to the agricultural economy, especially in Australia. In illustration [32] a culture of these insects bred on the inside of some cucumber peel may be seen.

On the surface of the water of ponds and pools one can see in the spring whole swarms of

32

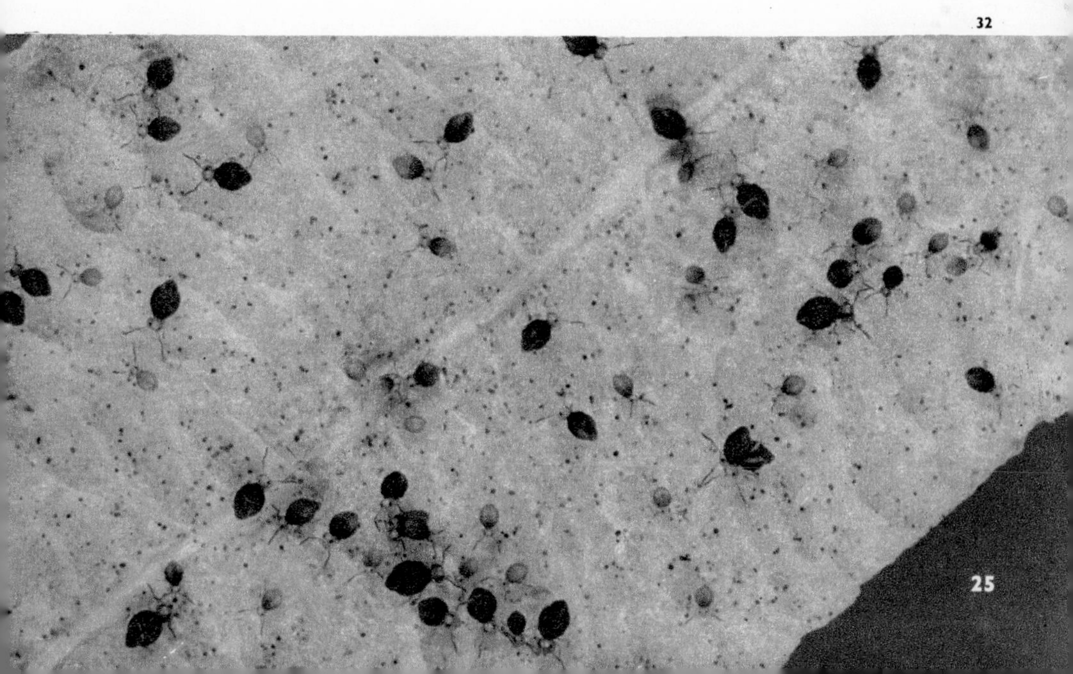

black-grey jumping creatures of a sombre appearance. They feed on putrefying vegetation or green water lentils: these are the **Water Springtails,** *Podura aquatica,* only 1 mm long [33, a photograph from central Europe]. These little creatures usually lie on the surface of the water, but can be picked up by the wind and carried over considerable distances. They are, therefore, like most related species, cosmopolitan.

In the second subclass, the **Half-insects,** *Protura* = *Anamerentoma,* there are about 70 species so far classified: minute beings which for the most part live deep under the earth. They are known in almost every corner of the globe, many researchers regarding them as the most primitive and ancient of all insects. In the fully-grown state they have a thorax section with three pairs of legs and 12 hind-quarter segments, the foremost of which may carry a further pair of legs. The half-insects are apparently predators, with piercing and sucking mouthparts. They hold on to their prey with the front pair of legs.

33

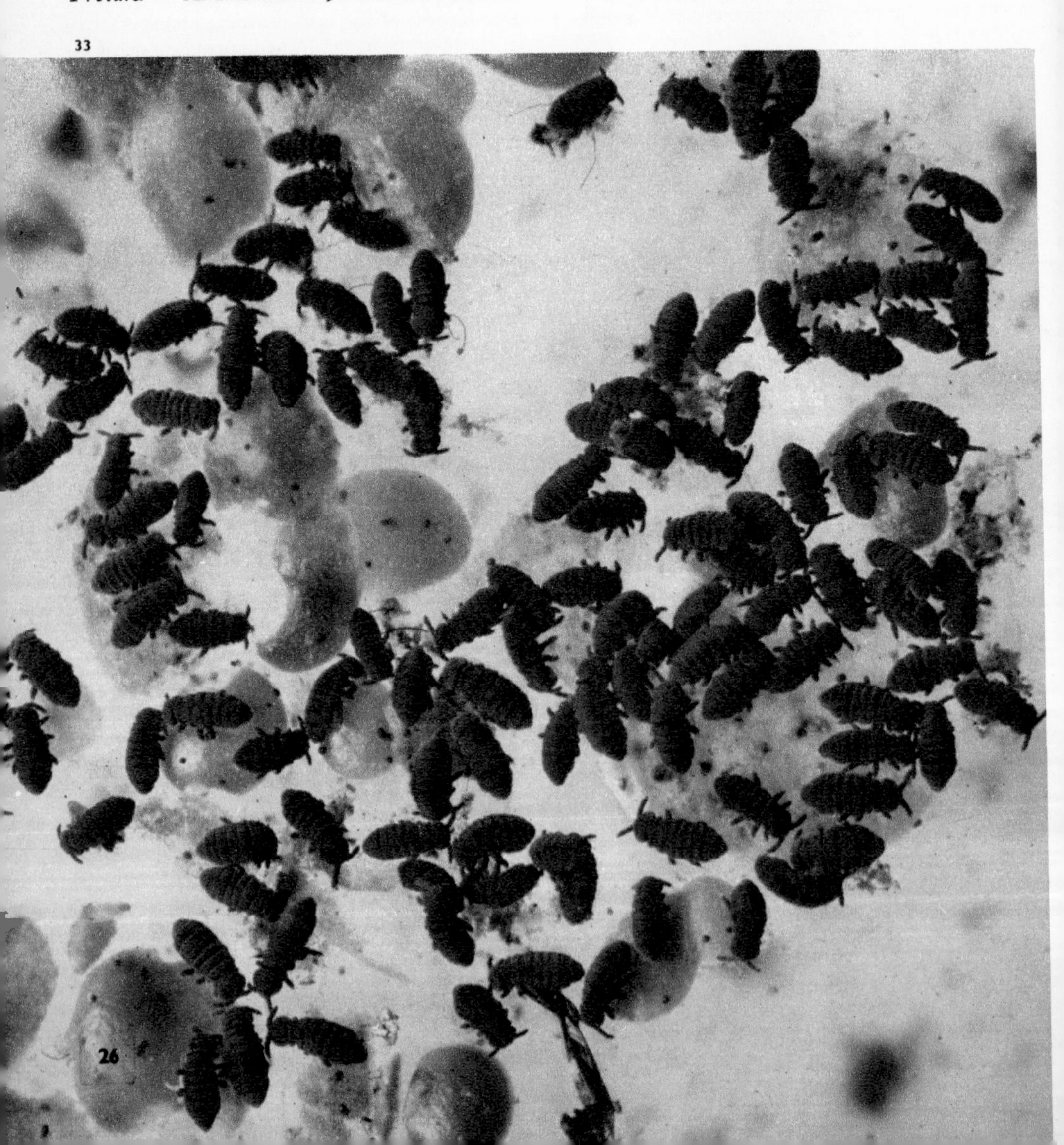

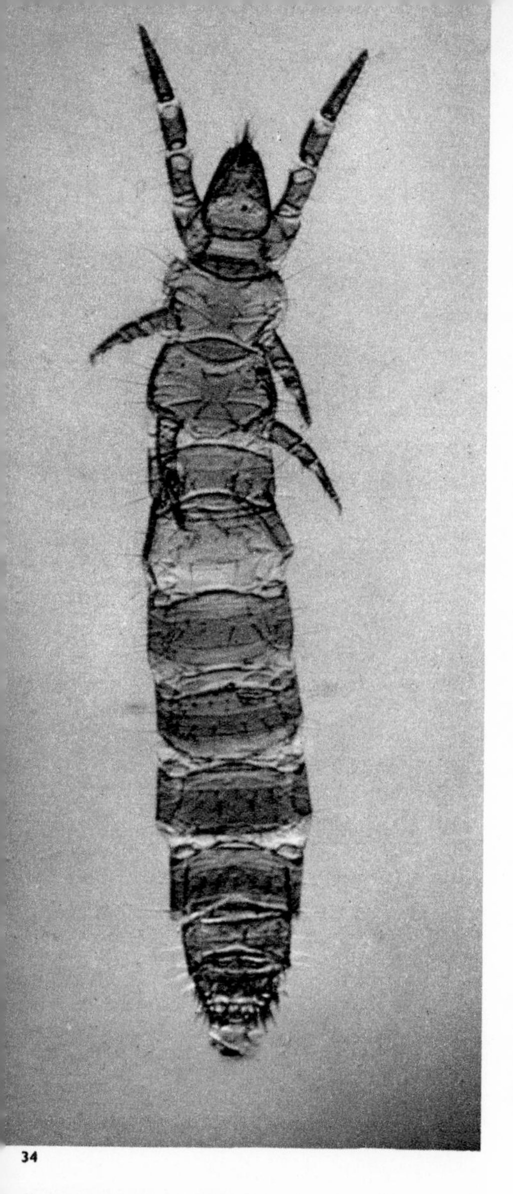

34

35

Acerentomon tuxeni [35], 1.7 to 2 mm in length, inhabits the deciduous forests of central Europe, deep in the earth, on the border between soil and rock stratum.

In the third subclass, the **Double-tails,** *Diplura,* there are about 500 species in the family *Japydidae,* distributed throughout the warmer lands of the world.

Catajapyx confusus [35] inhabits the Balkan peninsula and is spread from there up through central Europe, with southern Slovakia as its northern border. The creature measures about 7 mm and is predatory: it hunts in the earth for even tinier invertebrates. It holds its prey still with the little nippers which are attached to the end of the body.

28

The fourth sublcass, the **Bristletails,** *Thysanura,* comprises some 360 species of slim, fragile creatures which are covered with an easily stripped-off layer of small scales, usually of an inconspicuous colour. In contrast to the previous groups of primitive insects, which have simple eyes or are sightless, the bristletails possess efficient compound eyes. They are in the main small creatures—up to 1 cm in length, the biggest measuring about 2 cm. They feed on plants and have a natural life of two or three years. The suborder **Cliff** or **Coast Bristletails,** *Archaeognatha,* contains species which make their habitat in cliffs, rubble and rocky coasts, where they feed on algae as the tide goes out. In mountainous districts (up to 4,000 m) they live on lichens. At the end of their bodies they have three thread-like tail bristles, the longest of which, the middle one, is called a filament. These tail appendages resemble the forked springs of the springtails, but they are not used for jumping. If they are in flight, they jump by using their legs and increase their speed with the aid of their elastic hindquarters. **Lepismachilis notata** [36] is a species inhabiting mountains and foothills in the warmer parts of central Europe. It sometimes measures slightly more than 1 cm. Its variously coloured scales form a symmetrical, saddle-shaped design on the upper side.

In all these primitive insects, males and females can be distinguished. They reproduce through eggs. The young of the cliff bristletail go through various metamorphoses during their development. They shed their skins and, although they are overwhelmingly herbivorous, eat part of the shed skin.

In **Dilta hybernica** [37] the antennae are shorter than the body: it is slightly over 1 cm long. The scales and limbs are blue-grey. This species is extremely common in western Europe, where clearly it lived through the pleistocene glaciation. In northern and north central Europe, the most northerly areas of its distribution, only females are found, which reproduce females without fertilisation (parthenogenesis).

37

38

The second suborder of **Bristletails** are the **Zygentoma.** The **Silverfish** or **Sugar Guest,** *Lepisma saccharina* [38], is about 1 cm long. It is covered with glossy silver-coloured scales, has long antennae and three tail bristles. It reproduces by eggs, which are laid in batches of up to 20.

The silverfish is distributed throughout the world. In warmer regions it is freely found in nature; in colder zones—for example, in central Europe—it is closely associated with human habitations, where it lives a furtive life in dark, damp corners, in pantries, storerooms, wardrobes, bookshelves and libraries. It feeds on tiny bits of rubbish, dust, paper, cloth, leather, paste, flour and sugar. Although it does no actual damage, its presence is unwelcome. It whisks along at great speed on six slender legs and comes out only at night, so that very often we don't even realise that it is in the house. Its life is comparatively long. It only reaches maturity after three years, goes on to live up to eight years, during which time it sheds its skin several times. In warehouses in the tropics some species do considerable damage, while others have adapted themselves to a subterranean life in ant hills and termitaria.

Another species, the **Ovenfish** or **Firebrat,** *Thermobia domestica* [39], is somewhat larger and mottled. The body itself, without appendages, is up to 14 mm long and the tail bristles are curved outwards. It is comparatively rare and prefers to dwell in warm places such as boiler-rooms, bakehouses, hot-houses and places behind ovens, hearths and central heating.

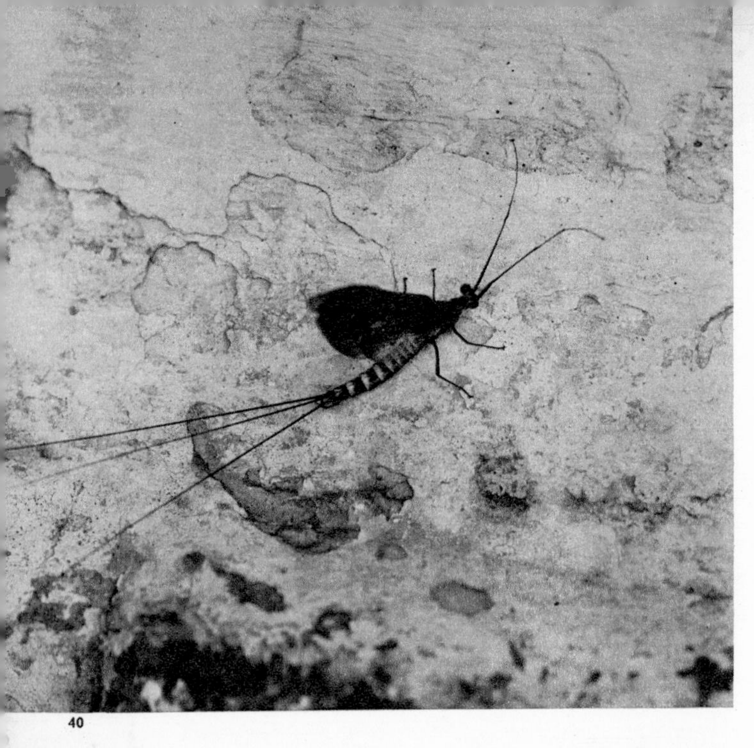

After the four subclasses of insects in which winglessness is a primary feature *Apterygota,* which are relatively poor in numbers of species, the overwhelming majority of known **winged insects** fall into the fifth subclass *Pterygota.* In this subclass belong all the insects which in the mature form are winged, as well as those which due to their mode of life (parasitism, subterranean habitat, etc.) have lost their wings as a secondary development.

The most primitive and the first superorder is that of the **Mayflies,** *Ephemeroidea.* The order of **Mayflies,** *Ephemerida* consists of small, slender

40

41

III The Crab Spider, *Thomisus albus,* lying in wait for its prey. Central Europe.

IV Some tropical Ticks:
Top left: *Amblyomma hebraeum*. Male. Central Africa.
Top right: *Rhipicephalus p. pulchellus*. Male. East Africa.
Centre left: *Amblyomma variegatum*. Male. Uganda.
Centre right: *Amblyomma splendidum*. Male. Cameroons.
Bottom left: *Hyalomma rutipes glabrum*. Male. Kenya.
Bottom right: *Rhipicephalus masseyi*. Female. East Africa.

insects with an extended body, degenerate mouthparts and two pairs of delicate, lace-like wings, which are laid side by side when at rest. On the head there are short antennae and a pair of large compound eyes. The whole digestive system is filled with air and at the end of the body there are two or three long, hair-like tail bristles, which apparently serve to aid navigation while the insect is in flight. The mayflies develop from larvae which live for one to four years in water and breathe with the aid of tracheal gills. They feed on plant and animal plankton on the bottom of streams, rivers and ponds where, depending on the species, they live either under stones or swim about freely among water plants. The larvae shed their skins about twenty times and from the last stage, called nymphs, emerge the fully-grown mayflies. On warm evenings they fly towards the light, unite into vast swarms, fly dancing into the sky and mate in flight. After this, the female lays a great quantity of eggs in the water. The life of many species in the mature state often lasts literally only a few hours or even minutes, though other species may live as long as several days. The mayfly larvae form an important fish-food. Mature insects are devoured by trout, dragon-flies, water birds, bats and other insectivores. About 600 species of mayfly are known.
Ephemera danica [40] is one of the largest central European species. The female measures up to 24 mm, has spotted wings and a characteristically coloured body, on the end of which there are three tail bristles. They can be found in central Europe in May and June on plants in the neighbourhood of water. Illustration [41] shows the larva of **Cloëon dipterum,** an equally common European species. The larva is about 2 cm long and the mature insect (the imago) approximately 1 cm. The larvae may occasionally be found living in extremely small pools or anywhere where water has gathered. In this species, too, the hind pair of wings is not developed, and so there are only two wings. On the banks of rivers in central Europe in late summer may be found in abundance **Potamanthus luteus** [42]. The female is about 13 mm long. The larvae live under stones in large rivers.

42

43

The first order of the second superorder, the **Stoneflies,** *Perloidea* is the **Stonefly** proper or **Bank-fly,** *Plecoptera.* It contains about 1,400 species so far described, from almost every part of the world, especially from the temperate zones and in particular the palae-arctic and neo-arctic. The larvae develop in fresh water. The stoneflies are small or of medium size; only a very few species achieve a length of 10 cm. The nymph of **Nemoura flexuosa** [43] measures about 2 cm and can live in stagnant water with a low oxygen content, whereas the stoneflies in general prefer swift-flowing mountain streams rich in oxygen. In this respect, they are very particular about the purity of the water. As a result of the present-day pollution of rivers they have virtually disappeared from the places where they were present in large numbers a few years ago and where they formed the staple diet of the trout.

In central Europe, the adult form [44] of *Nemoura flexuosa* emerges in the spring and summer. It is coloured brown. These stoneflies may be found by day motionless on plants and stones in the neighbourhood of water. They are nocturnal insects.

45

46

In streams and rivers in mountainous and hilly areas, the larvae of **Chloroperla torrentium** [46] may be found in the spring. The adult insects [45] are grey-green and about 1 cm long. They emerge in May, June and July, are very swift running as well as flying and are distributed throughout central Europe, wherever the running water of streams and rivers has not been too heavily polluted.

Perla burmeisteriana [47] is one of the largest central European species of bank-fly, being almost 3 cm long. The larva [48] lives under stones in clear streams. It leads a predatory existence and feeds chiefly on mayfly larvae and other water insects. The view has been propounded that the fully-grown stonefly needs very little nourishment, some species none at all. On the other hand, it has been established that they do drink water. *Perla burmeisteriana* is much prized by anglers as bait for trout. The stoneflies are very primitive insects, which have changed very little since their first occurrence in the Permian.

47

48

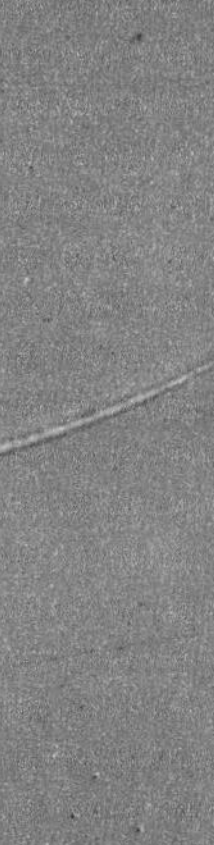

49

50

The third superorder, the **Dragonflies** and **Demoiselle-** or **Damsel-flies,** *Libelluloidea* contains the **Dragonflies,** *Odonata,* about 3,600 species of for the most part larger insects, which are distributed mainly in tropical regions.

As far as their history and development are concerned they are extremely primitive insects. The earliest fossil forms of dragonflies are found in the Permian and many of them, especially those from the Jurassic, closely resemble recent forms.

51

52

The suborder **Lesser Dragonflies** or **Damselflies,** *Zygoptera,* contains two central European lesser dragonflies, brilliant, dark-blue creatures which in the summer flit to and fro over the overgrown banks of running water, hunting their prey. The females of the two species are indistinguishable, and the males are distinguished by the colouring of their wings. The **Banded Agrion,** *Agrion splendens* [49], has expressive blue bands on its wings. The wings of the male **Lake Demoiselle-fly** or **Demoiselle Agrion,** *Agrion virgo* [50], are blue over almost their entire bodies; only the ends are lighter. The former lives in Europe, in the nearer parts of Asia and North Africa; the latter is even more widely distributed and occurs very far north. Both species are found high in the mountains.

The nymphs of the genus *Lestes* [51] are about 35 mm long and possess three tapering caudal gills at the end of their bodies.

The nymphs of the genus *Agrion* [52] have, on

the other hand, rounded caudal gills. Illustration [53] shows the larvae of the **Common Coenagrion Demoiselle-fly,** *Coenagrion puellum* [53], which is about 25 mm long. In spring and summer the adult insect flies to and fro above pools, places where water has collected and reedy ponds. Its flight is curiously uncertain, often settling on plants or stones which rise above the surface of the water. The range of distribution of this species stretches over Europe and Asia, extending southwards as far as North Africa.

53

All members of the suborder *Zygoptera* fold their wings together when at rest, more or less parallel to the fore-and-aft axis of the body; some fold them perpendicularly above their bodies, like the butterflies. The larvae of the slender dragonfly are generally green with irregular-shaped bands on the legs and tail appendages. This colouring makes them almost invisible as they hunt their prey among the water plants. They spot their victims, small water insects, from a considerable distance with their large eyes. They then shoot out their mask (the extended lower lip) and seize it with strong, claw-like hooks. With astonishing speed the prey is conveyed into the mouth and broken down by powerful jaws. The larvae of the various species of the genus **Agrion** [52, 55] differ from those of the genus **Lestes** [51] in that they have a shorter mask. In addition, a distinction can be observed in the arrangement of the branchial basket in the tracheal caudal gills. In the genus *Lestes* the tracheae branch out almost perpendicularly from the central axis of the gills; in *Agrion* they run obliquely towards the end and divide into extremely delicate bush-like ramifications.
Illustration [54] shows the imago of the **Scarce Green Lestes,** *Lestes dryas.* It may be seen in late summer hunting its prey over still water and, indeed, often in places far removed from any water. It inhabits many parts of

54

55

56

57

central Europe, temperate areas in Asia and North America.

The suborder **Great Dragonflies,** *Anisoptera,* comprises larger, stronger insects and the most accomplished fliers it is possible to conceive. In the rest position they spread their wings out flat. Only during emergence from the nymph and for a short time afterwards are they folded together. Then the insect spreads them out, waits until they become hard and swings itself into the air. The maximum wingspan achieved by present-day *Anisoptera* is about 10 cm; however, earlier species now extinct, which lived in the Carboniferous and Permian, achieved giant proportions. **Meganeura monyi** is known to have had a span of 70 cm, being the largest known insect which has ever existed on earth.

After a life of four years under water, during which the larva of the **Common Aeschna,** *Aeschna juncea* [56], hunts, grows and passes through several instars, the moment comes at the beginning of summer for the nymph to cease to take food and leave the water. The skin splits down the middle of the back and the whitish, temporarily shapeless body of the future demoiselle-fly appears. It writhes its way out in a struggle lasting almost two hours [57]. The body gains shape and colour, the chitin hardens gradually in the air and finally sets. Not until the next day is this demoiselle-fly ready for its whirring flight [58].

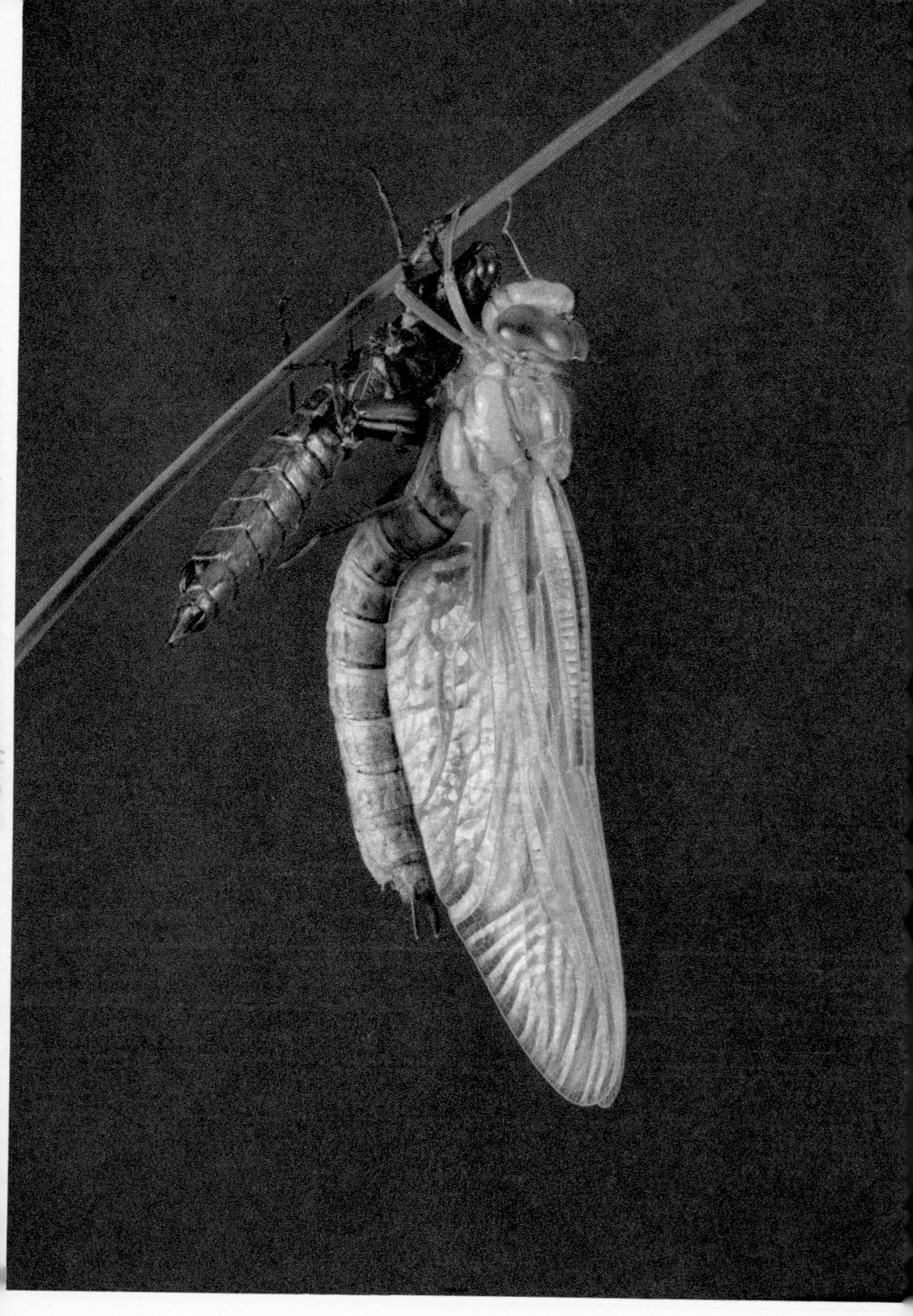

59

The **Common Aeschna,** *Aeschna juncea* [59], grows to a length of up to 8 cm, with a wingspan of 10 cm. Whereas the zygopteran dragonflies only live for one or two weeks, the anisopterans can hunt over forests, clearings, paths, bogs, pools, ponds, streams and rivers for four months, if they can avoid falling prey to som enemy. It is impossible to conceive a summe day without them.

Even larger is the **Emperor Dragonfly,** *Ana imperator* [60]. Its span is up to 11 cm and is rarer than *Aeschna.* Its range of distributio

60

extends over North Africa, Mediterranean area and the warmer parts of central Europe and central Asia. It only needs one year to complete its development. The nymphs leave the water during the night and climb high up into the tops of trees, where the imago emerges on the following morning.

Another attractive dragonfly, which is still present in comparatively large numbers in various places, is the **Broad Bodied Libellula,** *Libellula depressa*. It emerges as early as the beginning of May and has a span of some 8 cm.

61

62

It has a strikingly flattened and broadened body; the colouring of each sex is totally different. The females [61, 62] are honey-brown; the males have hind quarters with an upper covering of grey-blue. These dragonflies can be seen sitting in large numbers on the surface of small ponds, where they lie in wait for their prey. Like all dragonflies they catch their prey, which mostly consists of smaller flying insects, on the wing. They seize their victims securely with the aid of their front and middle legs, which are set with spikes. The larvae of the broad bodied libellula [63] lives for two years in dried-up ponds and grows to a length of about 25 mm. In contrast to the larvae of the anisopterans, which have long, thin bodies, the bodies of these and certain other genera of dragonflies are shorter and often greatly flattened. They usually have hairs and in addition are frequently covered with algae, so that under water they are often very well camouflaged. Dragonfly larvae breathe with the aid of their rectal canal, which is equipped with a number of little lamellae, serving as gills. They take in a quantity of water which can then be expelled with some force, developing a powerful thrust, enabling the creature to move out of danger with great speed. The larvae of the larger dragonflies feed in the water on weaker members of their own kind, as well as on the larvae of other insects, worms, small vertebrates, tadpoles and fish fry. For

64

this last, however, they make full reparation, since they themselves are a particular delicacy for large fish.

The **Foot-spinners,** *Embioidea,* form the fourth independent superorder in the class *Insecta,* together with the related order of **Embioids,** *Embioidia.* It is not particularly numerous, containing less than 100 species. The foot-spinners are frail little, warmth-loving creatures with cylindrical bodies and a thickened tarsus on the front pair of legs. With the aid of the spinning bristle of these legs they prepare their little tunnels—tiny, silk-like tubes, in which they live, associating together in considerable numbers. Most species are natives of the tropics and live a secret life on the ground under heaps of vegetable remains, much like earwigs. If they increase to sufficient numbers they can cause considerable damage in places where food is stored. The females and the larvae are wingless. The female of **Haploembia grassi** [64], from southern Bulgaria, is about 1 cm long.

Of the fifth superorder, the **Straight-winged Insects,** *Orthopteroidea,* the largest order is that of the **Grasshoppers,** *Saltatoria.* The **Hothouse Grasshopper,** *Tachycines asynamorus* [65], belongs to the superfamily the **Crickets,** *Gryllacridoidae.* This wingless insect is about 2 cm long, with long antennae and strong hind legs for jumping. Unlike most crickets, it does not chirp. Although it was originally a native of the tropics, it can nowadays be found in hot-houses throughout the

Va Waisted Scolopender, *Scolopendra morsitans*. West Africa. 12½ in. long.

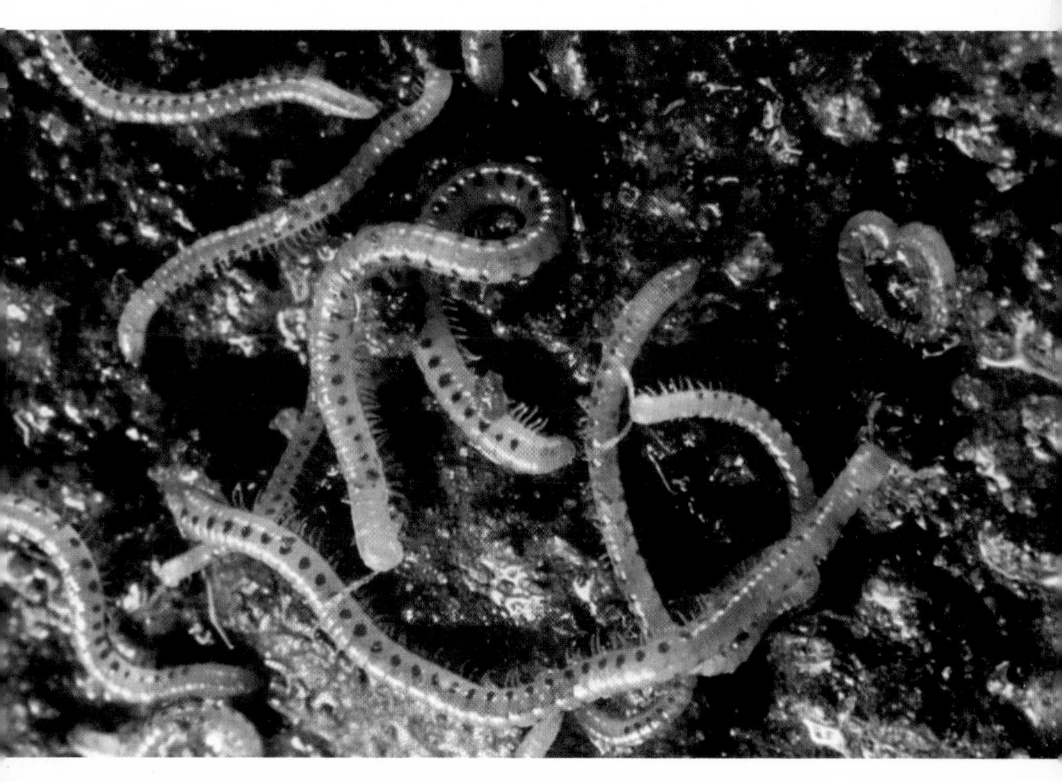

Vb Spotted Millipede, *Blaniulus guttulatus*. Central Europe. 1 cm. long.

VI Common Aeschna, *Aeschna juncea*. Imago emerging from the nymph. Central Europe.

65

world. It hides timidly by day in dark corners, coming out at night to hunt greenfly and other insects. It has, incidentally, been established that it eagerly drinks milk. By nibbling at young seedlings, it effects great damage in greenhouses. Our illustration shows a female with her ovipositor.

In southern Europe and western Asia lives a large, wingless, bronze-coloured cricket, which measures about 8 cm: **Bradyporus dasypus** [66, female]. In dry, barren places it moves slowly, feeding on desert plants and animal remains. Although it has a ferocious appearance, it is in fact a harmless insect.

66

67

68

In the stony deserts of North Africa, the interesting species **Eugaster guyoni** [67] lives a similar type of existence. It belongs to the same family as *Bradyporus,* the *Ephippigeridae.* It is an unforgettable sight, when the desert cools quickly in the evening at sunset after the heat of the day and hundreds of big, cricket-like creatures, about 4 cm long, crawl out from their hiding places. They move slowly and we can watch how they consume low plant growth. They are black-brown, but the growths and spurs on the shield are bright red. They defend themselves by squirting a liquid from openings along the side of their body.

One family in the suborder **Long-horned Grasshoppers,** *Ensifera,* is that of the **Leaf Grasshoppers,** *Tettigonoidea.* The green colouring and the venation of their wings give them an extraordinarily good camouflage, so that they can become completely invisible in leafy surroundings. Throughout the tropical zone we encounter a vast number of species in which this mimesis goes so far that we never cease to be amazed. Hundreds of medium-sized and large species cannot be distinguished from a leaf when they are quite still. The family *Pseudophyllidae* [68] is particularly well known for this phenomenon.

The creature has a span of 8 cm and simulates a dry leaf with its yellow-brown colour flecked with red. However, if it is disturbed into flight the luminous red-blue "eyes" on the rear pair of wings give a gruesome effect.

The species **Deroplatys trigonodera** [69], which is green and about the size of a walnut, resembles the burst-open fruit capsule of certain plants. It also affords a convenient example of the complete phytomimesis of a large number of insects. The same is also true of **Onomarchus cretaceus** [70, female], from Assam, which simulates either a greenish-yellow shell or a long leaf. In the tangles of boughs and in thickets there are countless numbers of them.

The family *Phylloporidae* lives in New Guinea and is rich in species. The largest is **Phyllopora grandis (speciosa).** It is green and bears over the thorax a large, angular, chitinous hood, with a projection over the wings [71, female]. The males

69

70

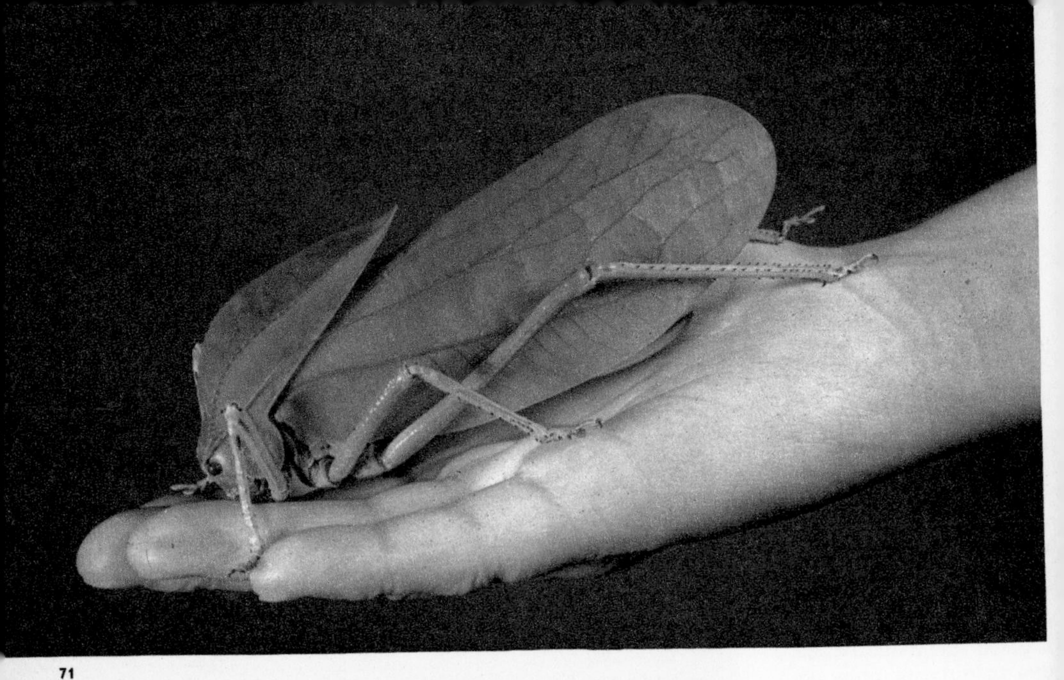

71

72

have no stridulatory organs; in other words this is a species which is unable to chirrup. The complete length, excluding the antennae, is usually more than 13 cm. *Phyllopora grandis* is therefore one of the giants among insects, and is the wildest ambition of most collectors to possess. As well as these, the life of this huge grasshopper is threatened by many other enemies: apes, lizards and other insect-eaters. The natives, too, regard them as a tremendous delicacy.

The auditory organs of the long-horned grasshoppers are situated on the tibia or shin of the fore legs. They consist of small depressions equipped with a skin or typanum full of nerve-endings, which are sensitive to sound waves and transmit them to the brain through the auditory nerve. Illustration [72] shows this organ from the front.

73

Throughout the whole of Europe, in North Africa and as far east as Amur is the range of the famous **Great Green Long-horned Grasshopper** or **Katydid,** *Tettigonia viridissima*. It is up to 3.5 cm long and coloured brilliant green with brownish or sometimes reddish wings. The female [73] possesses a long, sabre-like ovipositor. In the early stages of its life this grasshopper feeds on plants; the fully-grown insect, however, hunts for smaller insects, especially on the continent of Europe for the larvae of the potato beetle. They can be found among corn, in bushes and hedgerows, as well as on trees. Its unmistakable chirruping at night and its splendid flights of up to 100 metres are among the most pleasant features of summer.

The nymphs of the grasshoppers differ from

74

75

the adult insects chiefly in that their wings are not yet developed, and the positions of the wings seem to be reversed. Our picture [74] shows a male nymph shortly before the last time it sheds its skin and develops into an adult insect. The male long-horned grasshopper produces its chirp by rubbing a grooved vein of the left wing-cover over a rough-edged area on the right, thus causing it to vibrate [75].

In the higher plateaus of Europe and western

76

Asia the order in question is represented by the **Chirping Grasshopper,** *Tettigonia cantans* [76, male]. It measures about 2.8 cm. Its short wings are not even as high as the thighs of its hind legs. It lives in the lowlands of the north and can often be found near the earth, especially in potato fields and on tall wild herbs and grasses where, however, it is difficult to discern. Its chirping is interrupted by frequent pauses.

77

The **Steppe Saddle-backed Grasshopper,** *Ephippigera ephippiger* [77, female], inhabits southern Europe and the warmer plains of central Europe. It measures up to 3 cm; the top side of the thorax is arched into a distinct saddle-shape and the wings are markedly degenerate. Males, as well as females, are equipped with stridulators and can chirp. This grasshopper is a relic of the warmer post-glacial period.

Illustration [78] shows the male of the familiar **Field Cricket,** *Liogryllus campestris;* it is up

78

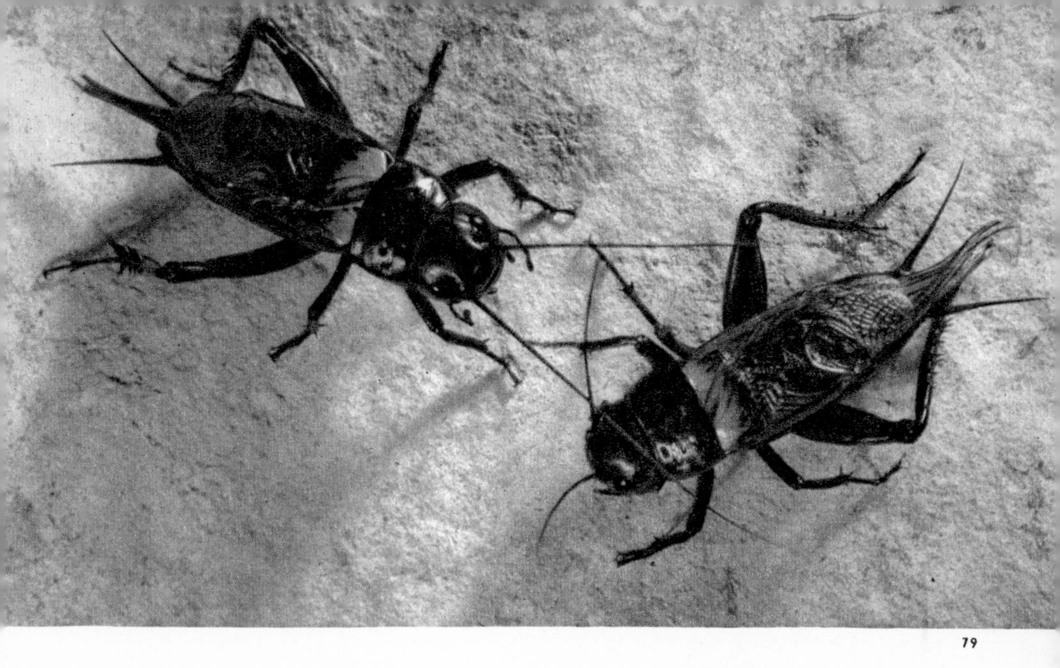

79

ɔ 2.6 cm long and coloured brilliant black vith reddish legs and yellow-brown wing-:overs. The field cricket is an inhabitant of entral and southern Europe, western Asia and ıorthern Africa. Its soft chirruping is heard inging out during May and June, principally on sandy, warm hillsides, on the fringes of woods, on dry slopes, meadows and in sunny forest clearings. This shy creature buries itself in a tunnel up to 30 cm long in the earth. It feeds on a plant and vegetable diet, but could not be described as a pest.

80

81

Gryllus bimaculatus, on the other hand, does great damage to the agricultural economy of many places. It is somewhat slimmer and weaker, with a conspicuous spot on each wing-cover. It is distributed throughout the whole of southern Europe, the warmer parts of Asia as far as India and China and all over West Africa. Its stridulation is comparatively loud [79, two males; 80, larva].

The **House** or **Hearth Cricket,** *Gryllus domesticus,* possesses a slender body up to 2 cm long, coloured dark yellow with a darker pattern. It has accustomed itself to live in close proximity to man, stays close to his habitations and loves warmth. It prefers dark corners at the backs of fireplaces and chimneys, warm cellars, stables and boiler houses, where it can be assured of a more or less constant temperature. It feeds on vegetable and animal waste, leads a secret, nocturnal life and betrays its presence only through its night-time concert. The house cricket is found scattered through central Europe, being more common in southern Europe. It has been imported into North America. In North Africa it occurs naturally outside human settlements and sometimes lives communally with cockroaches. The concert of the males is pleasant and comparatively weak. The females of all crickets are equipped with a long ovipositor, with the aid of which they deposit elongated eggs in spongy and damp substrata. The best place for their culture is damp earth. Picture [81] shows a female in the process of laying her eggs. The house cricket has lost its ability to fly, but is nevertheless very nimble, well able to leap some 25 cm. It forms an excellent meal for caged reptiles, insectivorous birds and mammals. Predatory insects also hunt crickets. They enjoy great popularity as laboratory animals.

The **Mole-cricket,** *Gryllotalpa vulgaris,* is one of the largest European crickets. There are 45 known species with world-wide distribution; this one, however, is restricted to Europe, western Asia and North Africa. It is completely adapted to its underground way of life, digging and burrowing. The front pair of legs have been transformed into a pair of powerful shovels, resembling the front legs of a mole. The eggs, 200 to 300 of them, are laid in an underground breeding chamber or nest, and from them the larvae grow, needing 2 to 2½ years to complete their development. In autumn they transform themselves into the adult insects, which hibernate and then mate in the following spring. At evening and during the night the male gives forth its whirring stridulation. On warm evenings they fly cumbersomely about, searching for females. Their large, transparent wings, constantly outstretched, horizontal and finely fluttering, carry the comparatively heavy body, with the hind quarters hanging perpendicular to the earth, very slowly and low over the earth. The life of the adult mole-cricket only lasts a year. It dies after it has made provisions for the continuation of the species. Picture [82] shows the larva just before the last moult; picture [83] gives a side-view of the fully developed insect. The mole-cricket measures up to 5 cm, feeds principally on wireworms, cockchafer grubs and other larvae in the earth, as well as pursuing beetles and earthworms. It does some damage to crops, since it undermines the roots of young plants in much the same way as does the mole, except that it also gnaws them off. The mole is as inexorable an enemy to this insect, as is the hoopoe. I once had under observation a pair of hoopoes in the nest which fed their young exclusively on mole-crickets.

82

83

84

85

The suborder **Short-horned Grasshoppers,** *Caelifera,* with the superfamily **Field Grasshoppers,** *Acridioidea,* numbers considerably more than 10,000 species—that is, the vast majority of all known *Orthoptera.* The short-horned grasshoppers can be easily recognised by their short antennae and the short ovipositor, which cannot normally be seen. They stridulate by rubbing veins located on the inside thigh of the hind legs over the surface of the wing-covers. The hearing organs are at the sides of the first body segment of the hind quarters. They are voracious plant-eaters. Some species unite and fly in huge swarms **(locusts).** Where they descend the ground is covered foot-deep with insects which consume everything green within an extraordinarily short time. This suborder of the grasshoppers is distributed throughout the world, but is especially common in the drier parts of the tropical and temperate zones. It is coloured a nondescript brown, green or yellow, usually with a symmetrical pattern. The rear wings, however, often have an impressive, lively colouring which almost certainly is designed to have an intimidating effect on its enemies.

One member of the family *Oedipodida* is the attractive **Blue-winged Waste-land Grasshopper,** *Oedipoda coerulscens* [84]. The fore wings or wing-covers are brown-grey and decorated with two black stripes or fascia; the rear wings are usually coloured light-blue for up to half their area, occasionally misted over with yellow or pink. The waste-land grasshoppers measure up to 2.8 cm; the males are smaller than the females. We can find them towards the end of summer in the warmer parts of central Europe, in southern Europe, North Africa and eastwards into south-west Asia; they inhabit heaths, dry hillsides, quarries and sandpits and are completely harmless.

In the more mountainous regions of central Europe live two species with beautiful red wings: the **Red-winged Waste-land Grasshopper,** *Oedipoda germanica,* and the **Red-winged Scraping Grasshopper,** *Psophus strididulus.* To the same family belongs also the notorious **Locust,** *Locusta migratoria* [86], which we have already mentioned. It is on average about 5.5 cm long and lives in several species and two ecological phases in south-west Asia, Africa and southern Europe. The "*Solitaria*" phase develops among reeds and in

86

87

88

marshes. The "*Gregaria*" phase consists of the swarming together and migration of thousands of individuals. The earliest literatures carry accounts of the plague of swarms of locusts attacking the countryside. These attacks continued in Europe up to the end of the 19th century. Indeed, occasionally —though certainly not very often — smaller swarms still advance into central Europe from the East at the present day.

The handsome **Desert Locust,** *Schistocerca gregaria* [86, 87], belongs to the family *Catantopidae*. It has wings over 6 cm long and inhabits the warmer parts of Asia and central Africa where, like the true locusts, it forms huge swarms. Occasionally it reaches Europe and even England. There are other species of grasshoppers elsewhere in the world—such as in South America —which have a similar life pattern. Locusts fly at a speed of about 18 k.p.h. (11¼ m.p.h.). Their breeding places are chiefly in river deltas; they multiply rapidly till there are millions of them and, once the winged adult insects have emerged, they form a huge, rustling swarm, like some great cloud which even darkens the sun. Wherever they descend they spread terror and dismay and bring famine and suffering to the land.

The short-horned grasshoppers are good climbers, incredible jumpers and have tremendous endurance as fliers. Their specially adapted, rear legs, which are equipped with

89

ttle thorns, serve them well in penetrating ndergrowth, moving on the ground and in scaping from their enemies. These legs [88] ave strong muscular "thighs" and long, traight "shins", which are so excellently quipped for jumping that the insect can often over more than two hundred times its own ength in a single hop.

n interesting species which lives in the armest parts of central Europe, in southern nd western Europe and through central Asia far as western China is the **Towered** or **osed Grasshopper,** *Acrida hungarica* [89]. It measures about 7.5 cm, has a remarkable oblique, elongated head and long, flattened antennae. Its colour is green, brown or reddish, with various patterns. This charming insect prefers damp meadows and plains with bushes or long grass. It can fly very well, but does no damage. Other species of this genus, however, from tropical Asia, like the related South American genus *Truxalis,* are greatly feared in agricultural areas.

The family **Grouse-locusts,** *Tetrigidae,* contains many species whose habitat is chiefly in the tropics and subtropics. They only measure

at the most 1 or 2 cm. These dwarf locusts have an angular, compact body, a pair of powerful hind legs for jumping and short antennae. The wings generally extend beyond the length of the body. Many of the tropical species look very bizarre. They often have thorns on their bodies and can therefore easily be confused with some plant seeds. The breast parts are extended, flattened into a leaf shape, or ends with remarkable projecting horns. The central European species have a less extraordinary appearance. Nevertheless, in these the top of the breast parts is extended back to form a horn-like cover for the hind quarters. The wings are either atrophied or of a normal form.

The **Two-pointed Grouse-locust,** *Tetrix bipunctata* [90], measures about 1 cm; its rear wings are two or three times as long as the front

90

91

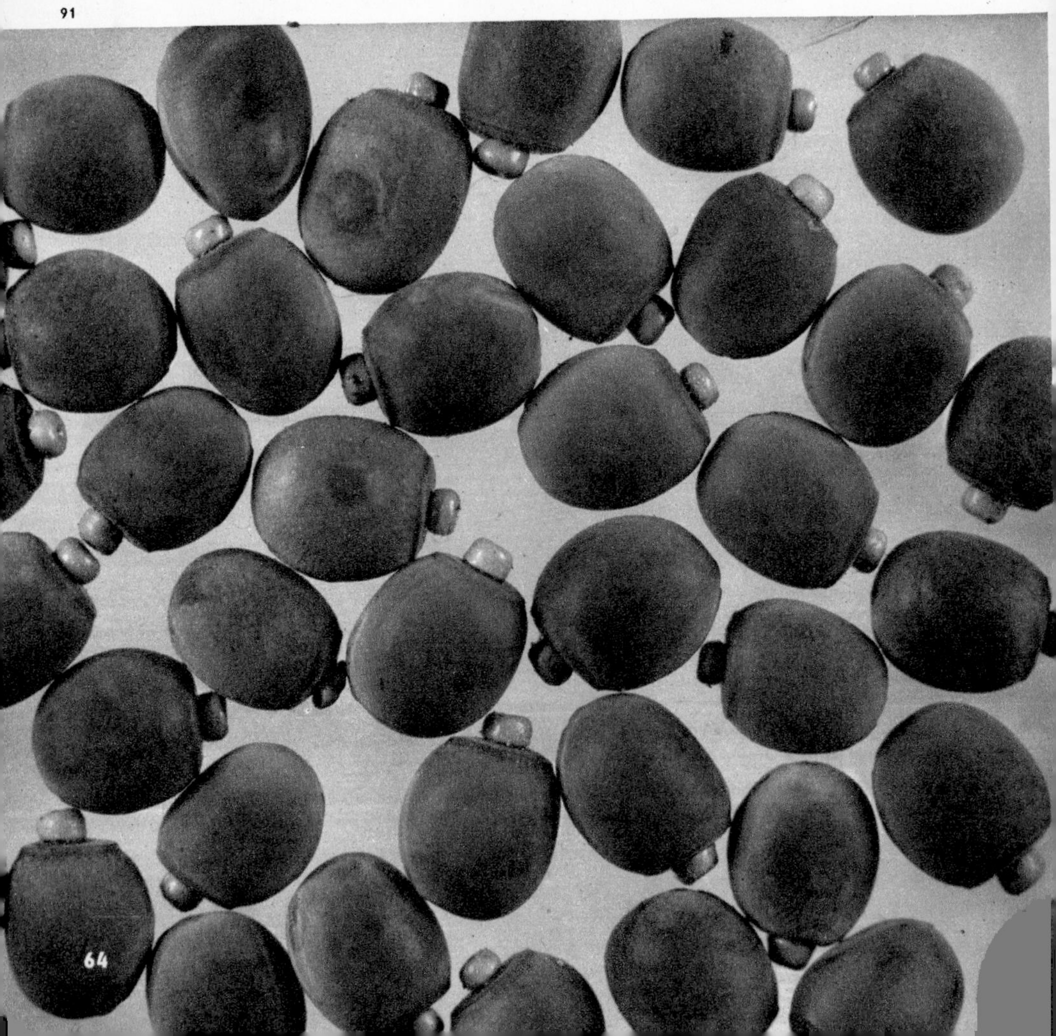

VII Common Aeschna, *Aeschna juncea*. About 15 minutes after emerging.

VIIIa Larva of the African Praying Mantid, *Sphodromantis lineola*. West Africa.

VIIIb *Sphodromantis lineola*. Adult female eating an insect.

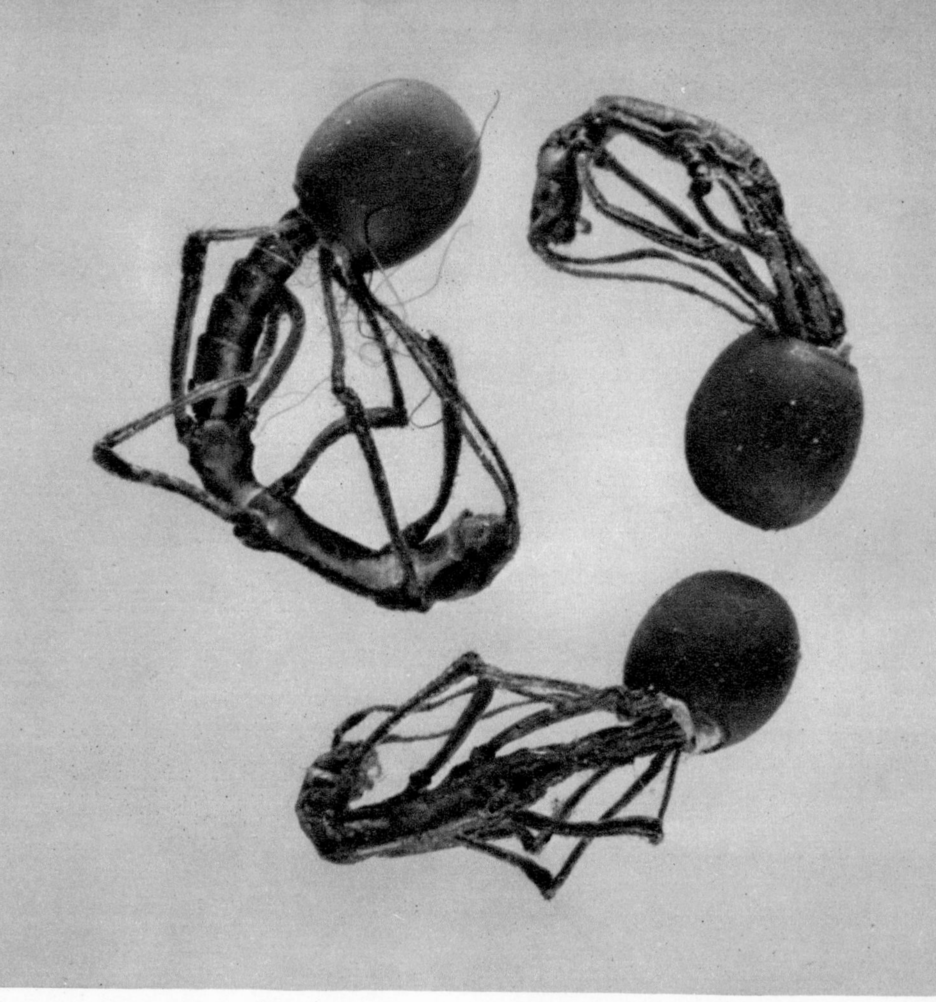

92

93

ving-cover. This species is found in most parts of central and southern Europe, in Asia and as far east as China. Its range of distribution stretches as far north as the Siberian Taiga. In the temperate zone this grouse-locust hibernates as an imago and appears early the following spring. It is a relic of the ice-age, and lives in forests as well as on the banks of rivers and streams. Sometimes it can also be found in dry and warm places, in areas not yet disturbed by cultivation.

The second order of orthoperoids, the **Stick and Leaf-insects,** *Phasmida,* include about ,000 species so far described, in some 300 families. Bushes, forest eaves and clearings with the luxuriant vegetation of the tropics, thorny shrubs in the savannah and bushy thickets in subtropical regions form the chief dwelling

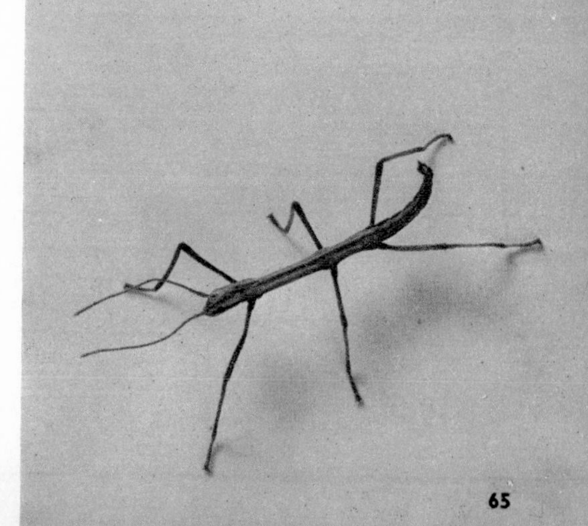

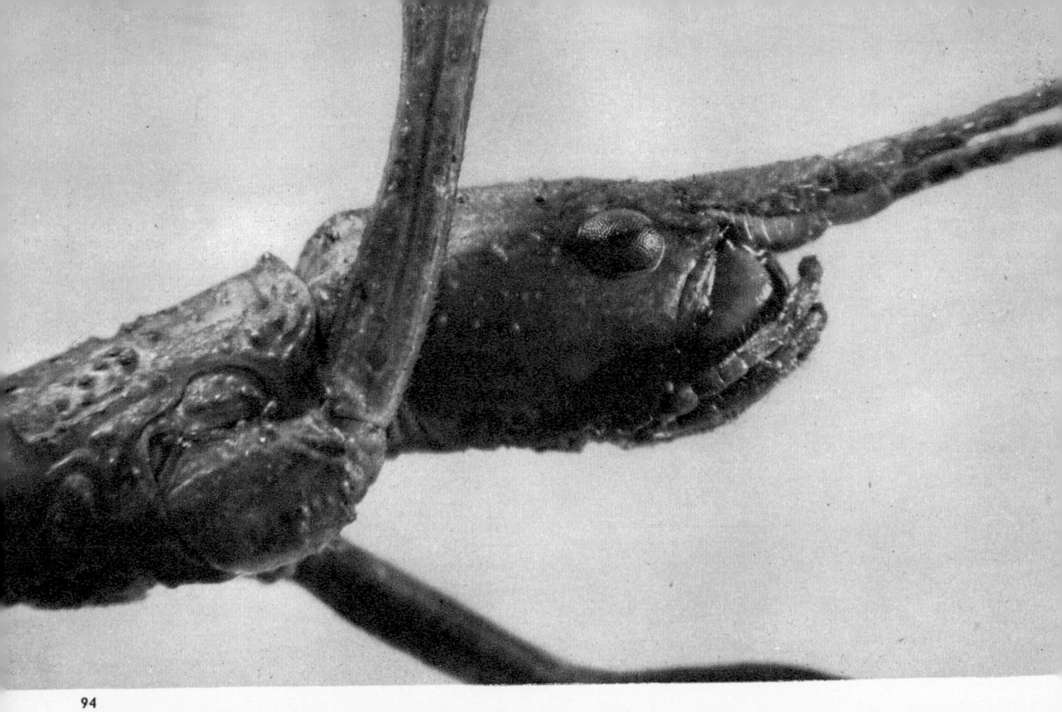

94

95

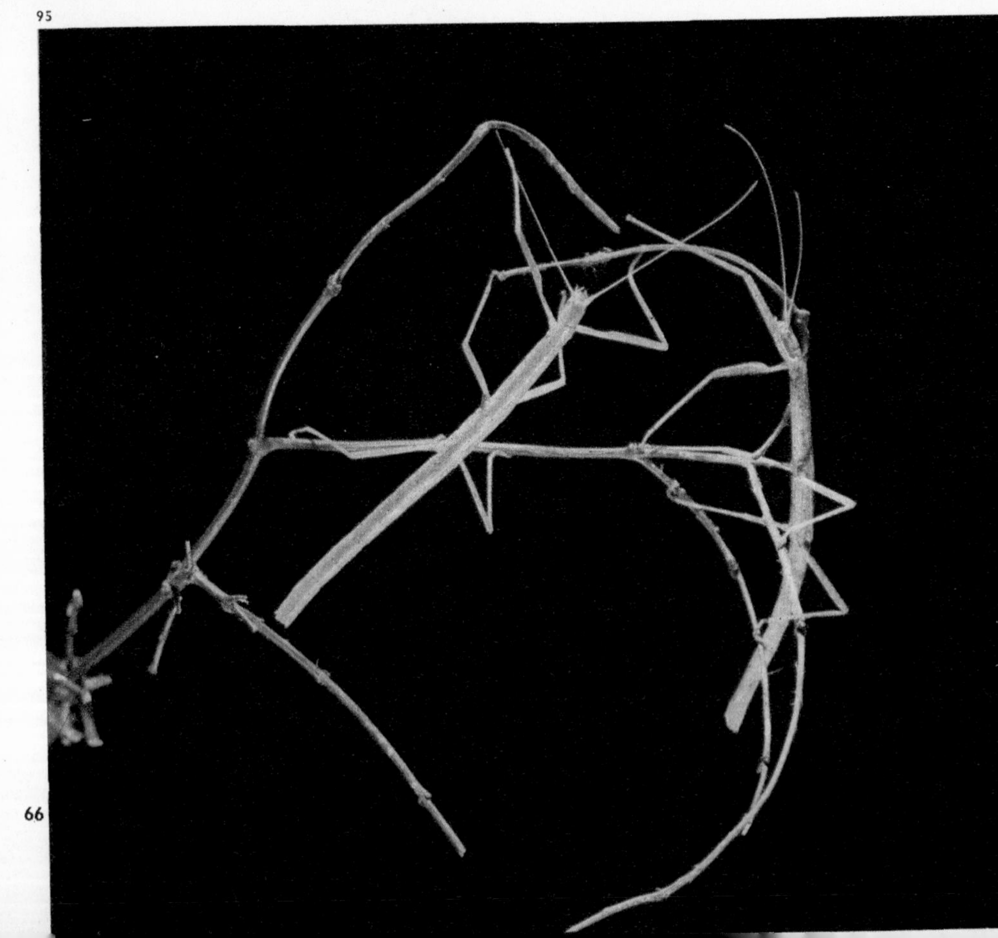

places of these remarkable plant-eating insects. They are distinguished by an astonishing phytomimesis, since their bodies take the form of a stick or leaf so completely that only a movement can give them away. Their behaviour completes the effect of their appearance: by day they usually take up a motionless, rigid position. One of these species is often kept in captivity and has become a laboratory animal, as it reproduces parthenogenetically. The **Common Stick-insect,** *Carausius morosus,* of the family *Lonchodidae* reproduces in captivity without difficulty. From the eggs [91] which are about 2 mm in size emerge, after a few months, nimble larvae [92, 93] about 1.5 cm long.

The fully grown stick-insect [95] is by contrast extremely reluctant to move, but if it is disturbed it is capable of great speed in escaping. It can be accurately identified by this characteristic, for it is as good as invisible otherwise, with its long, thin, motionless, stick-like body, the front legs stretched forward parallel to the axis of the body. However, it observes carefully with its large compound eyes [94] everything that is going on around it. The common stick-insect is a native of India; it is flightless and in the adult stage measures about 8.5 cm. In many laboratories only females are kept, which generally lay eggs, from which only females develop, without fertilisation—that is, parthenogenetically. Only very rarely—the proportion is something like 1,000 to 1—are males found. It has been calculated that this phenomenon accounts for wenty generations of virgin birth compared vith one sexual reproduction. The common stick-insect can be fed on various leaves, such as privet leaves, rose leaves and euonymus eaves. In winter, it will also take ivy leaves.

Some of the tropical species of stick-insects each considerable proportions. They are as hick as a finger and measure 20 to 25 or even s much as 30 cm. **Phoboeticus fruhstorferi,** vhich is a member of the same family as the ommon stick-insect, is certainly the longest resent-day insect, with a length of 30 cm.

n illustration [96] we show a female **Cyphorania gigas,** which is 17.5 cm long and elongs to the family *Phasmidae.* The males re smaller. They live in parts of Malaya.

96

97

Some species of stick-insects are wingless; others, on the other hand, have well-developed wings with which they are able to emit a rattling sound in flight. These wings serve to glide from tree to tree, but act more as a parachute than anything else. Our illustration [97] shows **Vetilia spec.,** which comes from New Guinea and achieves a length of 20 cm.

The bodies of some stick-insects are covered with thorn-like growths. **Obrimus asperrimus** comes from the family *Bacillidae* [98]. The male is 5.5 cm long. This species lives in Borneo (Kalimantan). The head, breast and legs are all covered with these thorns. **Euryacantha horrida** [99], from New Guinea, has thorns on the sides of the hind quarters and the rear limbs. It is 13 cm long.
Some insects in the order **leaf-insects** bear an astonishing resemblance to leaves, in both

98

99

form and colouring. Most of these belong to the family *Phyllidae,* which have greatly flattened and broadened bodies. A classic example of complete phytomimesis is the 8.5 cm long **Moving leaf,** *Phyllium siccifolium* [100], from Ceylon. The rear wings are atrophied and the wing-covers have developed a perfect resemblance to a leaf. The legs have broadened and taken on the shape of parts of leaves.

The body of **Extatosoma tiaratum** [101] from Australia and New Guinea looks as if it has been put together from broken bits of thorny twigs and dry leaves. It is about 12.5 cm long.

The third order, the **Earwigs,** *Dermaptera,* contains about 900 species, distributed throughout the hot and tropical zones. Some tropical species are up to 6 cm long and the smallest are only a few millimetres. They have an elongated, cylindrical body, a hard, shiny chitin exoskeleton and short, thread-like antennae. The mouthparts are adapted for biting as well as for sucking up liquid nourishment. The

100

101

102

103

rear wings are laid together in three folds and protected by short, scaly wing-covers. At the end of the body is a forceps composed of the cerci, which is used as a tool, and helps in the folding and unfolding of the wings. Does the earwig crawl into the human ear? No, it is a completely harmless creature, which raises and waves its dangerous-looking rear end whenever it is disturbed, but is quite unable to pierce the skin with it, nor even to deliver a scratch.

The male **Common Earwig,** *Forficula auricularia,* is usually larger than the female and measures up to 16 mm [102]. The females have also weaker and less strongly curved forceps [103]. The earwig is a nocturnal creature, feeding on greenfly and other small arthropods, but which will readily devour fallen fruit, the sweet sap of flowers and other plant matter. It is cosmopolitan, having been introduced into America, Australia and Africa from Europe, and leads a concealed existence in gardens, courtyards and fields as a mostly unseen companion of man.

A similar, but wingless species is the **Forest Earwig,** *Chelidurella acanthopygia* [104]. It is somewhat smaller, the male measuring up to 15 mm. It is found behind the bark of trees in forests in central Europe, in the wood itself and tree stumps. It hibernates as an imago.

104

In the sixth superorder, the **Cockroaches** and **Mantids,** *Blattoidea,* we will take as the first order the **Mantids,** *Mantodea.* Most of the 1,800 or so known species are tropical, the remainder live in all parts of the subtropical region. There are just a few species in the warmest parts of the temperate zone, as for example in central Europe the **Praying Mantid,** *Mantis religiosa* [105, 107, female]. In Germany it is found in the upper Rhine area, but its principal area of distribution is throughout the Mediterranean countries and Africa and Asia. It has been imported and has naturalised itself in North America. In view of its extremely mobile head, its raptorial front legs and the intense concentration with which it surveys its surroundings, this is an extremely interesting insect. It scares away enemies by raising its wings and making a noise like the hiss of a snake, by rubbing the rear part of its body. The fact that the female often eats her partner during or after copulation is no less curious. Only the males are able to fly; the females are usually filled to bursting with eggs throughout the summer. These are laid in a case (ootheca) [106] made of a frothy secretion which hardens to a firm, spongy substance, and they pass the winter glued to places where the praying mantid lives—on dry, warm hillsides, meadows and steppes.

105

106

108

When at rest, the raptorial front legs of the praying mantid are held folded together below her head. She seizes her pray by a lightning attack and, darting out her front legs, seizes the victim with great strength and holds it still while she eats it. The West African species **Sphodromantia lineola** is shown in illustration [108] and colour plates VIII and IX. The little southern European **Ameles abjecta** has a shortened body and is beautifully coloured [colour plate X]. The female measures at the most 2.6 cm.

The tropical species of mantids grow as long as 20 cm. They often take bizarre forms and lively, bright colours reminiscent of flowers. The larger species attack not only insects but

109

also small lizards, frogs and young birds and mammals.

In southern Europe lives the remarkable mantid **Empusa fasciata** [109, female], which measures up to 6.7 cm in length.

The second order, the **Cockroaches,** *Blattaria,* contains some 2,500 species. The largest measure about 12 cm, and the smallest—the American myrmecophile cockroaches (the ones which live in ant colonies)—only a few millimetres. The cockroaches have a comparatively mobile head with mouthparts designed for biting, a flattened body and legs equipped with thorn-like growths. Some tropical species are brightly coloured and look something like beetles. There are also aquatic species, and

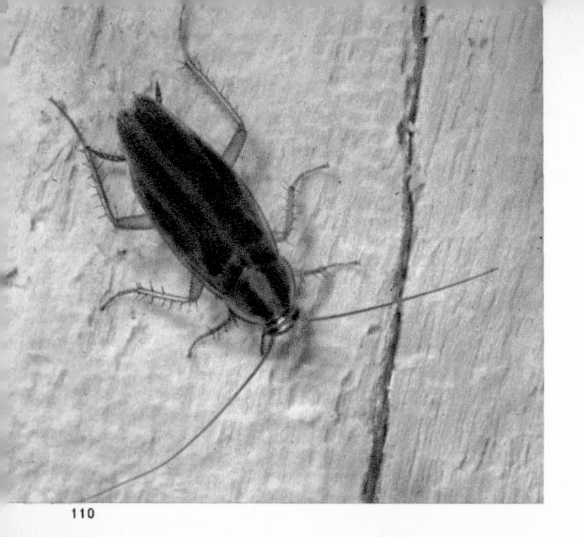

still others which can be found buried deep ir the desert sand.

The cosmopolitan **German Cockroach** *Blattela* or *Phyllodromia germanica* [110], is th smallest and one of the most unpopular. It i in addition a lover of warmth. In Germany thi cockroach is also called **French** or **Russian** while in the Soviet Union it is called th **Prussian Cockroach.** It is up to 13 mm long light brown in colour, and has two dark stripe on its shield. Both sexes have wings. It live in dark corners which can give it a constan high temperature and where it has access t food or rubbish.

110

111

112

One of the great cosmopolitan cockroaches, which also infest ships and mines, is the 4.5 cm long **American Cockroach,** *Periplaneta americana* [111]. In the picture we see the different stages of larvae and nymphs; in the middle is a fully developed insect. The females of this species are winged. A famous, but nowadays very rare species is the 26 mm long **Oriental Cockroach, Kitchen Cockroach** or **Kakerlak,** *Blatta orientalis* [112, male]. It is generally a shiny black-brown in colour; the wings of the male do not quite reach to the end of the body, and the females are wingless. It has a similar life pattern to that of the German cockroach.

Cockroaches lay their eggs in little packets (oothecae), which the female carries about with her until the larva hatch, or sometimes discard rather earlier. The ootheca of the oriental cockroach [113] is about 1 cm long, shiny brown and contains some 15 eggs. The larvae emerge as a rule after about two months.

113

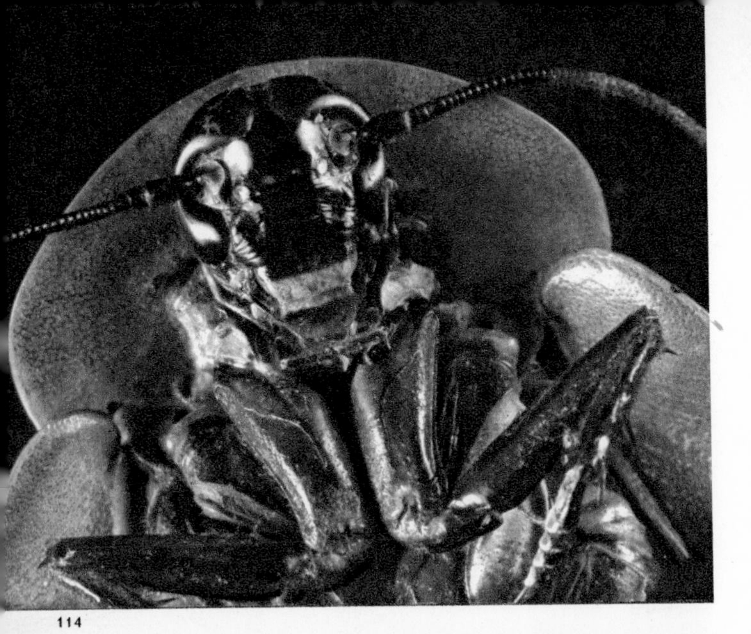

114

115

Some large species of cockroaches live in the tropical rain-forests of South and Central America. **Blaberus craniifer** [114, 115] is often kept in insectaria. Like the genus *Periplaneta,* this species breeds well in captivity. They are shy creatures with nocturnal habits: they hide by day and go looking for food—they are omnivorous—at night. They like freshwater, and reproduce by ovoviviparity. The larvae which emerge from the eggs are remarkable, resembling trilobites or other fossil animals [116]. They love warmth and live hidden in the earth or move about under rotting plant matter. Not until the last instar—that is, at the development into the fully grown insect — do they come out into the open and attach themselves to the bark of a tree. When they are excited, they emit a weak hissing sound and if picked up defend themselves and struggle with their thorny legs.

An even larger and more remarkable giant cockroach is **Gromphadorrhina portentosa** [117] of the family *Perisphaeriidae.* Our picture shows a male; it is generally larger than the female. This cockroach is about 8.5 cm long, brown-black, flightless, and has a remarkable bordering to the top of the thorax. It lives in Madagascar. If it is excited it attempts to frighten off its pursuer by a snake-like hissing.

Some tropical cockroaches have beautiful colouring. The light-green **Panchlora cubensis** [colour plate XI], about 2 cm long

116

with antennae half as long again, has occasionally turned up in Europe with bananas.

The third order of the superorder *Blattoidea* is the **Termites,** *Isoptera*. About 2,000 species of them have so far been described, nearly all of which are natives of the tropics, only a comparatively small number living in subtropical areas. These insects live socially in colonies and are divided into different forms, each of which has a quite distinct function. Termites form a polity, the central point of which is the queen.

The so-called workers [118] are generally characterised by a very small, frail body without a chitinous shield; they are wingless with a whitish colouring. Besides these there are the soldiers and the king. Initially, both the queen

117

80

and king are winged. They meet during swarming, when after the fifth moult of the larvae vast swarms of winged individuals break out of their underground colony. Then the male and female seek a suitable shelter or burrow their way with the aid of their mouthparts into the earth. They cast their wings, and only then does mating take place. After this the female begins to lay eggs, at first only in small numbers. Depending on the species, these are about 0.5 to 1 mm in size. Immediately after emerging from the eggs, the workers begin to build a nest, constructing corridors and chambers until eventually a giant termitarium emerges, in the building of which every new worker begins to help vigorously as soon as emerging from its egg. At the same time they feed on rotting fungi, the growing larvae as well as the queen and king. These royal prisoners gradually lose the ability to feed themselves and become totally reliant on their servants. Termites feed chiefly on wood, but will also consume vegetable pulp. With the assistance of symbiotic parasites living in their stomachs they are able to digest cellulose. Some termites cultivate fungi in special "fungus gardens". These, too, serve as food for the whole termite colony, which can consist of up to 10 million inhabitants with a highly complex social structure. After copulation has been repeated many times at regular intervals, the queen's body develops into

IX *Sphodromantis lineola*.

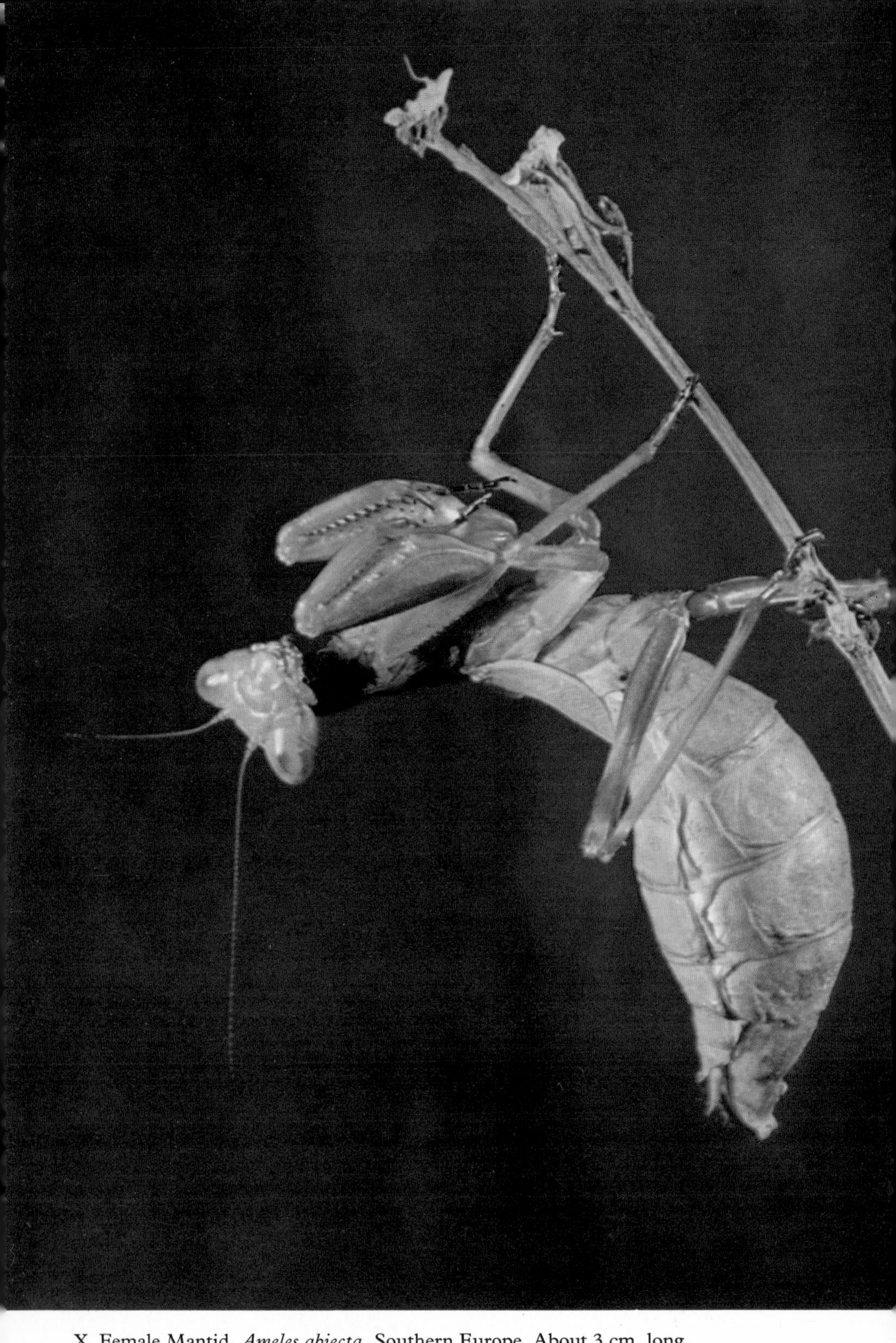

X Female Mantid, *Ameles abjecta*. Southern Europe. About 3 cm. long.

vastly swollen, shapeless nd almost motionless egg-aying machine, bringing nto the world during an verage life-time of ten ears many millions of ndividuals. It has not yet een established whether he so-called soldiers are atched from special eggs r if they are developed by neans of a special diet. 'hey are equipped with a emarkably large, glossy ark-brown head which as a stronger chitinous lating. As a rule, the head as long, powerful jaws nd is darker in colour nan the rest of the body. 'he task of the soldiers is defend the community, bove all in innumerable ncounters with aggressive nts which attempt to infil-ate the termitaria. These ests are, depending on ne species, either hidden nderground and invisible om the surface or take ne form of dome- or ower-shaped structures p to 7 m high, but can so be found attached as ll-shaped nests high in ees.

he **Light-shy Termite,** *eticulitermes lucifugus,* is n inhabitant of the Medi-rranean area. The work-s measure about 5 mm 18].

he second southern uropean species is the **ellow-necked Ter-nite,** *Callotermes flavi-llis.* The soldiers of this ecies [119] have large, lindrical heads and grow about 1 cm long.

he soldiers of **Odonto-rmes formosanus** 20], from southern hina, have extremely rge heads with an extra-dinarily powerful mus-lar system.

119

120

121

We have already mentioned that queen termites spend their whole lives once fertilised as batteries for the production of eggs. Their abdomens become enlarged, in some species to 3 cm in breadth and more than 10 cm in length. These queens are imprisoned in their

122

wn nests and become incapable of indepen-
ent movement. Only the peristaltic contrac-
on and dilation of the body and the slow
ovements of the limbs betray any sign of life.

The workers feed her, keep her body moist and attend to the eggs. The queen of the species **Termes bellicosus** [121, 122] measures about 6 cm.

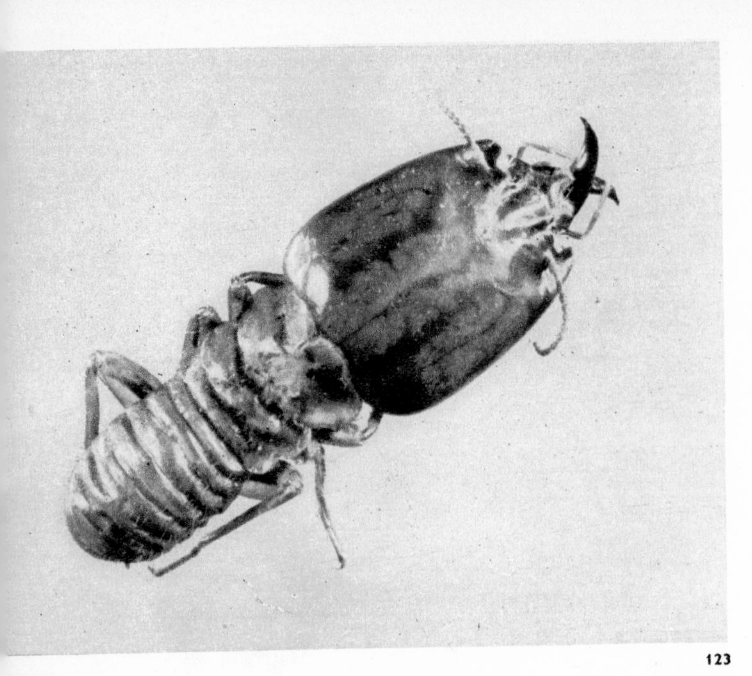

123

124

This species is one of the best-known of tropical Africa. They erect pyramidical structures about 4 m high, which are conspicuous, especially in light, half-desert forest and savannah areas. The defence of the nest is carried on by sightless soldiers [123], which in this species measure about 2.2 cm. The winged males [124] have a body of about the same length which is completely dwarfed by the wings. During swarming they can fly for distances of about 1 km. Most of them fall victim to various insectivores, and the natives also prize freshly roasted termites as a fine delicacy.

Nests which are underground are rarely noticed and can only be spotted by the little mounds of earth thrown up from them. Inside, however, they are completely organised, being built according to a pattern or design and containing entry and connecting corridors, a drainage system, ventilation shafts, living rooms and fungus gardens. The central point of the whole structure is the royal chamber. The workers collect wood and other vegetable matter, masticate it, gathering it overnight or by day in the shelter of the loamy pathways on the surface which they build for this purpose over very considerable distances.

The soldiers of some species dispense with biting and nipping and rely in battle on the effect of a nauseous secretion with which they spray their enemies.

These are called nasutes—

125

"nose-soldiers". In the place of the biting parts, which are atrophied, the hard heads of these creatures are prolonged into a conical, pointed projection which ends in a frontal gland, and has the appearance of the narrowed end of a syringe. The nasutes of the species **Eutermes parvulus** [125] from West Africa measure about 5 mm long.
In the inhabited parts of the tropics termites are a continual problem, as they frequently do an immense amount of damage to the wooden parts of buildings, although in nature they clearly perform an important part in the carbon-nitrogen cycle by this propensity for destroying wood.

The first order of the seventh superorder, the **Lichen Lice,** *Psocoidea,* consists of the **Book-**

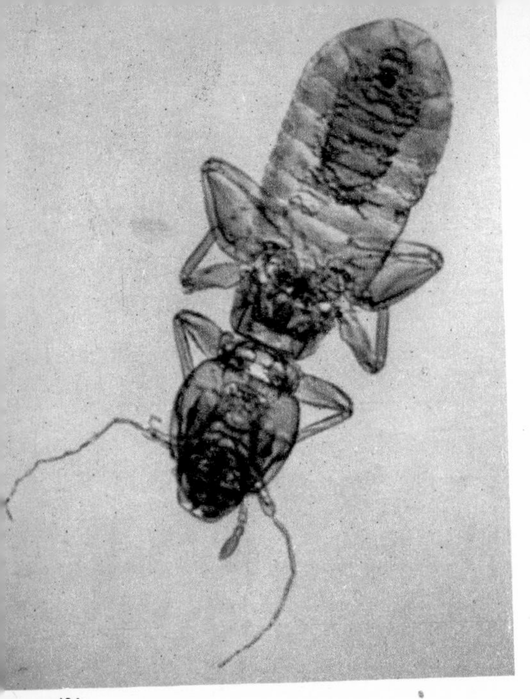

126

lice or **Dustlice,** *Psocoptera.* This contains approximately 1,000 species of tiny, soft, frail creatures, both winged and wingless, living chiefly in the tropics and subtropics. Only a comparatively small proportion of booklice live in the temperate zone, and then either in houses or on tree trunks, rotting wood, branches and leaves of trees and shrubs.

These tiny creatures feed on lichen, mould, algae, fungus mycelium and various fragments of decaying animal and vegetable matter. Their mouthparts are adapted for both biting and piercing. Indoors, booklice live on organic matter and invade larders and collections of insects. In libraries, they eat starch, glue and the paper of old books. Some species make their homes in birds' nests or those of mammals, or in anthills. They reproduce through eggs. Their transformation is incomplete. The example of a wingless psocid given here is the little **Liposcelis silvarum** [126]. It measures little more than 1 mm and is a tree-dweller, especially on bark and in nests.

The largest central European species is the winged **Psococerastes gibbosus** [127]. It is found mainly towards the end of summer, especially on pine trees in mountainous areas. It measures about 12 mm.

127

The second order of *Psocoidea* consists of the **Animal Lice,** *Phthiraptera.* It is divided into three suborders. The first that we shall mention is the **Biting Louse** or **Bird Louse,** *Mallophaga.* This suborder consists of about 2,500 species of flattened, wingless insects which live as parasites in the feathering of birds and the coats of animals. The smallest species measure about 1 mm, the largest about 1.4 cm.

The male **Neophilopterus tricolor** [128], which lives on the feathers of the black stork, measures about 3 mm.

Two males of **Lipeurus maculosus** are shown in illustration [129]. They are parasitic on pheasants and light brown in colour, with a long, extended body. On the head there is a semicircular projection. They measure 2 mm.

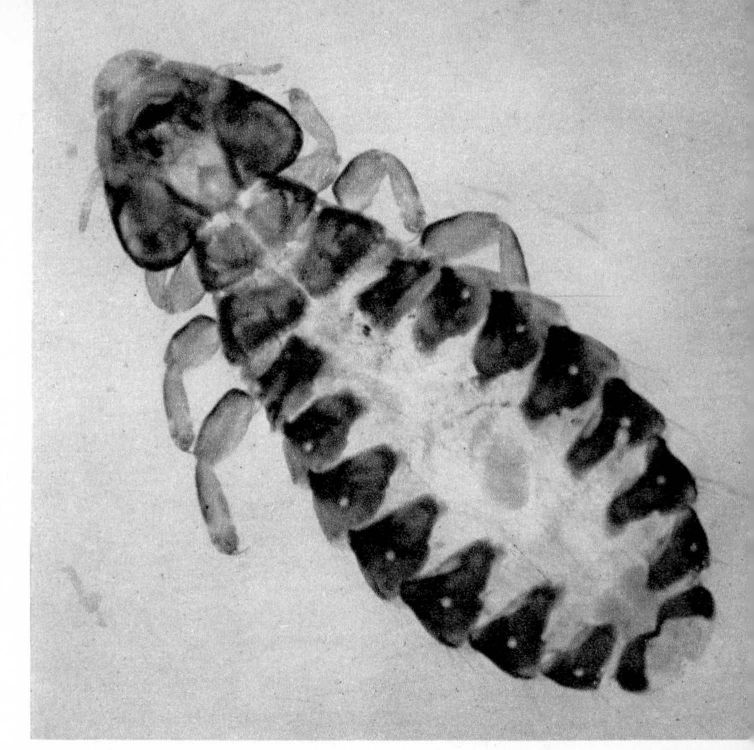

128

129

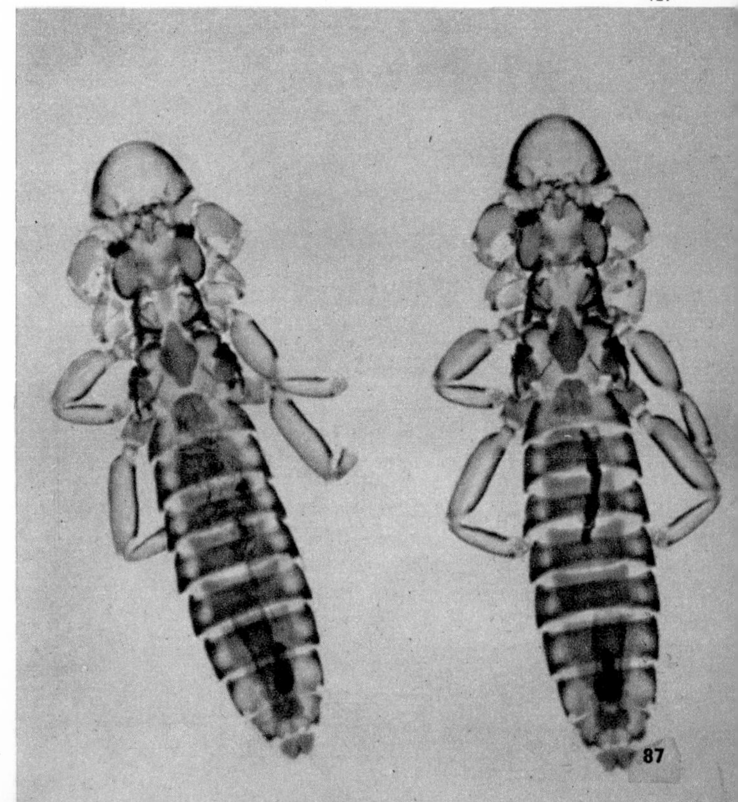

130

131

Frequently, thousands o bird lice live on the sam host. They feed on frag- ments of skin and scurf They eat new feather quil and some species depen exclusively on the blood o their host. They work thei way nimbly through th coat or plumage, creepin across the skin and tor- menting their host wit constant irritation so tha it often will no longer tak food and, indeed, some- times dies. Bird lice ar divided into sexes, an attach their eggs in a ro to the feathers. They can however, also reproduc parthenogenetically, s that males are compara- tively rare. Generally species are restricted t particular hosts, keepin strictly to them. The transfer themselves fron adult birds to the fledg- lings in the nest. I hav found living bird lice o the eggs in the nests o marsh birds. They do no infest man. In handling dead bird, however, it ca happen that in spite of al precautions they stray o to the warm body of th handler. I can remembe how I was once obliged t interrupt the dissection o a dark-brown monk vul- ture for this reason, al- though I was constantl renewing with a larg brush the sticky coating c glue round my arms. Afte a short time these bird lic had reached the hair of m neck. It is comparativel easy to rid oneself of ther by hair combing and pick- ing them off the skin, bu even if one did not do thi man is not a suitable host for them. To a considerable extent bird lice are restricted to only one species of host in respect of their food, temperature, humidity and other conditions of life.

Both males and females of **Ardeicola maculatus** [130], which lives on the central European black stork, are light brown and 4 and 5 mm long respectively.

Illustration [131] shows two females of **Craspedorrhynchus aquilinus** taken from the plumage of a central European golden eagle. They are brown-black, their bodies are remarkably short and measure 2.5 mm.

The profusion of forms of the various bird lice is astounding. Some species look as if they have huge eyes at the side of their head. However, nearly all are sightless or have strongly degenerate eyes.

The **Peacock Louse,** *Goniodes pavonis* [132], which, as its name suggests, infests the peacock, is about 3 mm long. Its head is broader than it is long and the so-called "sleep-horns" project sideways from it. In the specimen illustrated the digestive organs filled with food are clearly visible.

A very pretty species is **Koeniginirmus normifer** [133]. It is white with a dark-brown pattern on the head, breast and every segment of the abdomen. Both the females shown were found on seaweed. They are some 2 mm long.

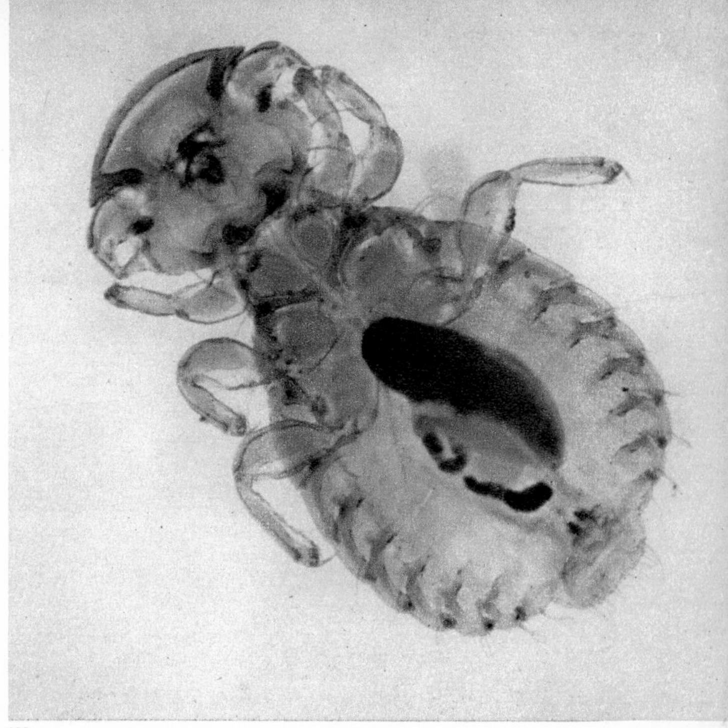

132

133

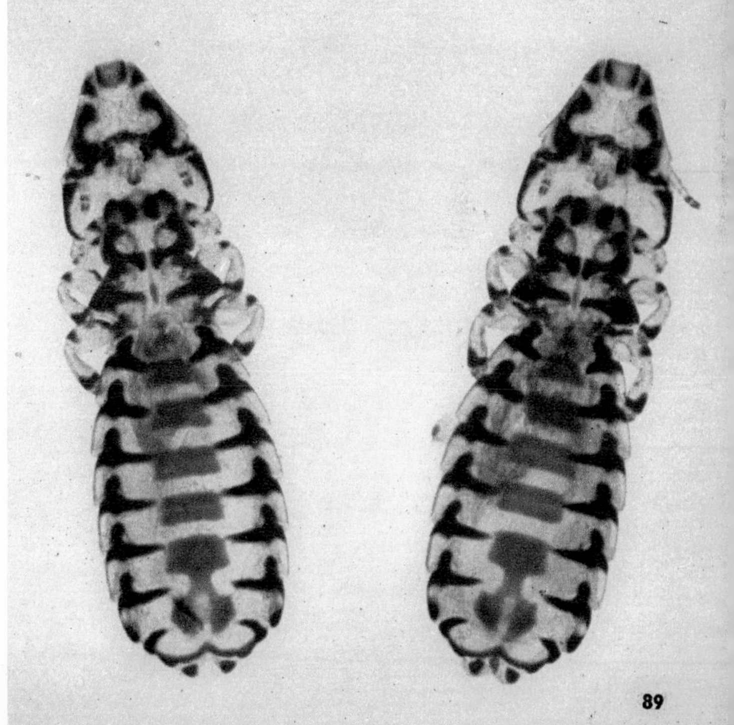

134

135

136

. very interesting principle, called Kellogg's ıw, states that the relationship of the bird lice orresponds to that of their hosts. As a result f this several surprising phylogenetic relation-nips between various orders and families of irds have been discovered. The species **.oeniginirmus punctatus** [134], which in-ests the Lach gull, resembles in form and attern the preceding, which lives on a related ird in the gull family, *Lariformes*. The uropean spoonbill is infested by the robust, mm long louse **Ibidoecus plataleae** [135]. 'he vast majority of *Mallophaga* live on birds. n the family *Trichodectidae* there are, how-ver, about 100 species of these insects which infest a great variety of mammals, although each species is closely bound to its corresponding host species. Only closely related mammals —for example, dogs and wolves—can carry the same species of parasite, in this case the **Dog Louse,** *Trichodectes canis.* Illustration [136] shows the parasite, from underneath, clinging to a dog's hair. This louse is rust-brown colour and infests in particular young dogs, preferring to take up residence on the neck and head. *Trichodectic mallophaga* can infest almost every species of domestic animal. Only cleanliness, frequent bathing and generous use of DDT can guarantee complete success in ridding a pet of these pests.

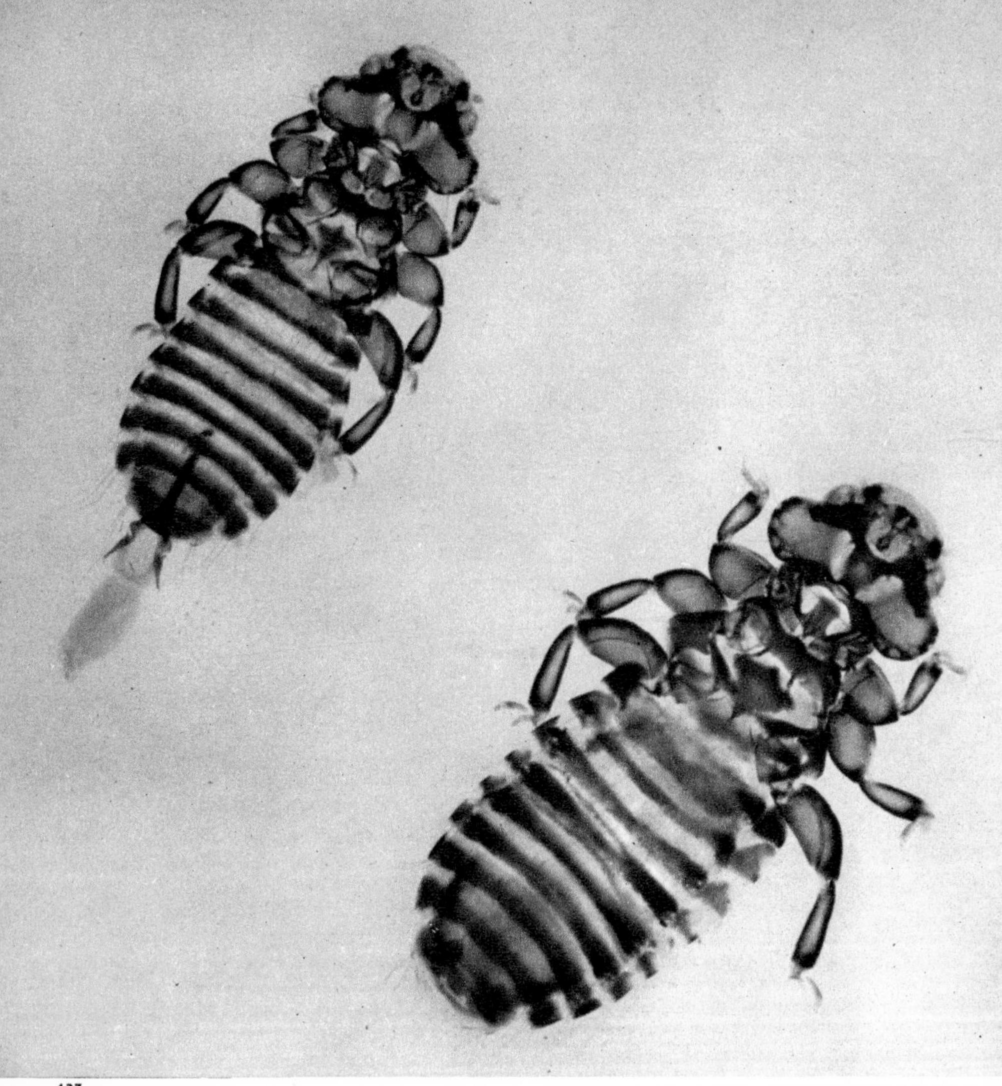

137

138

Many songbirds are infested by bird lice of th genus *Myrsidea*. The male and female **Myr sidea isotoma** in illustration [137] were take from a rook. The larger, the female, measure 2 cm.

Bird lice of the family *Ricinidae* possess at th front of their head a special organ for anchorin themselves and for sucking.

The remarkable **Ricinus elongatus** lives o blackbirds. It is found on the underside of th host and lays its eggs cemented to the brea plumage. The female shown [138] is 4.5 m long.

The species **Ricinus bombycillae** is ligh

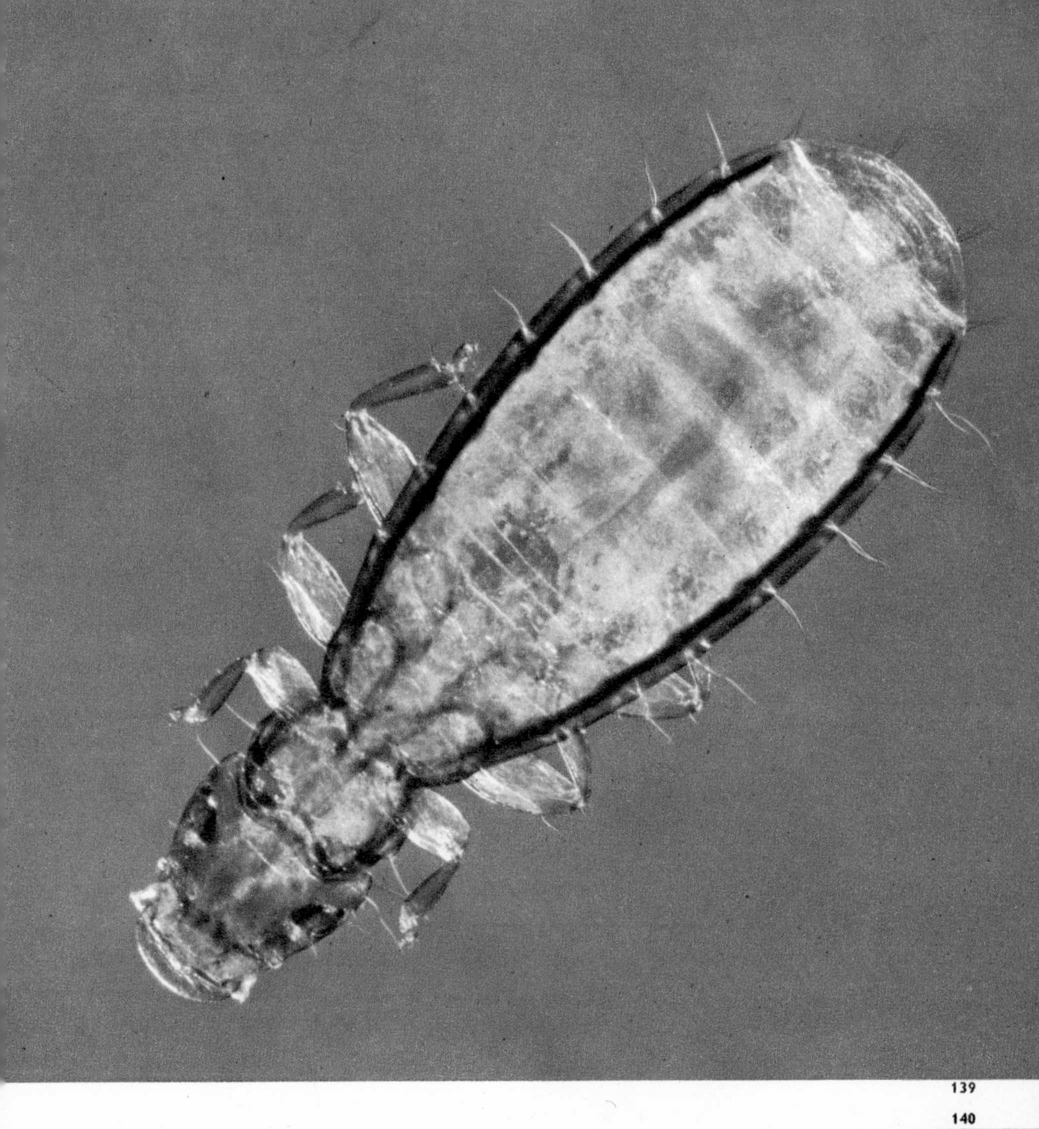

139

140

rown, about 4 mm long and sucks the blood f the waxwing, a lazy, attractively coloured ird. The female [139] measures about 4 mm. The second suborder, the **Elephant Lice,** *Rhynchophthirina* consists of only a single amily of remarkably specialised forms, which ive on both species of elephant and are dapted with their unique mouthparts to ierce the thick skin of the elephant. **Haema-omyzus elephantis sumatranus,** a parasite n the Indian elephant, measures about 2 mm xcluding its sucking tube or proboscis. It is oloured red, has a long tube with tiny biting arts and comparatively long legs. The bell-

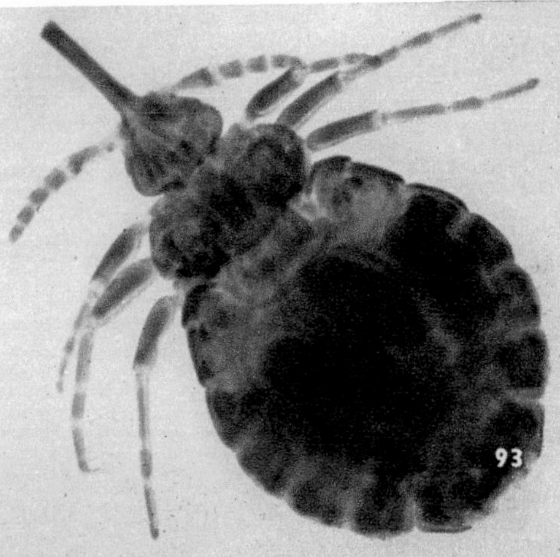

shaped eggs are cemented to the elephant's hair. Our picture [140] shows a female, full of eggs.

The third suborder, the **Sucking Lice** *Anoplura,* is distributed throughout the world in some 250 species. They live parasitically on a wide variety of mammals and, indeed, are present on pinnipeds, such as seals, which spend the greater part of their lives under water. They have very frail, transparent and flattened bodies, which, however, after feeding swell up and take a sack-like form, and prehensile legs with sharp claws which can be dug in. The openings of the tracheal respiratory system are situated on the upper surface of the body. The mouthparts are developed for piercing and sucking and the antennae are short. Most species are sightless. They feed on the blood of their host and attach their eggs (nits) to the hair. The larvae resemble the adults and the metamorphic process is incomplete.

Two species of sucking lice infest man—the **Body Louse,** *Pediculus humanus corporis,* which also includes the subspecies the **Head Louse,** *Pediculus humanus capitis,* and the **Crab Louse,** *Phthirus pubis.*

The head louse [141] generally attacks children but can also be found on adults. In times or circumstances where personal hygiene cannot be observed with the necessary attention (as in time of war, in refugee camps or prisons)

141

142

. used to be rife, but recently has become much ·ss significant. It attaches itself only to the ·ead and its nits to the hair. It measures about .5 mm long.

'he body louse [142] is much more dangerous. t is up to 4.5 mm long and of a darker colour. t lives chiefly on the front and back of the ·uman torso, laying its eggs in the hems and orders of clothes. It is a dangerous carrier of pidemic typhus, relapsing fever and five-day ever. Pediculosis (infestation with lice) was common phenomenon in ancient times and 1 the middle ages, and before the discovery of 1secticides extremely difficult to combat. As ecently as the First World War on one risoner 3,000 living lice were counted, and 1 1916 another was recorded as carrying 6,000.

'he larvae emerge from the nits after about to 7 days and after a further 14 days they each sexual maturity.

'he crab louse [143] is distinguished by a short, road body and more strongly developed claws n the prehensile legs. It lives on the hairy laces of the human body with the exception f the head. It is 1.5 mm long, grey-white to eddish in colour and attaches its nits to the airs of the host.

'he **Hog Louse,** *Haematopinus suis* [144], neasures up to 5 mm long. It is often present n the domestic swine, especially behind the ars.

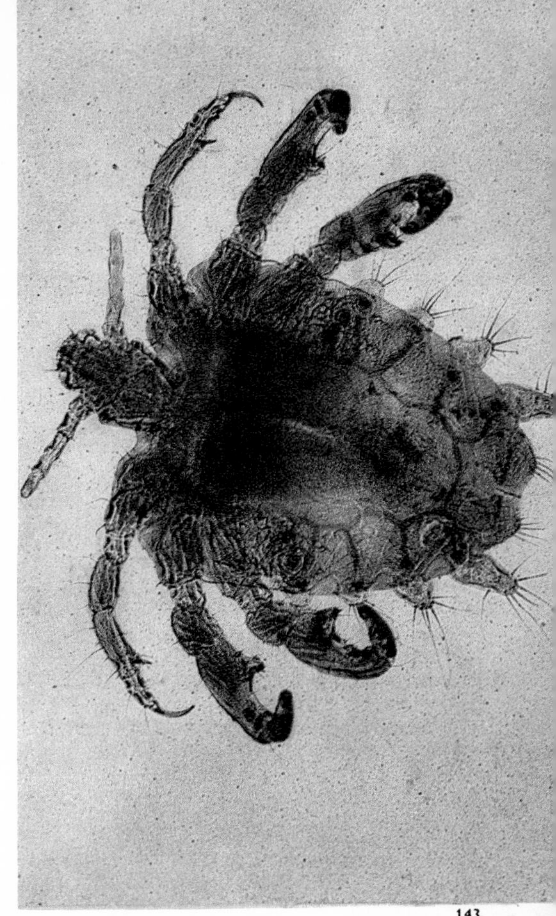

143

144

145

The eighth superorder, **Fringed-winged Flies,** *Thysanopteroidea,* has only one order. The **Thrips** or **Bladderfoots,** *Thysanoptera,* embraces roughly 1,500 species so far recorded, distributed throughout the world. These insects are generally 1 to 3 mm long; the largest of them, the tropical "giants", measure 13 mm. They are either wingless or have wings bearing remarkable fringe-like appendages. At the end of the extremely short legs are protrusible vesicles on the feet. Most thrips suck the sap of plants or consume fungi or fungispores. A

XIa Cockroach, *Panchlora cubensis*. Greater Antilles. 3 cm. long.

XIb Hog louse, *Haematopinus suis*. Central Europe. 4 mm. long.

XIIa Green Stink-bug, *Palomena prasina*. Europe. 11 to 14 mm. long.

XIIb Frog-hopper or Cuckoo-spit Insect, *Cercopis sanguinolenta*. Central Europe. 9 mm. long.

146

w species are also predatory on plant-lice. hey are much feared as plant pests, the more since they not only do great damage to the lants in our gardens and fields, but also are rriers of various plant diseases, especially ruses. **Frankliniella intonsa** [145, female] is about 2 mm long. It lives in August in the temperate zone of bulb flowers.

A dark body and brown wings are the characteristics of the familiar pest of peas and other leguminous plants, **Kakothrips robustus** [146], which is also about 2 mm long.

147

The ninth superorder, the **Plant-bugs,** *Hemipteroidea,* contains about 48 to 50,000 species so far described, divided into two orders: 1. **Bugs,** *Heteroptera* and 2. **Plant-suckers,** *Homoptera*.

There are about 25,000 species of heteropterans and most of them can be clearly distinguished from one another.

Besides the piercing-sucking mouthparts, they have a characteristic three-cornered breast-shield. The wings lie alongside this shield, and overlap each other. Some bugs have the wings hidden underneath the shield, which resembles the wing-covers of the beetles, but is not divided down the middle. The rear wings, which are always somewhat smaller and membranous, lie hidden under the forewings, which are developed normally and laid flat side by side, having a leather-like chitinisation at their roots ("half-cover wings"). Many species especially those from the tropics, are brightl coloured and sometimes have an almost metallic sheen. All terrestrial bugs emit a noxiou smell, which they also leave behind at thei source of food. Their stink-glands are situate at the sides of the rear breast-segment.

Among the most attractive species of centra and southern Europe is the red-and-blac **Striped Bug,** *Graphosoma italicum* [147], o the superfamily *Pentatomoidea*. It is 1 cm long harmless and lives at the end of summer o the blooms of umbelliferous plants. The shiel (scutellum) covers almost the whole of the rea of the body.

Sphaerocoris ocellatus [148], of the famil *Pentatomidae,* comes from East Africa and i equally attractive to the eye. It is about 1 cn long, yellow-brown in colour, with black eye having a red border and an attractive red desig on the shield which covers the whole of the rea part of the body.

The **Berry Bug,** *Dolycoris baccanum* [149 151], of the superfamily *Pentatomoidea,* i about 10 to 12 mm long and coloured yellow brown. The edge of the rear end of the bod has a light-and-dark chequered design. I sucks on various different plants, leaves an fruit in forest and countryside. It is particularl common on bilberries, which it despoils wit its smell. As well as being common in Europ and Asia, it is also common in North America

14

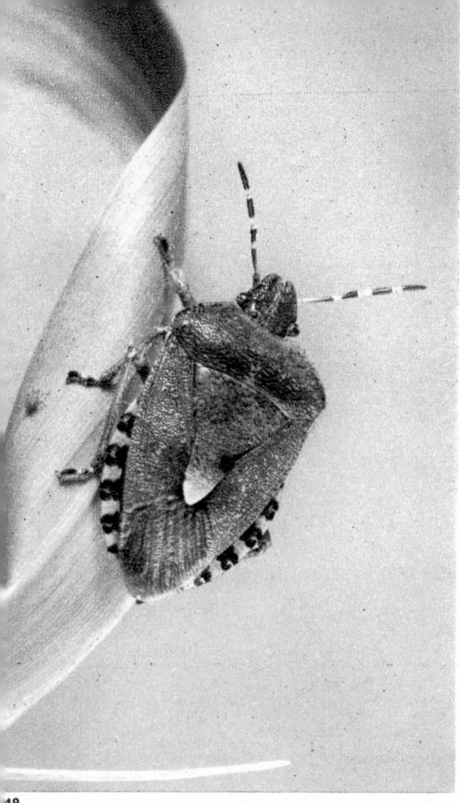

49

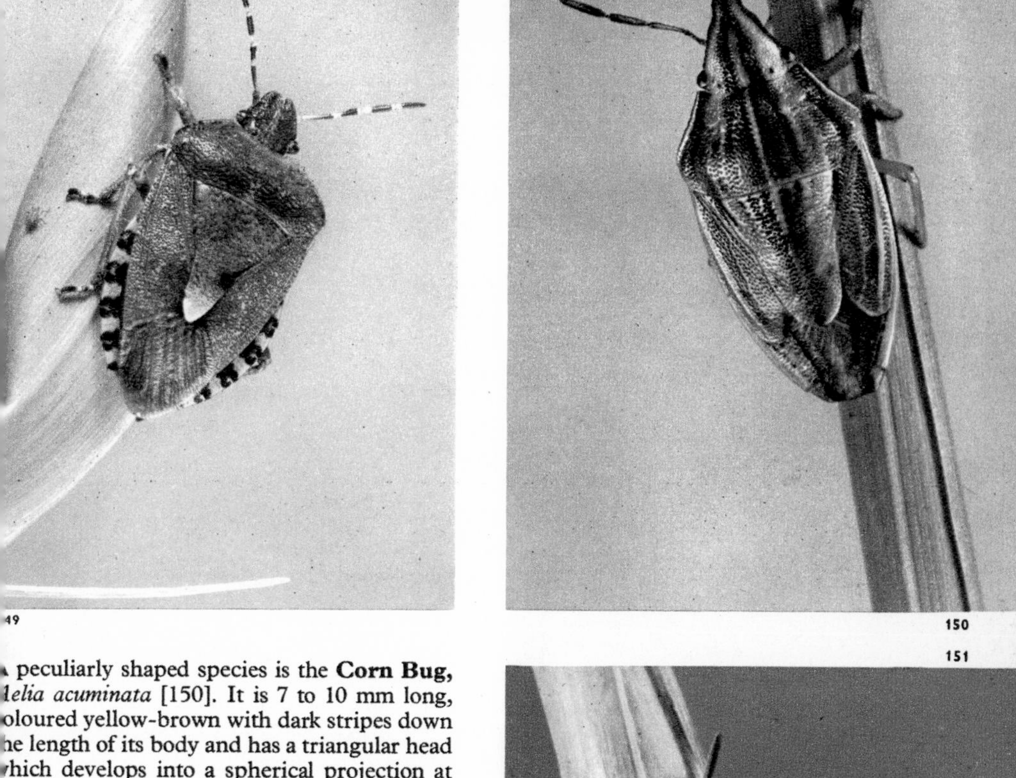

150

151

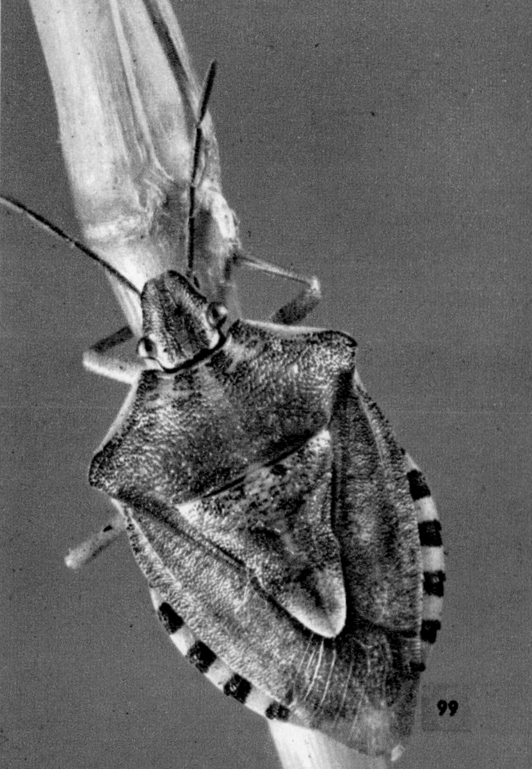

ı peculiarly shaped species is the **Corn Bug,** *lelia acuminata* [150]. It is 7 to 10 mm long, oloured yellow-brown with dark stripes down he length of its body and has a triangular head hich develops into a spherical projection at ıe front. The edge of its rear part is yellow. 'he black spots on its underside are arranged ı four rows. It lives on grasses and is a pest, ı that it damages standing corn by sucking ıe sap from the stalks of young shoots. It is resent in Europe, in the temperate parts of ısia and in North Africa.

'he **Green Stink-bug,** *Palomina prasina* :olour plate XII], is 12 to 14 mm long. 'hroughout the summer it is green, but hanges colour in the autumn, becoming rown, and survives the winter as an adult ısect. In the spring it recovers its green colour. : also lives in Europe, the temperate parts of sia, North Africa, and sucks the juice of ripe ilberries, which it prefers to anything else. 'he damage which it does, however, is com-aratively small, although it has an abominable mell.

152

1

From the family of **Leather Bugs,** *Coreidae,* we mention the **Seam Bug,** *Coreus/Syromastes marginatus* [152, 153], which may be found in central Europe. It is greyish-red with a bronze sheen on the wing-covers and measures 13 to 14 mm. It is easy to recognise because of the peculiar shape of its body. It is present from spring to autumn in heaths and fields, especially on sorrel. It has a pleasant scent, something like apples.

There are something like 2,000 species of seam bugs. Some of them are predatory, others feed on plant juices and many species are bo predators and suck the juice of plants. Mo of them have a thin leaf-like broadened proce on their rear legs, often of a striking colou **Leptoglossus phyllopus** [154]; from Nor America, is coloured brown with a white stri running across the middle of its body. It 2.5 cm long. Other genera of this family ha strikingly bent, strong, serrated rear leg **Anoplocnemis curvipes** [155], from Sen gal, is a brown coloured bug 2.7 cm in lengt It does a great deal of damage to the cultivatic

154

155

f sunflowers and cotton, as also in coffee lantations. Other species of this family, too, re much-feared pests in the agricultural conomy.

'he bugs of the family *Phymatidae* have a redatory mode of existence. Their front legs ave developed a raptorial form like those of e mantids and serve to seize small insects as rey. The body, which is extremely small, often kes a surprising shape, being equipped with llets and thorns. The family consists of about 00 species distributed chiefly in the tropical

and subtropical parts of the Americas and Orient. In central and southern Europe and in the neighbouring parts of Asia can be found the species **Phymata (Syrtis) crassipes** [156], which lives in warm, dry places. It is brown, about 9 mm long, and lies in wait on low plants for small flying insects, though it also hunts greenfly.

One of the most interesting genera of the family *Coreidae* is *Phyllomorpha*. The body of this plant-eating bug is flat and broad, shaped something like a leaf, the edges of which project in thin spikes. The only European species **Phyllomorpha laciniata** [157], lives in the Mediterranean area. This light grey-brown coloured bug is 1 cm long and lays its eggs on the backs of other members of the same genera males as well as females. There they remain held fast by the spikes on the body, until the emergence of the larvae. The bug produces a light chirping sound by rubbing its antennae together.

156

157

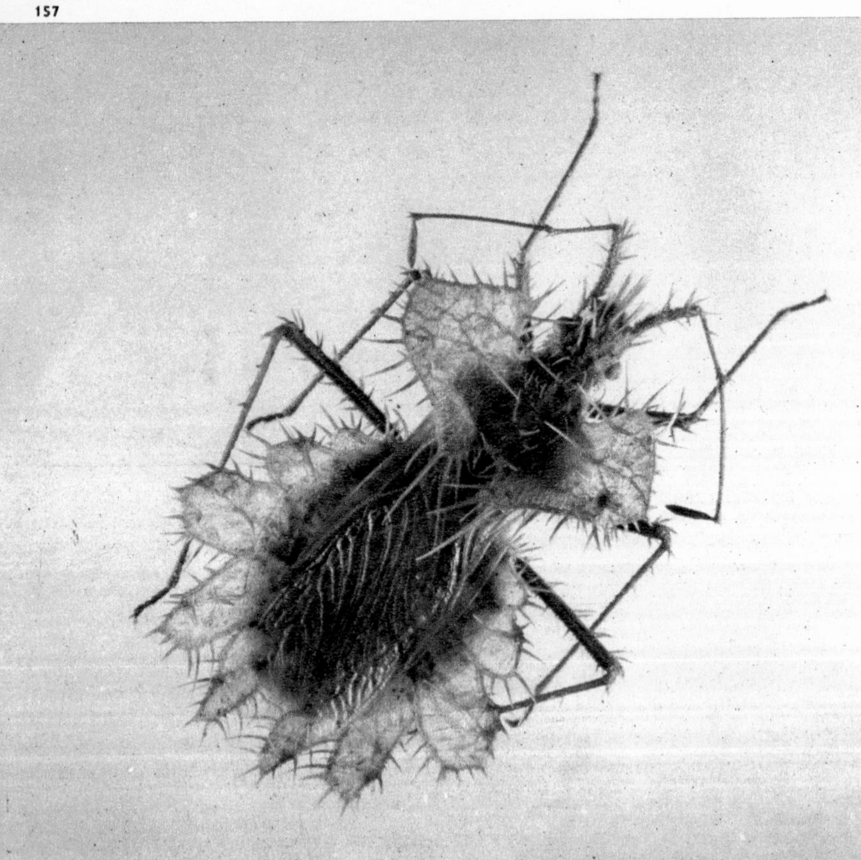

Another interesting family of bugs is that of the **Lacebugs,** *Tingidae.* They are distinguished from other bugs by their transparent upper wings, which have a lace-like structure on them. Many species also have a similar translucent edge to the side of their neck-shield. These little creatures only measure 2 to 5 mm. Some lacebugs live in the imago or adult state in plant galls. Their larvae leave sticky traces on the underside of leaves. Underneath them can be found well-known tree and plant pests. They are distributed in every part of the world, wherever they can find suitable living conditions.

Derephysia foliacea [158] has on its shield three veins like lengthwise ribs. It is about 3.5 mm long, very common in central Europe and found chiefly in grass. The **Lacebug,** *Galeatus maculatus* [159], is only 2.5 to a little over 3 mm long, bearing on its shield a small bladder-like growth. It is found in places where hawkweed and cinquefoil grow.

158

159

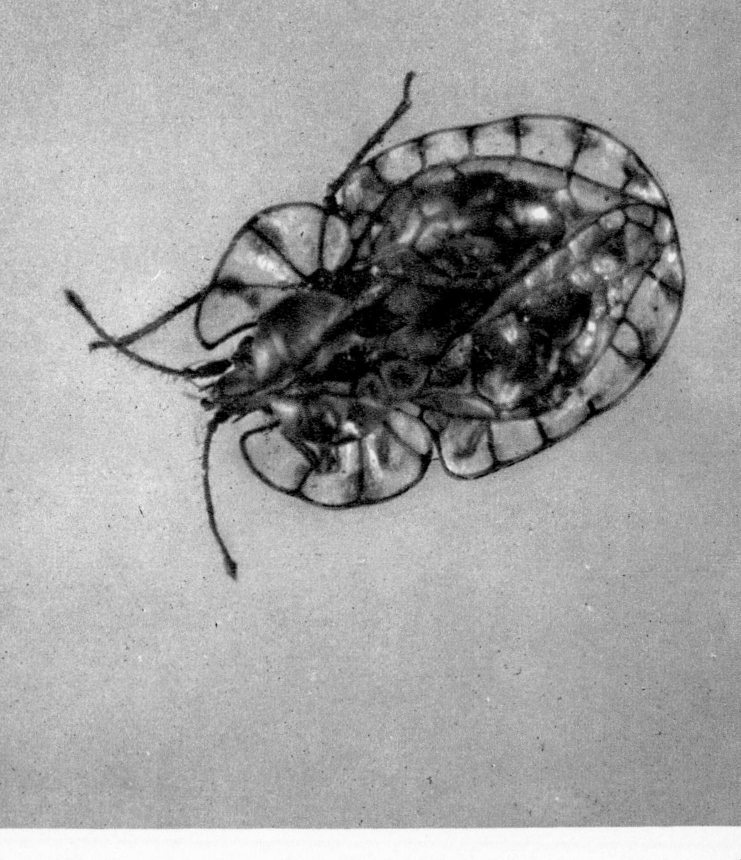

The family of **Firebugs,** *Pyrrhocoridae,* con
sists of wingless, fairly large insects, generall
of a striking blood-red colour. Their distribu
tion lies throughout the whole of Europe, a
far east as Siberia. They are also known i
North Africa, Central America and India
Some non-European species of firebugs ar
violent predators. The **Common Firebug**
Pyrrhocoris apterus, is well known due to it
presence in large numbers [160, 161]. It i
about 1 cm long, has a magnificent black-and
red colouring and is often found in swarms a
the foot of old lime and elm trees, especiall
when in blossom. In general, they are harmles
plant-suckers and carrion-eaters, though the
also live on various other organic remains
However, they can damage young trees i
plantations. They are also pests in vineyards
as they suck the juice of grapes.
The family of **Assassin Bugs,** *Reduviidae,* i
cosmopolitan, containing about 4,000 species
Many have mouthparts adapted for piercing
and are nocturnal in their way of life. The
attack small vertebrates as well as insects. Som
are even dangerous to man, especially those o
the large tropical species, which are carrier

160

161

1.

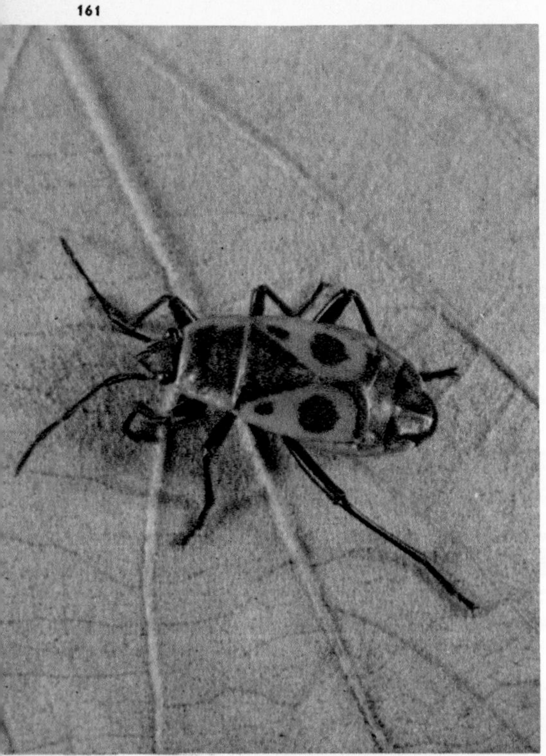

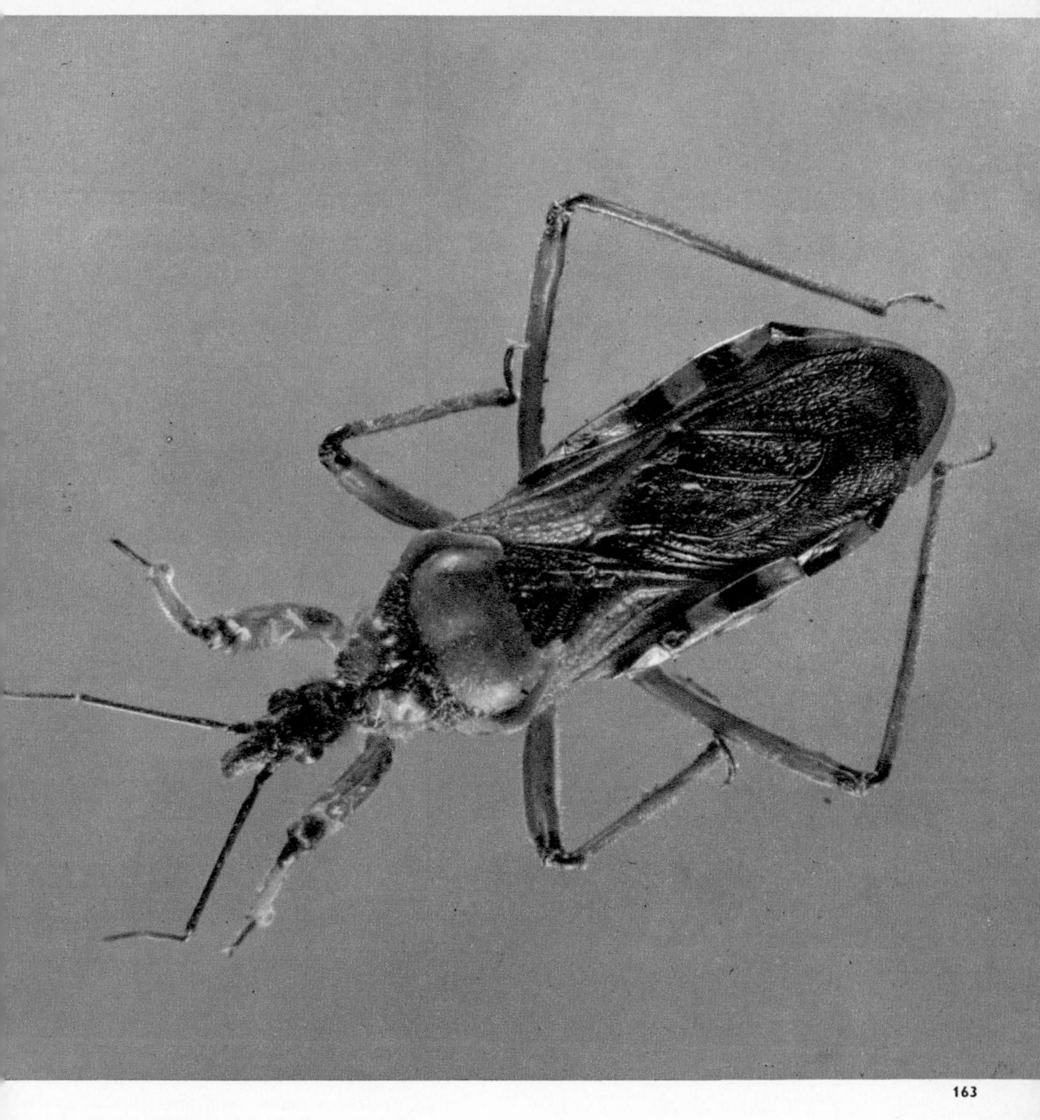

163

of the trypanosomiasis virus *Trypanosoma cruzei*. In Europe the commonest species is the **Flybug** or **Masked Bug,** *Reduvius personatus* [162]. It has also been imported into America. It measures between 15 and 16 mm, is coloured dark brown and lives in hollow trees, but also often invades houses. When disturbed it emits a shrill chirping. The nymphs, which hunt small insects in dark corners, camouflage themselves with dirt, which is picked up by the viscous hairs on their bodies. The bite of the flybug is quite painful and causes a swelling.

The **Red Assassin Bug,** *Rhinocoris iracundus* [163], is a powerful, red-brown predatory bug, 13 to 18 mm long, distributed throughout Europe. It usually takes up a position in flowers, where it lies in wait for small flying insects. Its bite is as painful as the sting of a wasp.

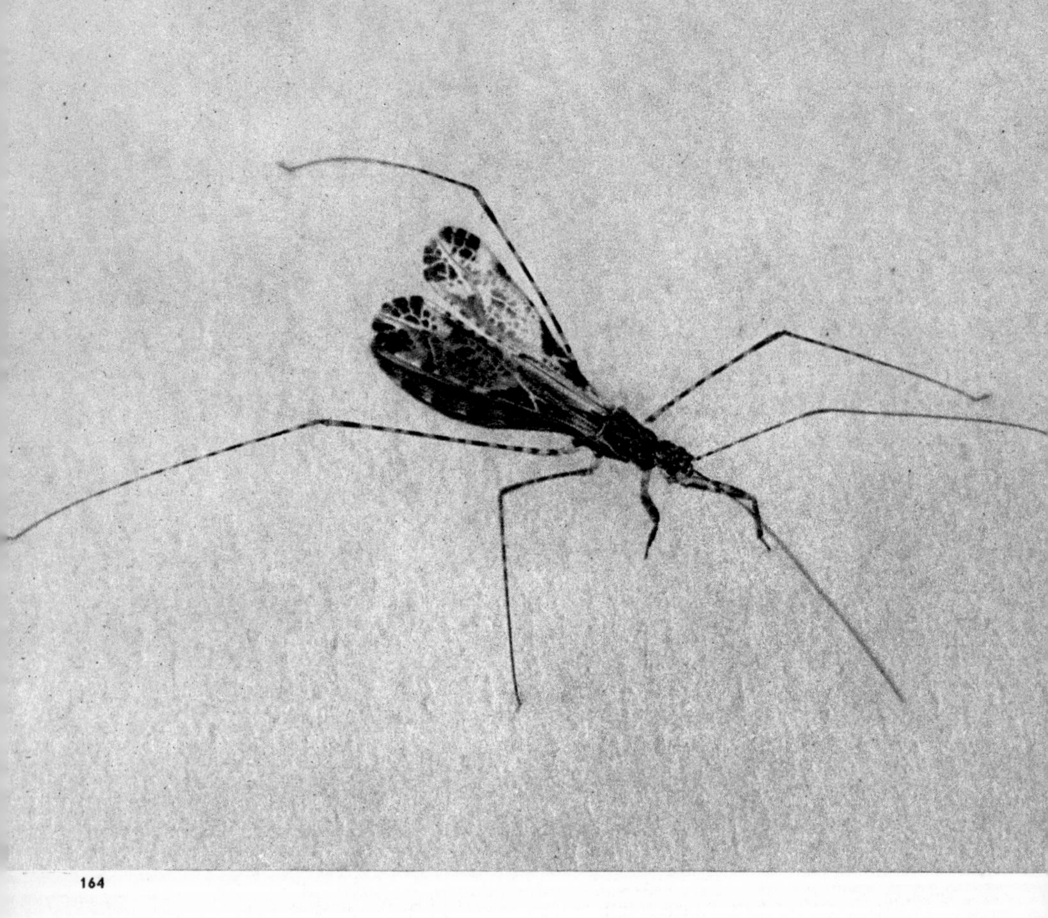

164

Empicoris vagabunda [164] measures about 7 mm in length, and an inconspicuous brown. It has a slender, elongated body and is often mistaken for a gnat. The front legs are prehensile, however, and by them this predator can be identified; it uses them to seize and hold flying insects. It lives in forests, haystacks and houses. It is also found in stables and toilets, where it acts as a pest-controller, attacking flies and gnats.

The **Bedbug** family *Cimicidae* is distributed throughout the world in about 30 species. All of them are parasitic on warm-blooded vertebrates.

The **Common Bedbug,** *Cimex lectularius,* grows to a length of between 3.5 and 8 mm, coloured brown and has a greatly flattened body with an extremely wide shield on the breast and an egg-shaped rear. It emits a characteristic bedbug smell, which betrays its presence. It finds its way into most human settlements, being distributed throughout the world, wherever man is to be found. During the day it hides away behind loose wallpaper and picture frames, in mattresses and bed-frames, so that at night it can attack human beings as they sleep and suck their blood, which forms their only food. Similarly, it afflicts all our domestic animals, including fowl, as well as mice and rats, and as a result it is a dangerous carrier of disease. The struggle against this parasite has made great steps forward since the development of more sophisticated insecticides. Our illustration [165] shows the head with the mouthparts,

165

166

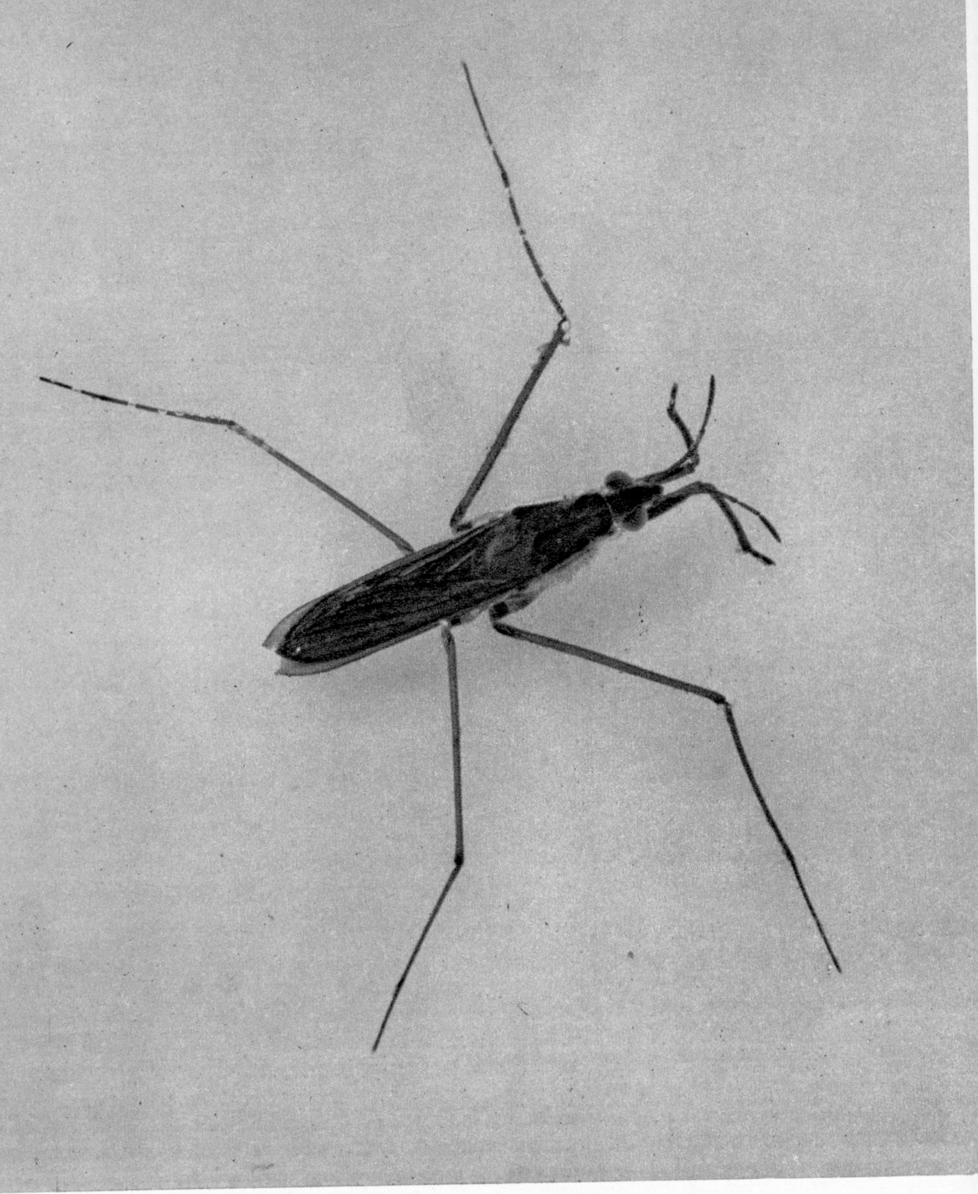

167

showing clearly the piercing and sucking "needle". In illustration [166] are shown an adult insect with two larvae.

The family of **Pond-skaters** or **Water-striders,** *Gerridae,* consists of slender, predatory bugs, which generally live in large numbers on ponds, lakes and other still or slow-flowing waters. They glide like skilful skaters over the surface of the water in powerful spurts on their long, slender legs. The under part of their hindquarters is set thick with tiny hairs, forming an air cushion. As they move

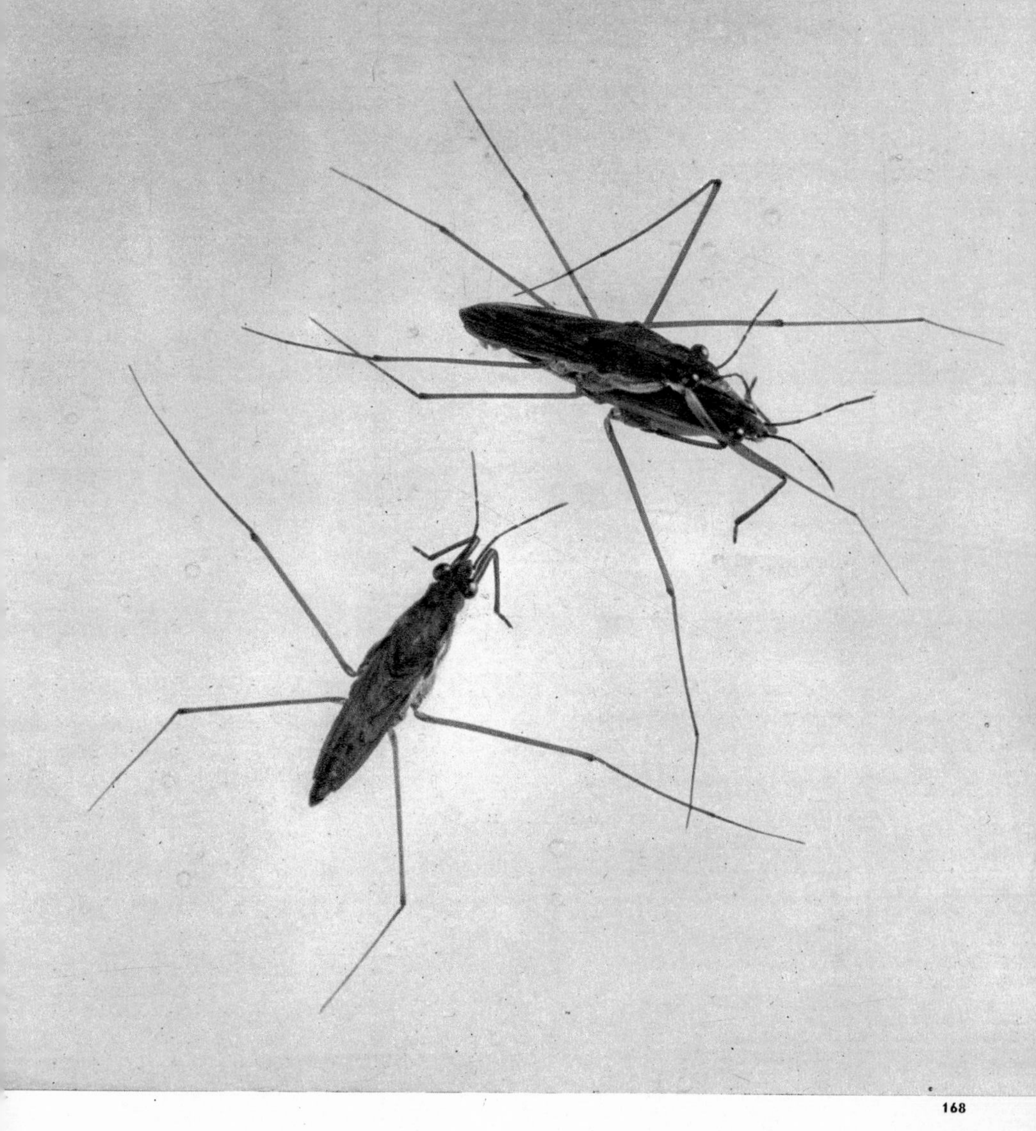

168

wiftly about they hunt every kind of insect which settles on the surface of the water. They will also suck the juices from the bodies of dead birds and small mammals floating on the surface of the water. **Gerris paludum** [167] measures between 14 and 16 mm.

The **Common Pond-skater,** *Gerris lacustris* [168], is somewhat smaller, being about 8 to 10 mm. It is particularly common on stagnant waters, but can also be found on the surface of slow-moving rivers.

The **Water-cricket,** *Velia saulii* [169], is

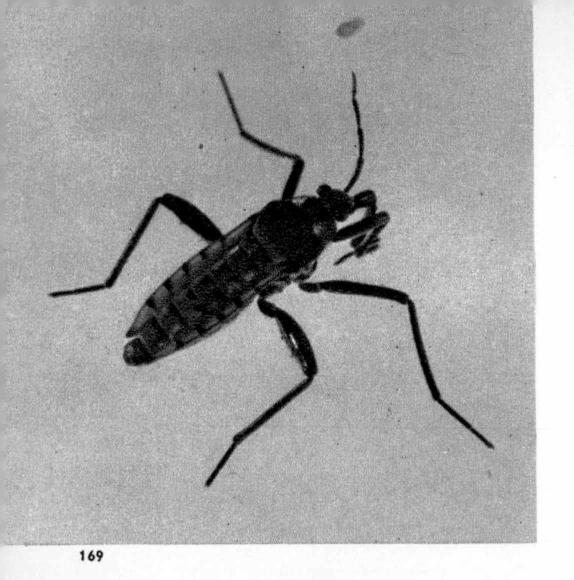

169

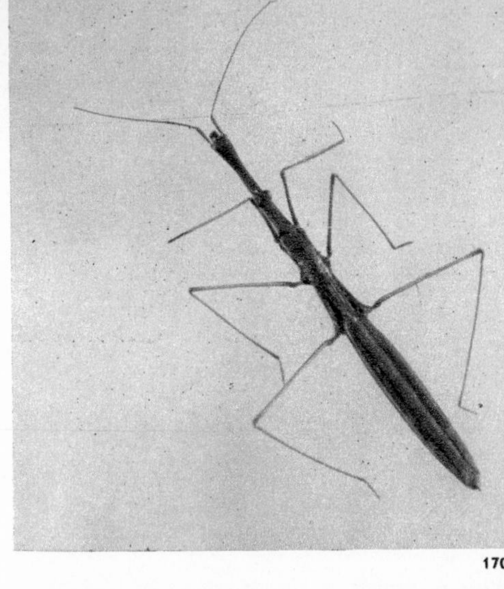

170

about 7 mm long. It is greyish-red with white spots. It runs rather than skates over the surface of the water. When hunting its prey it occasionally dives under the surface. It lives singly rather than gregariously on the slow-flowing streams throughout the palaearctic region.
A gregarious insect found on the edge of fresh-water swamps in central and southern Europe is the **Water-measurer** or **Water Gnat,** *Hydrometra stagnorum* [170]. It is 13 mm long, red-brown in colour and a predator, although it moves only slowly.
The **Saucer Bug,** *Naucoris/Ilyocoris cimicoides* [171, 172], measures about 15 mm, and is a beetle-like bug which hunts under water. It is very common in Europe. It delivers a painful sting and in some places this and

172

17

173

milar species have received the local name water bee".
he largest representatives of this order are the **iant Waterbugs,** *Belostomatidae.* Some outh American, Australian and Indian species hieve a length of well over 10 cm. As well as other insects they also attack tadpoles and small fish. In China they are prized as a great delicacy for the table. **Belostoma europaeum** [173], from Dalmatia, is 75 mm long and brown in colour.
The family of **Water Scorpions,** *Nepidae,* is

174

175

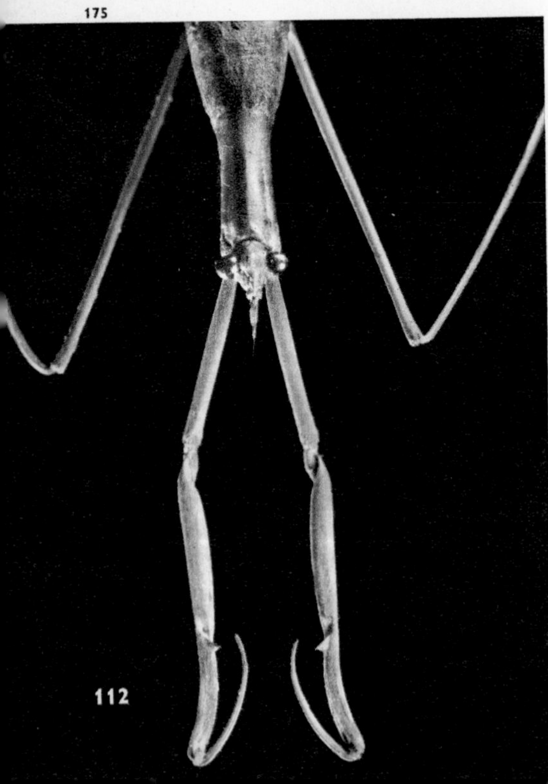

divided into two genera, which differ greatly in appearance: *Nepa* and *Ranatra*. Both are, however, an inconspicuous brown or greenish colour and have a long, double respiratory tube at the ends of their bodies. The front legs are adapted for seizing, and operate on the principle of a jack-knife. However, Nepa is short and flattened, whereas Ranatra has a long, needle-like body. Both are able to deliver a powerful and painful bite, of which man can also be the victim in wading through shallow, muddy water.

Familiar European species are the **Grey Water Scorpion,** *Nepa cinerea* [174], about 2 cm long, and the **Long-bodied Water Scorpion,** *Ranatra linearis* [175, 176], which is approximately 3 to 4 cm in length.

The Long-bodied water scorpion is distributed generally throughout central Europe, although it can seldom be observed, since its inconspicuous colour and body-shape make it very difficult for man to distinguish it from the water-plants among which it lives.

The **Backswimmer** or **Water Boatman,** *Notonecta glauca* [177], grows to a length of 16 mm. It is an inhabitant of large areas of water, not only in Europe but also throughout the Old and New Worlds. As its name suggests, it swims on its back. Its belly is yellowish, but often dark, and sometimes actually black. The flattened, greatly lengthened hind legs, which are beset with tiny but long hairs, are used as oars. It is a voracious predator and uses its powerful sting to kill not only insects but also small fishes. Its sting is also very painful to humans.

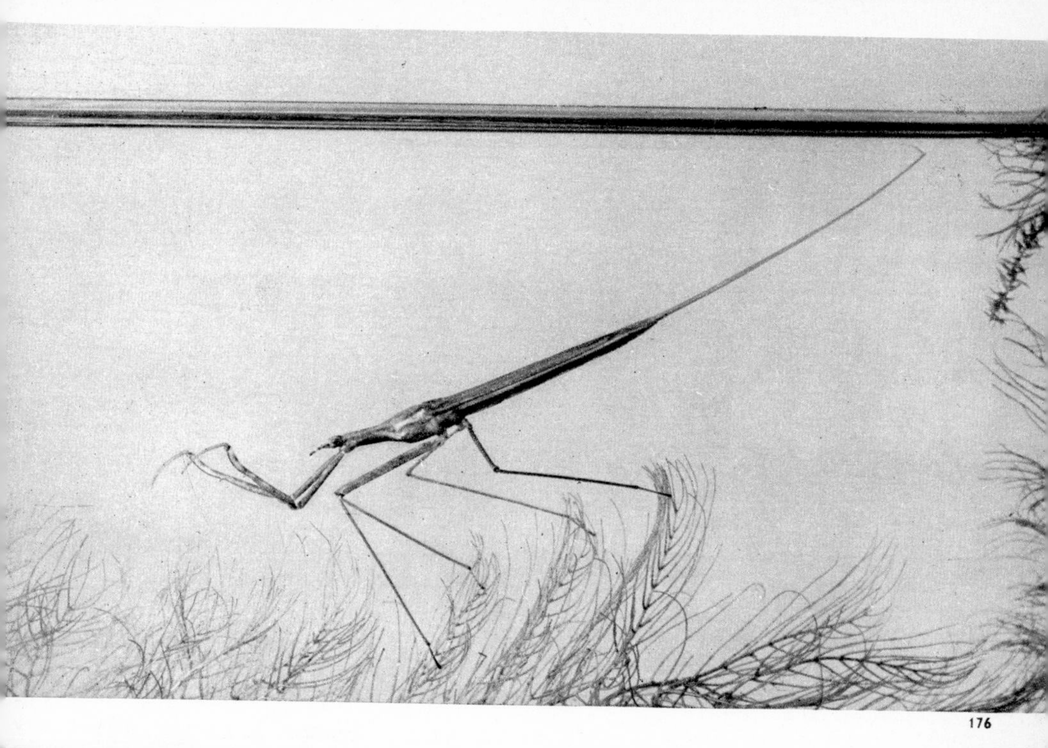

176

177

178

The **Lesser Water-boatman,** *Corixa punctata* [178], measures about 1.5 cm and is coloured brown. Its wing-covers are mottled and the shield irregularly striped. It is a member of the **Lesser Water-boatman** family *Corixidae,* which is distributed in about 300 species throughout the world in fresh and brackish water. The water-boatmen swim very swiftly, with a jerky motion, for which they make use of their rear legs, which are flattened into an oar-like shape and set with bristles. As a result of their ability to produce a clearly audible chirping sound, in some places they are called water cicadas. They have no bite or sting, but rake through the mud on the bottom with the shovel-like processes on their front legs and live on decaying vegetable and animal matter.

The widely distributed **Sigara balleni** [179] is about 8 mm long. It is one of the chirping members of the aquatic bugs and has the same habitat and way of life as *Corixa*.

Plea atomaria [180] is a small backswimmer, light brown in colour and about 3 mm long. It makes its habitat almost anywhere, and can even be found in puddles.

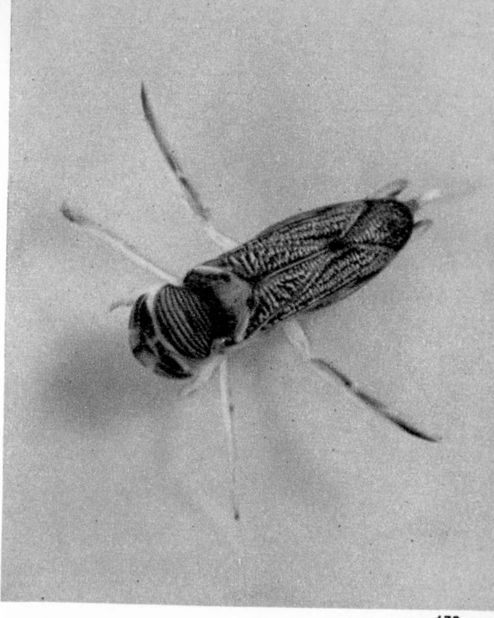

179

180

The second order, the **Plant-suckers,** *Homoptera,* consists of plant pests with mouthparts adapted for piercing and sucking. They are present in large numbers throughout the world. Among them are many which wreak great damage in agricultural and forested areas, as well as in gardens.

The first suborder, the **Cicadas,** *Cicadina,* contains the singing family *Cicadidae,* which is found chiefly in the tropics. It embraces more than 1,500 species, ranging from dwarfs only a few millimetres long to giants which are among the largest known insects. The largest species, from the Sundra Islands, have a span of 18 cm. The male cicadas emit, with the aid of two arched chitinous plates strengthened with fillets, a shrill, sustained note, which is particularly noisy during periods of bright sunlight and which the insect modulates by moving the rear end of its body from side to side. Many cicadas secrete wax. Of the few European species we mention the **Mountain Cicada,** *Cicadetta montana* [181]. It is about 2 cm long, has a wingspan of over 4.5 cm and lives in sunny clearings and on warm hillsides, specially in oak woods.

The **Manna Cicada,** *Cicada orni* [182], has a length of about 2.5 cm and a span of some 7 cm. The edges of its front wings are veined with a row of brown spots, which are thinly scattered over the whole surface of the wings. In central

181

182

183

Europe it can be found singly in very warm and sheltered forest clearings; in southern Europe, on the other hand, it is present in large numbers.

The **Lurker,** *Tibicina haematodes* [183], has a span of over 7 cm. It is black and the venations on its wings are by contrast a striking red or yellow-ochre. It is very common in the south,

184

185

1

but in central Europe is found only in very warm spots. The development of its larva [184] takes place underground and lasts for four years. It burrows its way slowly forward, feeding on the juice from plant roots.

The **Tree-hopper** family *Membracidae* consists of over 2,500 known species distributed throughout the whole world. They are, as a rule, only a few millimetres long, the largest measuring about 2 cm. Many of them, however, are able to leap considerable distanc with powerful leaps. Their breast-shield greatly extended and equipped with horn spikes and other bizarre growths. In th tropical species these forms are so striking magnified that today it is still not possible ascribe any actual function to them.

Hemikypha punctata [185], from Brazil, about 2 cm long and reddish-brown in colou with yellow spots. Tree-hoppers of the gen

187

188

Membracis [186] from Venezuela are about 1 cm long, dark brown with a white pattern. They have flattened sides. They are reminiscent of the fruit or seedpods of certain plants. The whole family *Membracidae* is named after this genus.
Another Brazilian example of this family, **Triquetra** [187], has its thorax continued forward in three sharp horn-like processes. Its shield, the continuation of which also covers the hindquarters, has an attractive, dimpled texture. Its camouflage in the undergrowth of plants is by virtue of this shield almost complete.
The female of the species **Umbonia spinosa** [188] is about 14 mm long and coloured green, with red borders around the edges of its body and on its thorn-like process, which in shape mimics the thorns of certain plants. It is a native of Central and South America.

189

One of the most adventurous shapes in the animal kingdom is surely that of the **Indian Tree-hopper,** *Hypsauchenia hardwigii* [189], from Sikkim. It is 8 mm long. It has a long, thorn-like process at the end of which are two banner-like growths.

190

191

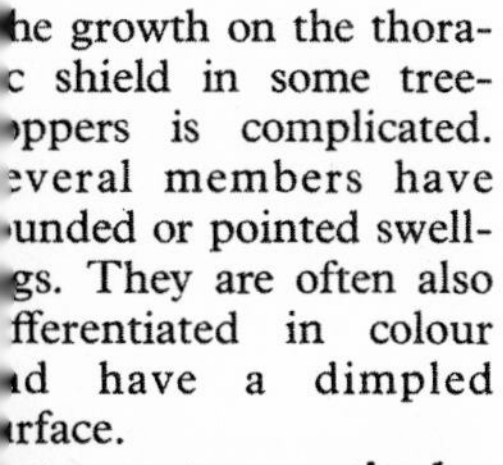

he growth on the thora-
c shield in some tree-
ppers is complicated.
everal members have
unded or pointed swell-
gs. They are often also
fferentiated in colour
d have a dimpled
rface.

**eteronotus reticula-
s** [190] combines a body
ngth of only 1 cm with an
tremely bizarre appear-
ce. It is light and dark
own in colour and lives
Brazil. In another Braz-
an species, **Bocydium
lobulare** [191], the
owth is extended into
ur ciliated balls and a
ng thorn. The whole
sect is 4 mm long.

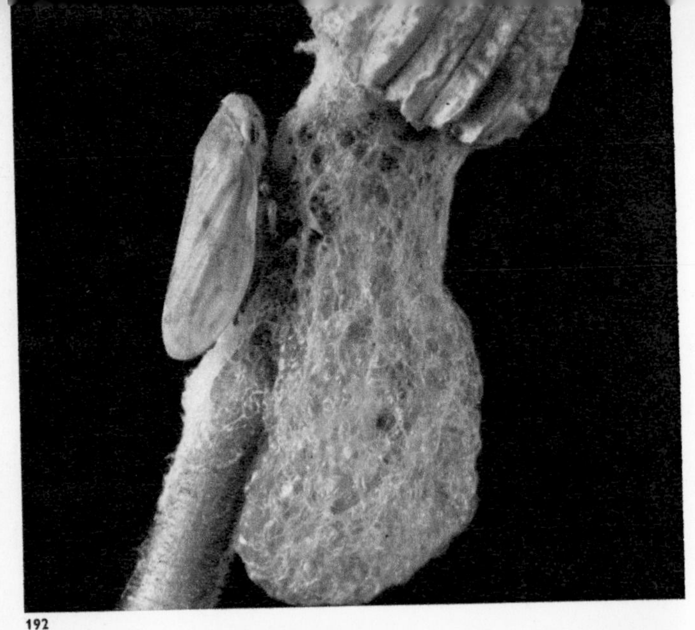

192

193

In the **Frog-hopper** or **Cuckoo-spit Insect** family *Cercopidae* there are some species whose larvae complete their development in a frothy envelope of their own making. The largest of them measures about 1.5 cm. They are present almost everywhere, in temperate as well as tropical zones, and there are about 3,000 species. They feed on plant juices. There are periodical increases in numbers, as is quite common among insects, and when this happens they effect considerable damage to agrarian economics.

The most widely distributed species in the Old World is the **Meadow Frog-hopper,** *Philaenus spumarius.* It is about 5 to 6 mm long and of very various colouring: often yellow, but black and mottled individuals are also found. The larvae of this species make frothy envelopes in which they live (the cuckoo-spit), attached especially to the cuckoo flower. Illustration [193 shows an adult insec (imago) just after emerg ing from the nymph. Par of the frothy secretion ha been removed in order t show the inner structure In illustration [192] we se another newly-emerge insect, which has not ye become fully coloured.

A somewhat larger species s the **Alder Frog-hopper,** *Aphrophora alni* [194]. It is about 10 mm ong and grey-brown with ighter oblique stripes. Its arvae live in the grass of lamp meadows; the adult nsect is found on bushes and trees, especially on villows and alders.

The **Leaf-hopper** family *Jassidae* contains over ,400 species. Most of them are less than 1 cm ong, and brightly coloured. Many of them warm easily and do great lamage to cultivated plants.

Empoasca flavescens [195] is not quite 4 mm ong. It is very common on bushes, also on forest and fruit trees. It does great damage to hops and vines, on the leaves of which it leaves light yellow-brown marks.

194

195

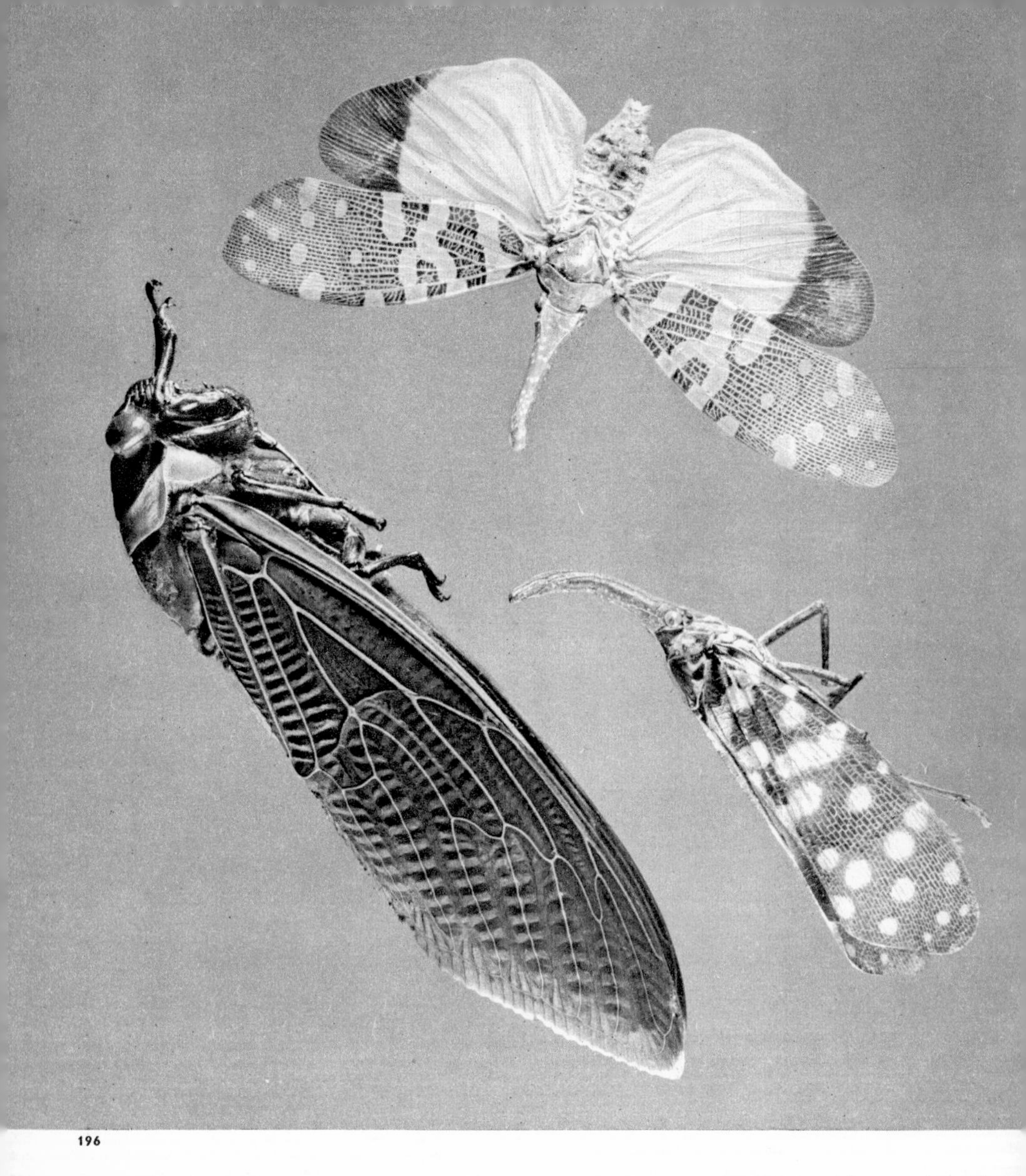

196

The tropical cicadas are often so gaily coloured that they rival in beauty and interest the butterflies. Under an inconspicuously coloured wing-cover they conceal radiantly coloured rear wings, so that when they unfold the effect of the contrast is all the more striking.
Illustration [196] shows on the left **Cicada speciosa** from New Guinea. It is dark green with a metallic gloss. The wings have a red venation and there is a yellow and red band across the breast. It measures 8 cm with wings. The specimen shown with extended wings is **Fulgora spinolae** from tropical South-East Asia. Its wing-covers are green with a yellow pattern; the hind wings are bright yellow with dark brown tips. Its span is 6.7 cm.
The family of **Lantern Flies,** *Fulgoridae* have an extremely long, narrow head with a

crescent-shaped thorn-like growth on top, which curves upwards and backwards. On the right, with folded wings, is **Fulgora maculata** from Ceylon. Its wing-covers are red with turquoise venation and white spots; the hind wings are light blue and black at the back. It has a span of 5.3 cm.

Another member of the lantern fly family is **Pyrops pauliani** [197]. It is about 6 cm long. At the sides of its hollow head-process it has a dark patterning.

The **European Lantern Fly,** *Dictyophora europaea* [198], is bright green, with transparent, green-veined wing-covers. The head is extended into a conical peak. Whereas it is very common in southern Europe, in central Europe it is only found on very warm grassy slopes. In central Asia it does considerable damage to melon plantations.

197

198

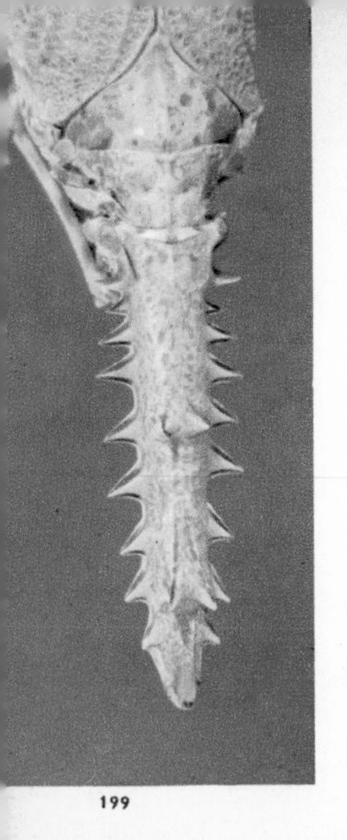

199

20

Cathedra serrata [199] carries on its forehead a most adventurous growth in the form of a thorny branch of exactly the same length as all the rest of the body, in all measuring 7 cm. It lives in tropical South America. The various large **Lantern flies** in that part of the world are distinguished from each other by the different bladder-like growths on their heads. Formerly, it used to be believed that these were shining lanterns, and it was from this mistaken notion that the family was named. Most of them possess striking designs on their hind wings in the form of "pheasant's eyes" which can be clearly seen on the wings when extended [200].

The function of the growth on the head of the

200

204

203

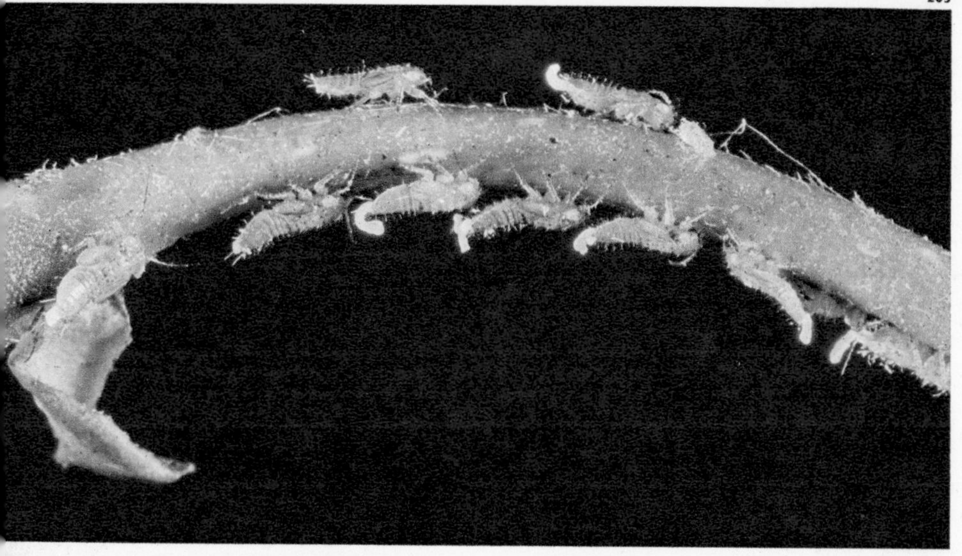

ntern Fly, *Laternaria laternaria* [201], is wise unknown. It has a span of 9 cm.

e second suborder, the **Jumping Plant e,** *Psyllina,* consists of about 1,000 species sap-sucking insects between 2 and 5 mm g. The adult forms are winged and resemble cicadas, except that they have long antennae. They are usually very active, their movements being a combination of leaping and flying. The larvae of some species secrete large quantities of wax. Many of them are feared as plant pests, as for example the **Apple Plant Louse,** *Psylla mali.*

The **Alder Plant Louse,** *Psylla alni* [202,

205

imago], is between 4 and 5.5 mm long. Its larvae stay hidden under an exudation of white wax which can be found on branches and twigs of alder [203]. Picture [204] shows the larvae.

The third suborder, the **White Flies,** *Aleuroidina,* consists of tiny sap-sucking insects with two pairs of wings which are dusted wi a white powdery wax. The largest of them not achieve a length of 3 mm. Only about 2 species of white fly are so far known. They a mostly natives of hot countries, only a f species being found in the temperate zon The larvae as well as the adult insects atta

206

207

ıe leaves of plants, such as the celandine or ıaple, sucking the sap. After the third instar f the larva the insect adopts an inactive state he pupal stage). During this period the evelopment of the imaginal or adult insect is ɔmpleted. The pupal case of **Aleurochiton ɔmplanatus** [206] is found on maple leaves.

It measures about 2 mm. **Trialeurodes vaporarium** [205, 207], a very harmful species, can breed in vast numbers in greenhouses as well as in the open. It is a native of Central America, but has been imported into Europe with orchids. It measures about 1.15 mm.

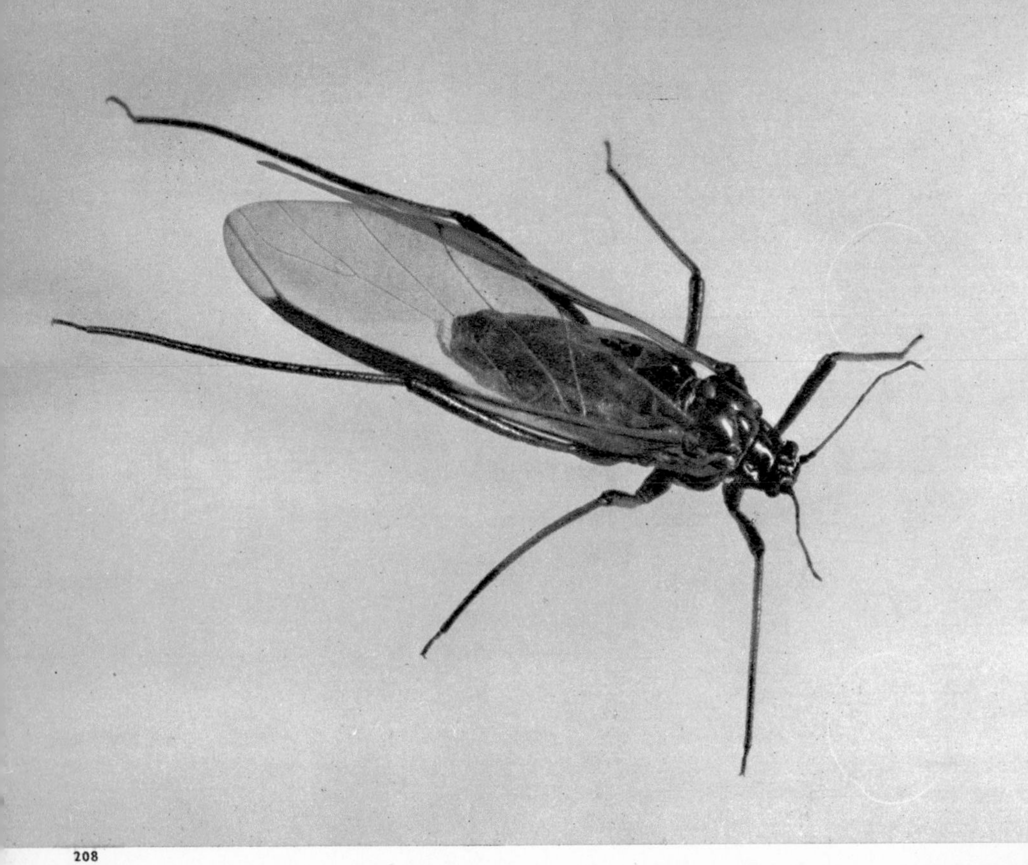

208

209

210

The fourth suborder, the **Greenfly** and **Plant Lice,** *Aphidina,* contains some 3,000 species of extremely harmful plant pests. They have oft bodies, long legs and measure between).2 and 6 mm. Very few species are as long as : mm. They can be found on almost every pecies of plant in the world. Their developnent is extremely complicated, also varying rom species to species. The alternation beween sexual and parthenogenetic, or winged nd wingless generations, is frequently closely elated to the life-cycle of the host plant.)uring a single year, as many as eight generaons can develop under favourable conditions. 'hey reproduce in enormous quantities and the damage which these tiny sap-suckers can effect is extraordinarily great. They are also feared as carriers of virus diseases.

Cinaria piceae [208] has a body 5 mm long, and with the inclusion of the wings measures over 7 mm. It lives on pine trees, in certain generations on the roots, but otherwise on the trunk and branches.

In illustration [209] we see a wingless generation of parthenogenetically reproducing females of the species **Callaphis juglandis** on the upper surface of a nut tree. They are about 3 mm long and yellow-green.

Brachycaudus lychnidis [210] has a dark colouring and is spherical in shape.

211

Our illustration [211] shows a culture of **Yellow Plant Lice,** *Paramyzus heraclei,* on the stem of cow parsnip, *Heracleum sphondylium.* They suck the sap from the leaves, which afterwards become flecked with spots and begin to curl.

Most plant lice are wingless and female. This, the most common form, called the fundatrix, emerges in the spring from fertilised, overwintered eggs. It brings forth a whole generation of progeny viviparously and parthenogenetically. These offspring grow into wingless females called fundatrigeniae, which resemble the fundatrix, live on the same host plant and can give birth to living young parthenogenetically. In a favourable year u

212

to eight generations of such asexually reproduced females can be gone through, as a result of which the host plant is overwhelmed by hoards of these attackers.

The **Plum Greenfly,** *Hyalopterus pruni* [212], has two lighter stripes running lengthwise over its green body. It has red eyes and is covered in white powdered wax. The illustration shows the summer form, which lives on reeds up until the autumn. Then it changes host, driven by the need to develop further. After a few parthenogenetic generations, winged females occur, which then leave the primary host (in this case the sloe, plum or apricot) for a different plant (in this case reeds). There they continue to reproduce until the

last parthenogenetic generation (sexuparae), from which both males and females are born.

Dactynotus (Uromelan) jacae [213, 215], from central Europe, attacks in the main the cornflower genus, *Centaurea scabiosa, C. rhenana,* the stems of which in June are covered in wingless females. The plant lice are dark brown with a golden sheen. They generally suck with their heads pointing downwards.

The wingless females of the **Cabbage Plant Louse,** *Brevicoryne brassicae* [214], are about 2 mm long, grey-green in colour and dusted with powdered wax. They attack, especially in late summer, the crucifers of the genus *Brassica,* among which are numbered some of our most important vegetables. This species is cosmopolitan in distribution. In warm countries the wingless females spend the winter on the plants.

213

214

216

217

The **Pear-root Louse** or **Elm Louse,** *Schizoneura lanuginosa,* measures in its wingless female form 2 mm. It is dark in colour, but powdered with wax and covered in long, delicate hairs. It is the occasion of the formation on elm trees of galls up to the size of a potato [217]. These are originally green, but become red and finally harden and become brown. Inside them live and reproduce several generations of parthenogenetic plant lice. At the bottom of the gall the honey dew collects, which the plant lice emit drop by drop and which is saturated with a sweet solution [219]. In August the galls open and the last generation of winged sexupares leave the primary host for the roots of a pear tree. There, a bisexual generation arises which after mating gives rise to fertilised eggs which overwinter on pear tree roots.

An important pest of apple trees is **Woolly Aphis,** *Eriosoma lanigerum.* Originally a native of America, it spread throughout the world during the second half of the last century, and can be found today wherever there are apple trees. It likes warmth and is therefore unable to survive in places where there are regular heavy frosts. In Europe it reproduces only parthenogenetically. It lives in colonies underneath a protective covering of white fluffy wax [218]. If we press one of these waxy forms a blood-red fluid comes out, which originates from the body of the plant louse (hence the nick-name "blood louse"). It does not migrate, but attaches itself exclusively to the trunk, base and branches of an apple tree. The tree is so weakened by the parasitism of these insects that it attempts to protect itself by the formation of excrescences, gnarled thickenings and scabby surfaces, which burst and look like cancer growths. These then become vulnerable to infection.

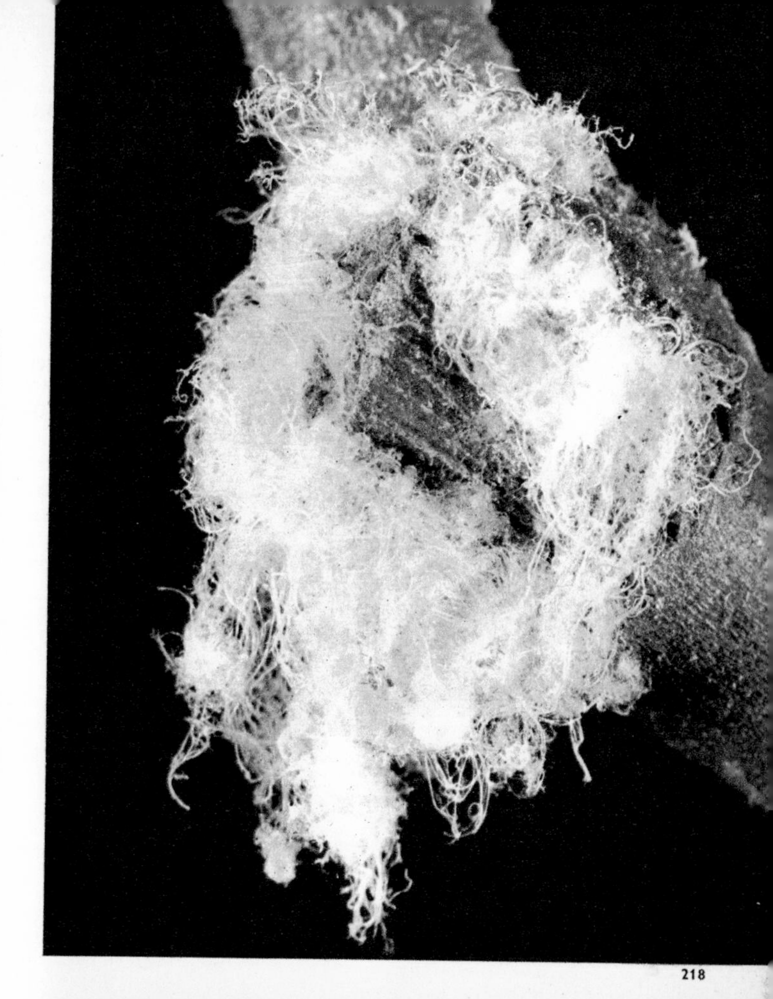

218

219

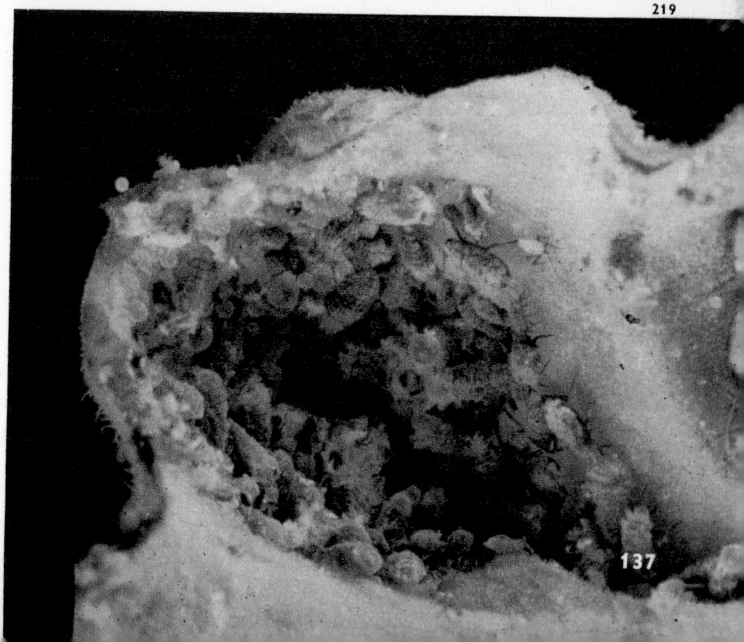

The **Plant Louse,** *Schizoneura ulmi,* attacks **Mountain Elm,** *Ulmus glabra,* **Smooth-leaved Elm,** *U. carpinifolia,* and, as secondary hosts, redcurrant and gooseberry roots. Its presence is betrayed by the leaves which become pale and roll up into conical or cigar-like shapes [220], in which only the one half rolls up, towards the central vein of the leaf. In this hiding place live and reproduce the yellow-brown, viviparous females which are about 1.5 mm long. At the beginning of June the winged generation of females is fully developed and flies to the roots of redcurrant or gooseberry bushes. In the autumn the fully-grown generation of sexuparae returns to the elm. The damage to gooseberry and red-currant cultures is generally very great.

Pemphigus filaginis forms on the leaves of the Lombardy and black poplars green galls, which swell up on the central vein of the leaf

220

221

21]. The fundatrigeniae migrate to the ɛms of **Cudweed,** *Filago* and *Gnaphalium*.
ɛmphigus bursarius is similarly parasitic ı the white and black poplars. Its galls can found on the thickened stems of leaves; ey are bag-shaped and have a small groove opening on the side [222]. At the end of ıne the winged females emerge from the ılls and make their way to the roots of lettuce, ıbbage or certain cornflowers. In late August, w generations of plant lice wing their way ıck to the poplars, where the females lay eir eggs for overwintering.
here is a further kind of gall to be found on e poplar, which causes the leaf-stem to roll into a spiral [223]. This is formed by **ɛmphigus spirothecae,** which remains on e poplar during the whole cycle of its velopment and does not migrate to other ants.

222

223

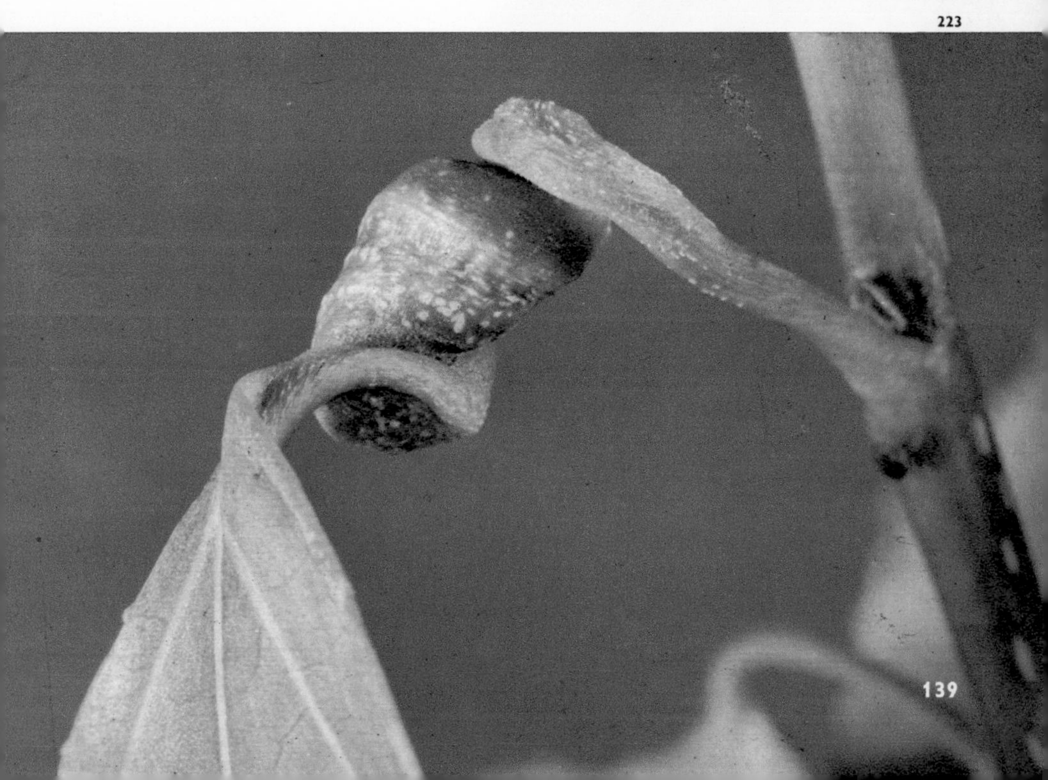

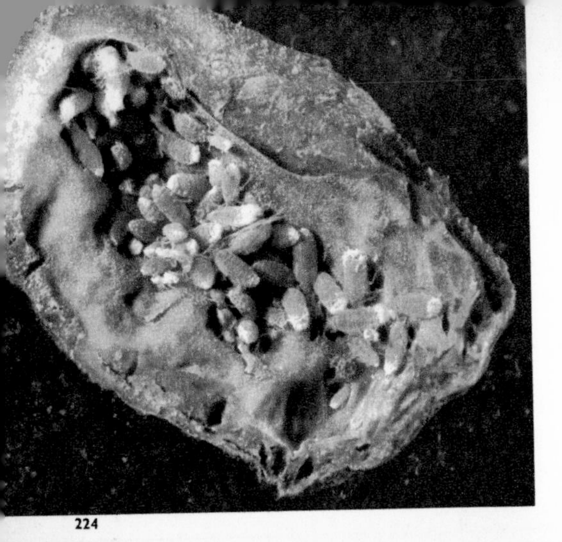

Sometimes, galls can also be found on young shoots. A cross-section of the poplar plant-louse gall reveals nymphs in the process of growing into winged forms [224]. In central Europe, the galls reach this stage at the end of November.

The *Adelgidae* family contains some very widely distributed species of conifer pests. The primary host of all *Adelgidae* is the spruce. The **Spruce Gall Louse,** *Sacchiphantes* [*chermes*] *abietis,* forms fleshy galls at the bases of young spruce shoots, which at first are green and look like small cones. Our illustration [225] shows a cross-section through a young, living gall, with plant lice in its chambers. During the characteristic periods of mass reproduction plant lice galls can be seen on

224

225

226

lmost every new shoot, and a tree so afflicted ecomes stunted in its growth. Illustration 226] shows an open, dried-up gall which has already been abandoned by the plant lice. The spruce gall louse is common everywhere where there are spruce forests, and has also

227

228

been carried across the Atlantic to America.
The fifth suborder, the **Scale Insects,** *Coccina,* contains over 4,000 species of these highly specialised plant-suckers. Their size varies between barely 1 mm up to 1.5 cm, and the longest of them achieve 3 cm. On account of their extraordinary powers of mass reproduction they are among the most widely feared pests of cultivated plants. They are frequently so highly adapted to their life on the bark and leaves of trees and plants that some of them lose the power of moving away. A particularly common member of the family **Cane Scales,** *Ortheziidae,* in Europe and Asia is the **Nettle Scale,** *Orthezia urticae* [227, 228]. The adult female is about 4 mm long but is completely covered with a wax excretion extending over its rear so that altogether it can measure up to 1 cm. in length. The rear part of this wax shield is hollow and serves as a receptacle for the eggs. The female is black but has a very striking chalk-white pattern on

230

r back.

ıe female **Coffee Mealy Bug,** *Pseudococcus onidum,* a member of the family *Pseudo-cidae,* is very active even as an adult, mbing nimbly about the host plant. It has a aracteristic form and structure [230], measur-; up to 6 mm and being covered with a white yellowish wax excretion. It is cosmopolitan, t in colder climates it is confined to green-uses, where it is parasitical upon the leaves and stems of succulent tropical plants. Each female gives birth to up to 300 successors. The males are lemon-yellow, with white wings.

The female of the **Common Scale Insect,** *Parthenolecanium corni,* is 4 to 6 mm long and a glossy chestnut-brown [229]. The adult female attaches itself to various deciduous trees and shrubs, such as plum, acacia, elm, maple and gooseberry. This parasite is one of the most virulent enemies of fruitgrowers.

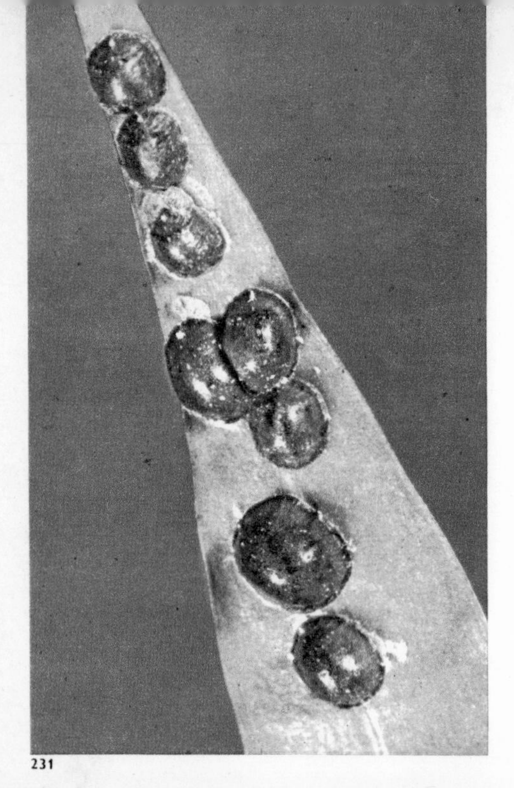

A well-known pest which attacks mainl ornamental plants in greenhouses and indoo is the **Hemispherical Scale,** *Saissetia hem sphaerica* [231, on a sago palm], which is member of the **Cupped Scale** family *Lec niidae.* The shield of the female has th form of an irregular hemisphere up to 4 m across.

The female **Orange Scale,** *Coccus hesperidu* [232], has a shallow, asymmetrical body up 4 mm long. It is generally of variegated colou ing—greenish, brownish, often with dar spots. Its area of distribution is all war countries, where it attacks mainly citrus fru trees, though it is also a parasite of greenhous and indoor plants. It is able to reproduce vast numbers if left unchecked.

231

232

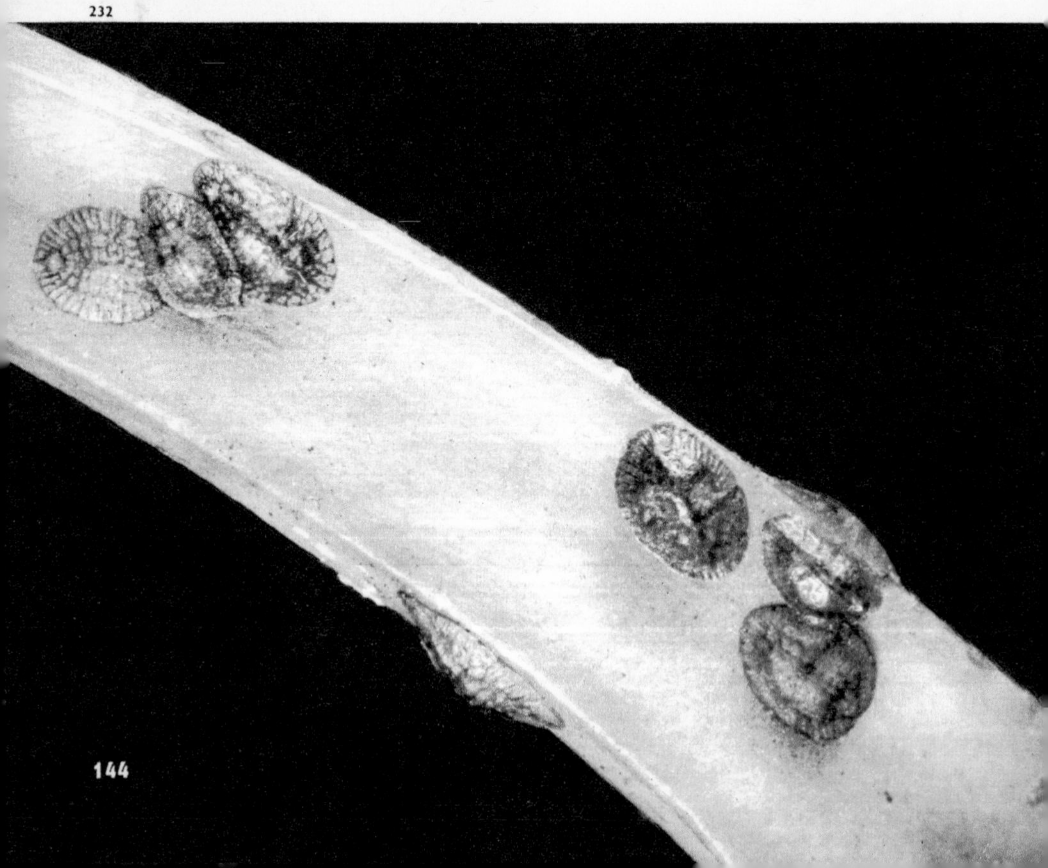

XIII Coffee Mealy Bug, *Pseudococcus adonidum*.

XIV Adult larva of the Pine-needle Sawfly, *Acantholyda hieroglyphica,* on the top of a pine sapling.

233

The **Olive Scale,** *Saissetia oleae* [233], is dark brown. It bears an H-shaped pattern on its arched outer shield. It attacks citrus trees, oleanders and other subtropical and tropical plants. It has been introduced into Europe, being found chiefly in large greenhouses.

One member of the **Armoured Scale** family *Diaspedidae* is the **San José Scale,** *Quadraspidiotus* [*-Aspidiotus*] *perniciosus*. It is yellowish-green, 1 to 2 mm in diameter and lives under a circular shield which at first is white, but later becomes dark grey. Apparently, it originated in North China and has spread as far as Europe since 1926, via California. It is very adaptable and hardy and can even stand frost. Since it reproduces at an incredible rate (a single female can bring up to 30 million offspring into the world in a single year), it is a

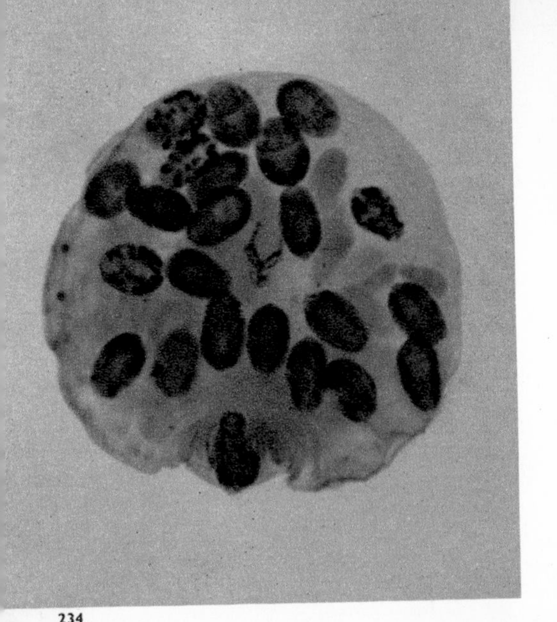

serious menace to European fruit growing in spite of all the counter-measures being taken. Illustration [234] shows a female with the offspring clearly visible.

Female **Oleander Scales,** *Aspidiotus hederae,* are shown in illustration [235] on an oleander leaf. They are about 2.5 mm in size and coloured light yellow. On top of them are the moulted skins of the first larvae. This species is a harmful pest throughout the world.

The **Mussel Scale,** *Mytilococcus—Lepidosaphes ulmi* [236], forms a mussel-like or bow- or S-shaped shield over its back. It can often be found in large numbers on the bark of deciduous and coniferous trees. It is a native of Europe, but today it can be found almost throughout the world and lives everywhere where fruit trees are cultivated. It is a harmful pest of apple and pear trees, which frequently kills off whole branches at a time. Illustration [237] shows the circular shields

234

235

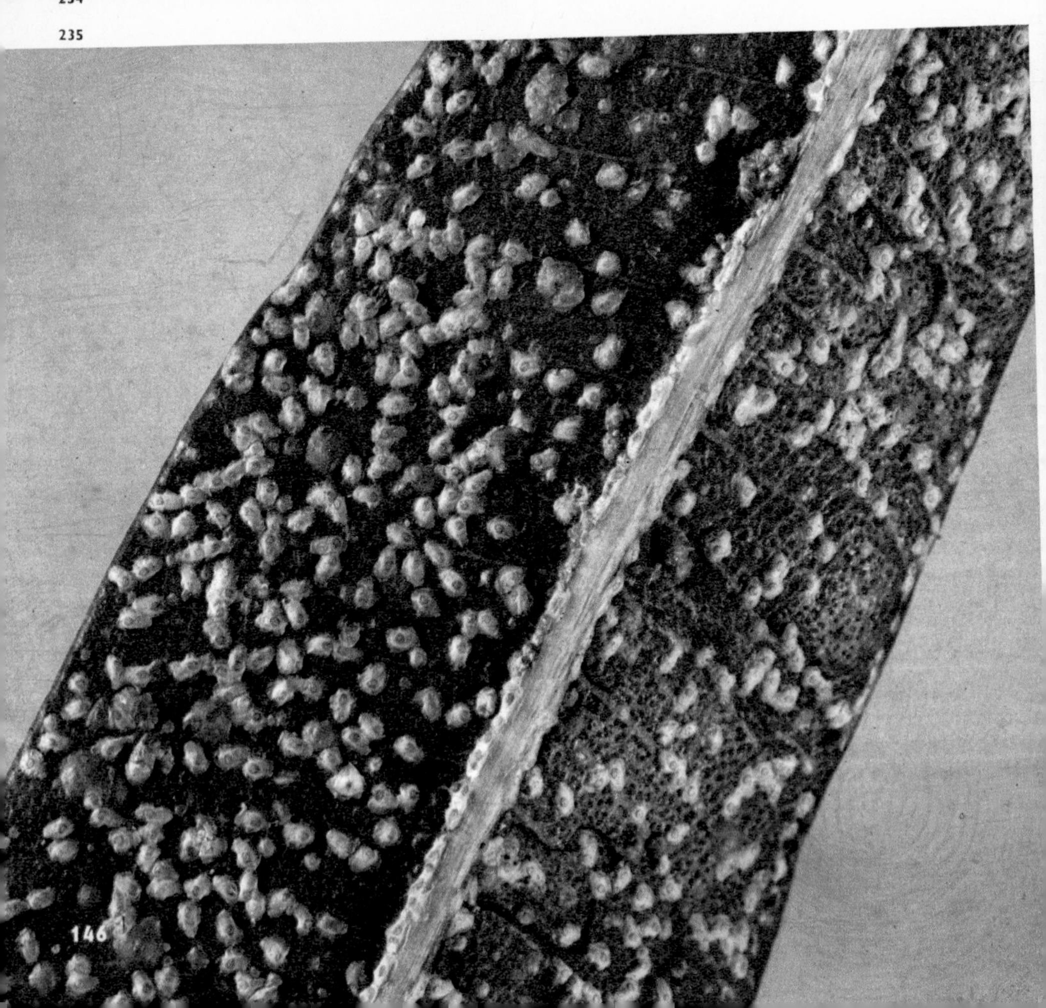

over the back of the female scale insect **Chrysomphalus ficus** on the peel of a grapefruit. They are 1.7 mm long and coloured brown with a red centre. This pest is distributed throughout Egypt and the Near East, Florida and Cuba.

Another scale insect found on citrus trees is **Mytilococcus becki.** The mussel-shaped shells of the females are about 2.5 mm long [238].

Various species of scale insects were once of industrial and commercial significance. The excretions of the female or indeed her ground body were used in the preparation of varnishes, shellac, artificial colourings, medicines, cosmetics and various foodstuffs. The development of chemical industries has, however, rendered the scale insect obsolete as a source of raw-materials; it remains, instead, a constant problem as an agricultural pest throughout the world.

236

237

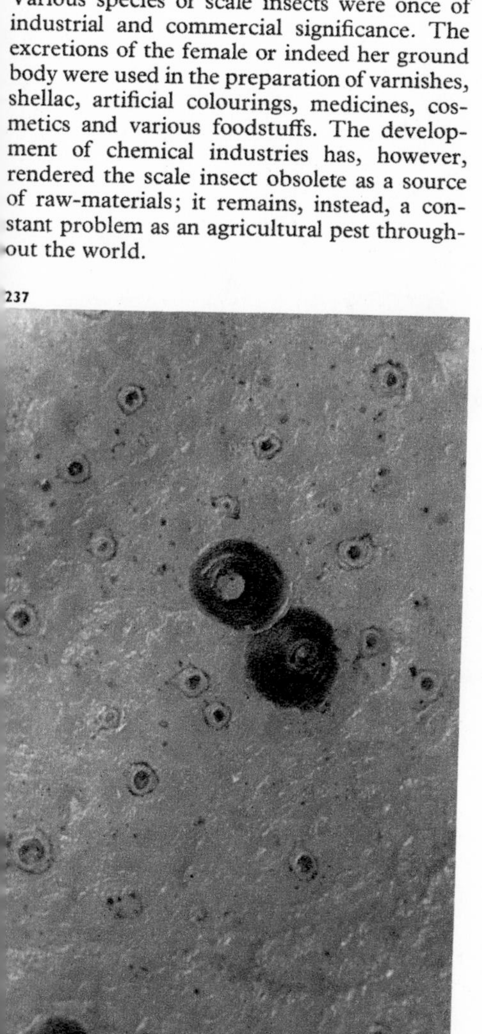

238

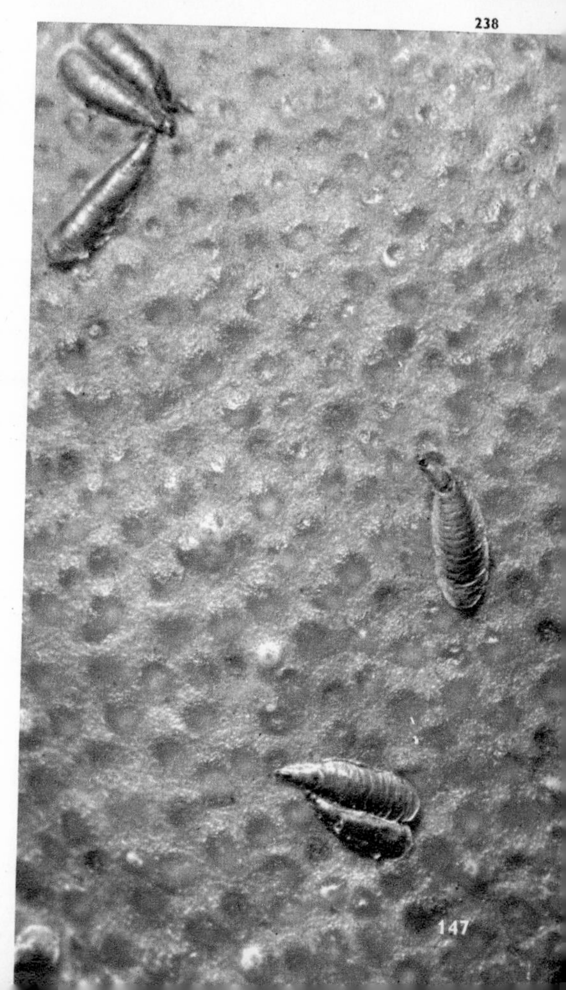

239

The only order of the tenth superorder, the **Hymenopteroidea,** surviving in our era is the order of **Membrane-winged Insects,** *Hymenoptera*.

To date, approximately 300,000 species of hymenoptera have been described, but the actual number has been estimated at about one million. The smallest of them measure only a few tenths of a millimetre, while the largest are several centimetres long. The hymenoptera are the most highly developed of all insects, and they undergo a complete metamorphosis.

They have one pair of compound eyes, three ocelli, comparatively long antennae and two pairs of transparent, membranous wings, which the rear pair is the smaller. They a moved in unison, and in flight the wings each side are hooked on to one another by series of little chitinous hooks, called hamu The mouthparts are generally of the biti type, though some suck only nectar. Several the order are plant pests, the remainder le various kinds of predatory existences.

The first suborder, the **Sawflies,** *Symphy* contains the oldest archaic group of hyme optera. Their abdomens are partially amalg mated with the thoracic segment. Almost species are herbivorous and many are pests agricultural and forest economies.

240

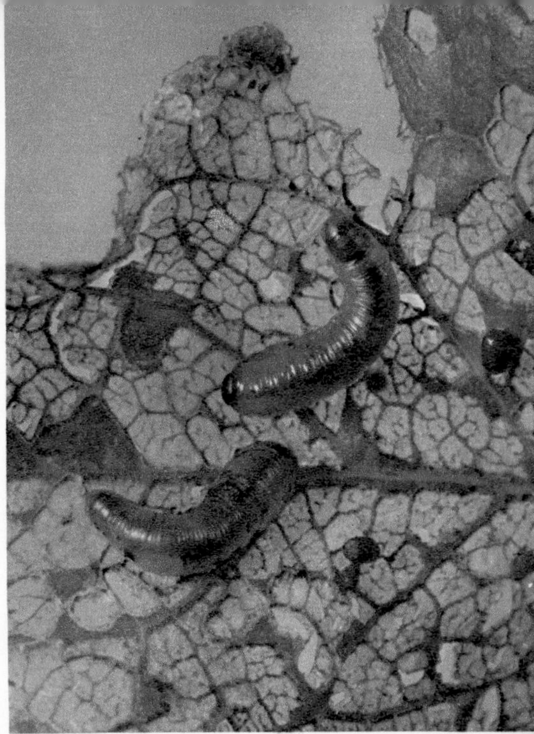

241

The **Rose Sawfly,** *Argo rosae* [240, female], grows to a length of 7 to 10 mm and is green. The larvae [239] feed on rose leaves. It inhabits Europe, the Near East and Siberia.

The **Pear** or **Cherry Sawfly,** *Eriocampa limacina,* attacks cherry and pear trees. Its larvae [241] resemble little black slugs and smell of ink. It is cosmopolitan.

The larvae of the sawfly family, *Tenthredinidae,* also have this general appearance, as illustration [242] shows.

242

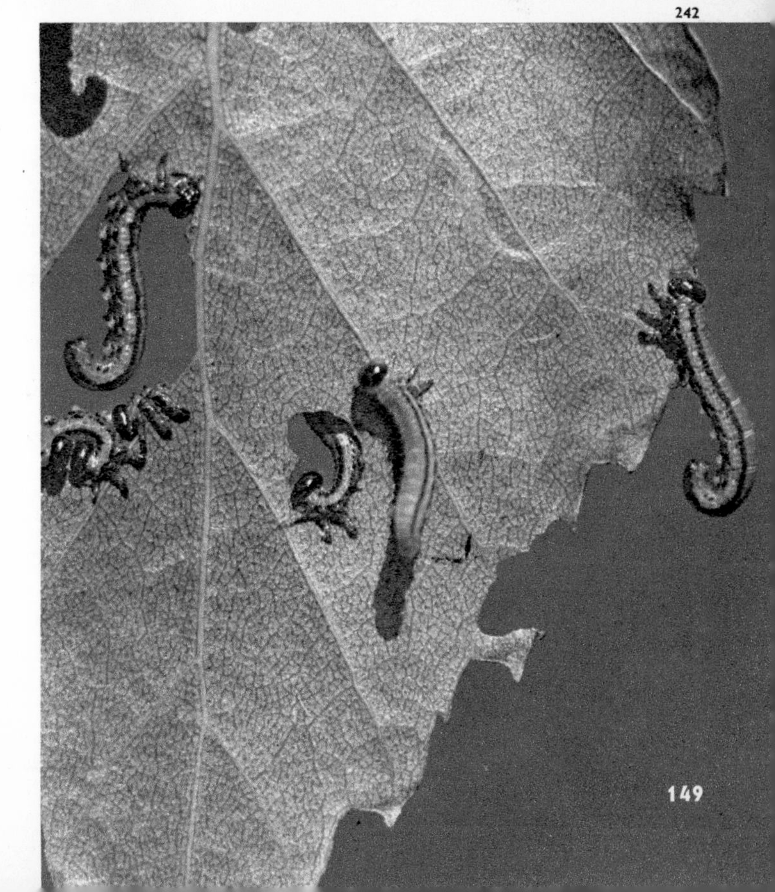

243

The **Birch-leaf Sawfly,** *Climbex femorata,* has short antennae ending in a knob, a characteristic of the **Knob-horn** or **Club-horn** family. The adult measures about 15 to 20 mm [244]. On the left is a male, at the top a female, and on the right the cask-shaped cocoon. The larva [243] feeds on birch leaves.

The larvae of the **Giant Wood-wasp,** *Sire gigas,* of the **Wood Wasp** or **Horn-tail** famil develop inside the wood. The female has syringe-like ovipositor with which it deposi its eggs preferably in the wood of conifers, bı also in ash and poplar. The female [24 achieves a size of up to 4 cm.

244

245

246

247

248

Րhe **Common Wood Wasp,** *Paururus juven-us,* is about 3 cm long. Its antennae are range up to half-way and its body a shimmer-ng blue-black. The male has a brownish-red tripe down his back. Illustration [246] shows pair, the female on the right. The larvae levelop in large numbers in the wood of onifers.

;irex noctilio measures about 3 cm. The ›ody is a metallic blue-black, with the antennae ompletely black. Illustration [248] shows a emale. These larvae live in conifer wood.

Րhe **Pine Wood Wasp,** *Xeris spectrum,* is bout 2.5 cm long and coloured black. The ides of the thorax are lined with two whitish tripes. The ovipositor of the female [247] is s long again as the rest of the body. The arvae develop principally in fir wood, but also n other conifers and even in oak. Occasionally hey are transported with the wood into new ›uildings and crawl, to the astonishment of he inhabitants, out of the ceiling beams, valls or floor.

Tremex fuscicornis measures, with the inclusion of its ovipositor, about 3.5 cm. It has comparatively short antennae and has a dark colouring; the rear part of the abdomen has yellow flashes. The wings are yellow. The larvae live principally in deciduous trees. Illustrations [249 and 250] show two females.

49

250

251

252

The second suborde **Parasitic Wasps** an **Gall Wasps,** *Terebrante* comprises a great num ber of species of hymen optera, whose abdomen generally only joined the thorax by means of thin rod. The females hav a sharp ovipositor, *terebr* The larvae of the **Para sitic Ichneumon Wasp** *Ichneumonidae,* develo generally in the egg larvae, pupae or imagin of other insects, in whic the female lays her eggs– the high point of sophist cation in the insect worl being provisioning for th succeeding generation. B and by the larvae start devour their host fro within, and in this wa kill it. The group of tru **Gall Wasps,** *Cynipoide* contains mainly plan eating species. They hav their development in gal which they cause to gro on plants and trees. Th gall and ichneumon wasp are divided into four larg superfamilies: *Ichneu monoidea, Chalcidoide Cynipoidea* and *Proctotr poidea.*

The **Ichneumon Flie** *Ichneumonidae,* consist about 10,000 species medium-sized insects wit an extended, strongl chitinised body and lon thin, extremely active an tennae. The females hav an ovipositor—usuall very long—at the tail-en of their bodies, by mea of which they can bore way even into solid wood i order to insert their eg into the larvae of wooc eating insects *xylophage* The species with a sho ovipositor lay their eg in caterpillars living o the surface. The larvae on emerging from the egg attack at first the fatty parts and the less important organs of their host, and finally towards the end of their development, consume it completely, so that inevitably it pays for its enforced hospitality with its life. Many species of these *Ichneumonidae* specialise in choosing as hosts agricultural and forest pests. They are, therefore, extremely important allies in the fight against these pests. Most ichneumons and related families live in the northern parts of temperate zones; in the south, they increase in exact proportion to the increase in the butterfly population.

In Europe and Asia lives **Xorides filiformis,** which is a good example of specialisation in one species of host. Its larvae consume the larvae of little **Red Capricorn Beetles,** *Pyrrhidium sanguineum.* This black ichneumon is about 18 mm long; its antennae have a white pattern and the abdomen is red [251, 252]. Its larvae [253] pupate in the larval chamber of the capricorn beetle in a web-like case made out of the body of the latter [253].

Illustration [254] shows two of the largest ichneumons from central Europe. Above is **Ephialtes imperator;** excluding the ovipositor it measures 32 mm and is black with red-brown legs. It develops in the body of the larvae of the larger capricorn beetle, *Ergates faber*. Below is **Rhyssa persuasoria;** including the ovipositor it measures up to 8 cm and is coloured black with yellowish-white spots and edges. It develops in the larvae of the wood wasp.

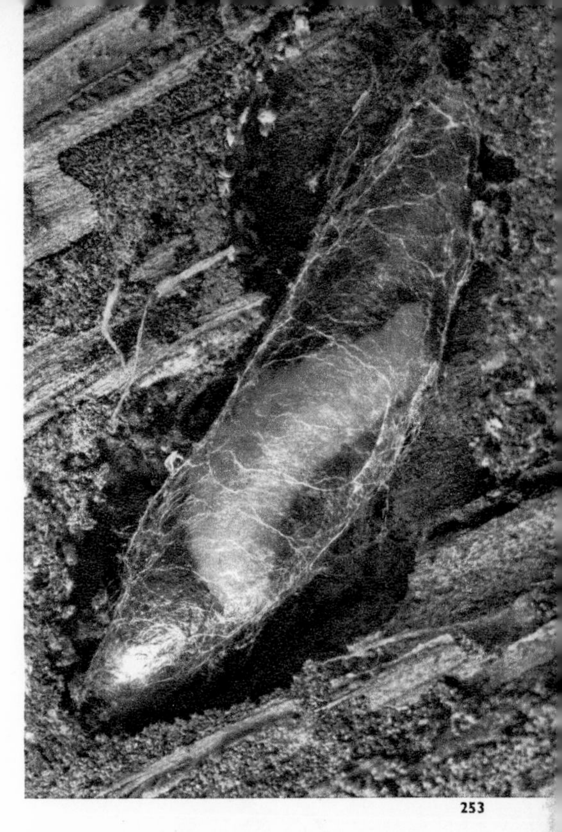

253

254

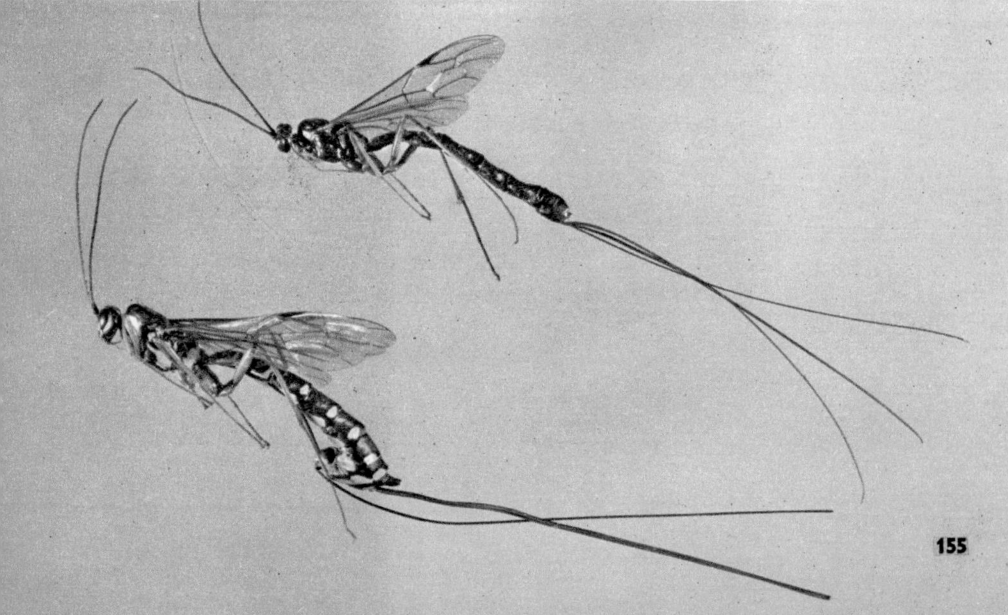

255

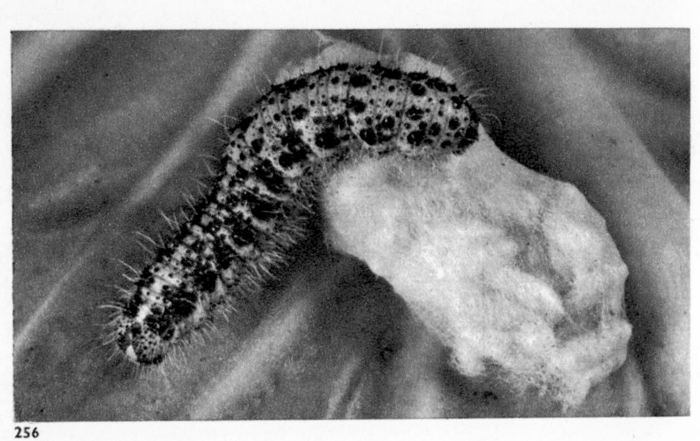

256

257

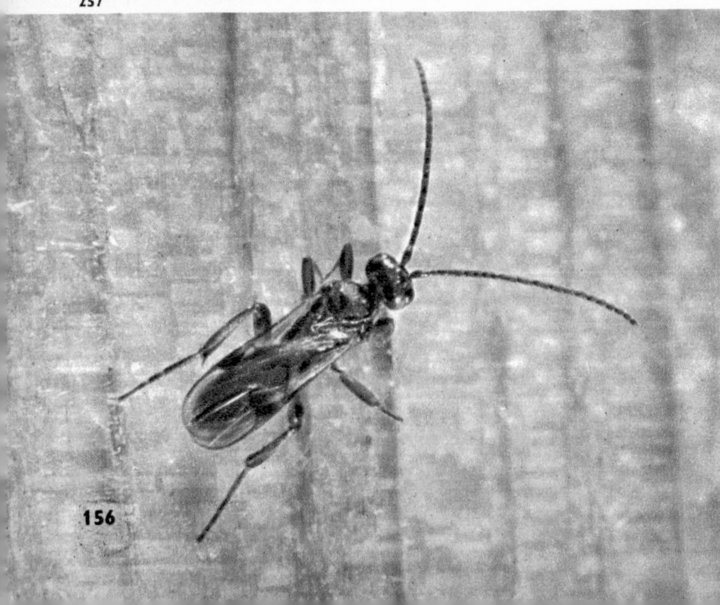

The male of the specie: **Metopius croceicorni:** [255] is 18 mm long; it ha black and yellow rings, a yellow pattern on its head and long, red-brown an tennae. The larvae con sume butterfly larvae and emerge as imagines fron the pupae of the latter The adult ichneumon feeds on sweet plant juices The **Garbage Ichneu mons,** *Braconidae* measure only a few milli metres. Generally they develop parasitically in butterfly caterpillars, in which they lay a larg number of eggs. The tiny white, maggot-like larva bore a way through th skin of the dying hos caterpillar once they ar fully developed, and as a rule immediately spin littl barrel-like cocoons fo themselves which remain attached to the host in littl groups. Some of thes ichneumons develop para sitically in aphids an are therefore importan allies in the biological wa on plant pests. A larg number of other braconid are parasitic on pupae and adult insects, especially th beetle.

A well-known and at th same time typical specie is **Apanteles [Micro gaster] glomeratu** [256], parasitic on th cabbage white butterfly In autumn in centra Europe one sees larg numbers of little yellov cocoons on cabbage leave (so-called "caterpilla eggs"), which contain th larvae of these ichneu mons and which ar attached to the body c their pupated prey, th caterpillar of the cabbag white butterfly. The adul

258

259

braconids measure only a few millimetres. They are dark in colour, have long antennae and a short, pointed abdomen.

The **Microgaster** [257] illustrated derives from a larva developed in the caterpillar of the **Tiger Moth,** *Arctia aulica.* The larvae of this genus are generally parasitic on the caterpillar of the **Lesser Swallow-tail,** *Cerura bifida,* and pupate a few minutes after leaving the body of their host. The latter lives for some three days after this and then perishes.

The cocoons of the **Microgaster** [259] are seen here attached to the dead body of a **Poplar Hawk-moth** caterpillar, *Amorpha populi.* Although there are many species of Microgaster, they are comparatively little known.

260

26

The **Gall Wasps,** *Cynipidae,* measure only a few millimetres. They are, by and large, glossy black or dark brown, with bodies flattened at their sides. Many species of gall wasps undergo a complicated developmental cycle with alternating generations between bisexual and parthenogenetic forms. The larvae of the gall wasps cause the formation of galls on plants, each species having a characteristic form.

262

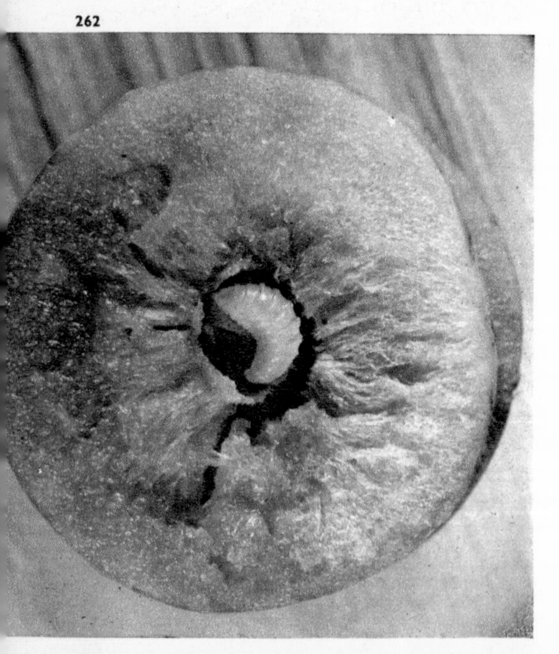

The **Common Oak Gall Wasp,** *Cynips* [*Diplolepis*] *quercus—folii* [260], is 4 mm long and glossy black. It causes by the insertion of its ovipositor the growth of the familiar "oak-apples" [261] on the under sides of oak leaves, which at first are yellow, but in autumn become tinged with red. These have a diameter of between 1 and 2 cm, with a chamber inside and a sponge-like structure; it is possible to squeeze juice out of them. The cross-section in illustration [262] shows a larva about 5 mm long in its central chamber. From the small, lentil-shaped galls [261] beside the oak-apples emerge females of the species **Neuroterus quercus-baccarum.**

Andricus fecundator [263] forms galls on oaks, in the form of hop cones, popularly called "oak roses".

Several species of gall wasps specialise in the oak. Illustration [264] shows galls of three species: on the right, **Cynips quercus—calicis,** which develops on the sides of acorns. The hard, cap-like growth almost completely covers the acorn. In the centre is the gall of **Cynips hungarica,** which is found in the warmer countries of Europe. It forms hard spherical growths with sharp knobs and low ridges on the branches and twigs of the oak; they have a diameter of up to 4.5 cm. On the left is **Adleria [Cynips] caput-medusiae,** which also grows on the sides of the twigs and branches of the oak. The galls are multi-branched, hard and measure 5 cm in diameter.

In former times, galls had considerable economic value as providers of tannin and inks.

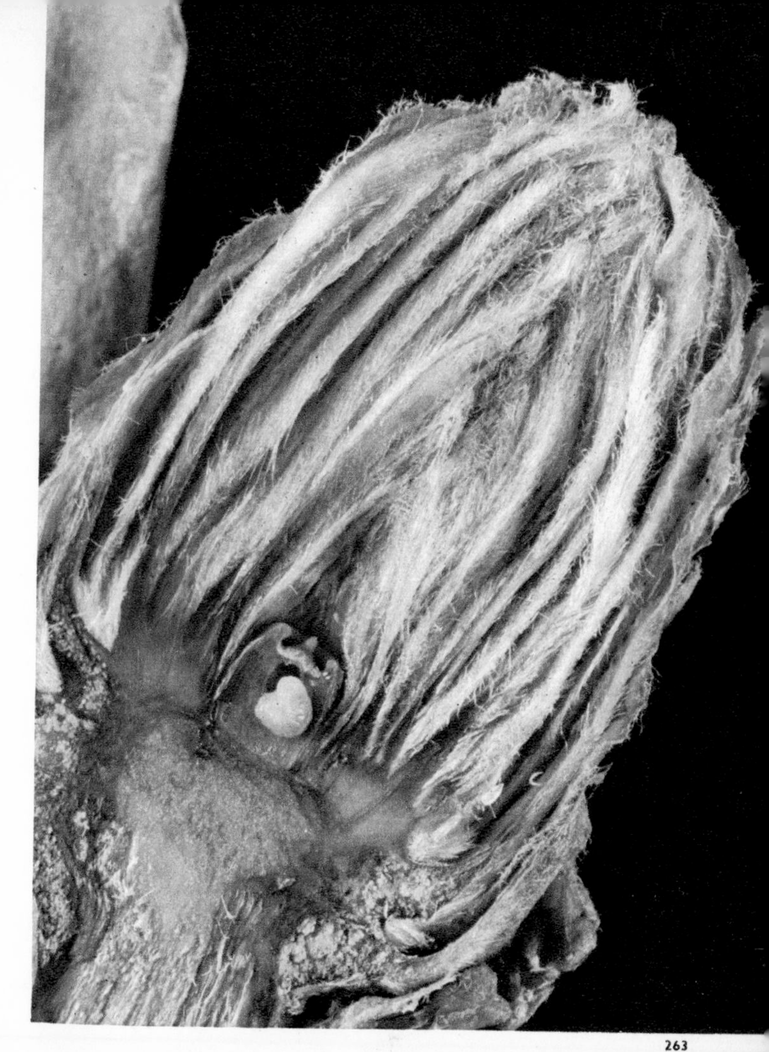

263

264

At the base of the trunk of young and sometimes older oak trees it is frequently possible to find little groups of angular-spherical galls, about 7 mm in diameter, of a remarkable structure. These are caused by the **Oak-root Gall Wasp,** *Andricus testaceipes,* which can even cause the death of young trees with these galls. Illustration [265] shows galls which have been opened and the larva removed by the woodpecker.

The familiar moss-like rose galls of climbing roses, in the vernacular also known as "rosekings" or "sleepy apples", [267] are the home of the larvae of the **Common Rose Gall Wasp,** *Diplolepis* [or *Rhodites*] *rosae.* Inside the matted fibrous growth are a number of sealed chambers, in which the larvae live and feed on the fabric of the gall. These gall wasps

265

266

XV Female Gold Wasp, *Hedychrum nobile*. Central Europe. Up to 1 cm. long.

XVIa Earth Bumblebee, *Bombus terrestis*. Central Europe.

XVIb 'Rose-kings' caused by the Common R• Gall Wasp, *Diplolepis rosae*. Central Eu

generally reproduce asexually, and males are very seldom encountered.

The females [266] are up to 4 mm long; parts of their abdomens and legs are yellowish-red. The bodies are black.

The very numerous *Myrmaridae* family consists of tiny little wasps which lay their eggs in those of other insects. **Camptoptera papaveris** [268] is only about 0.5 mm long, with feathery wings with long lashes.

Some myrmarids are aquatic and lay their eggs in those of other water insects. Thus, the species **Prestwischia aquatica** [269] has its development parasitically in the eggs of large water bugs and sometimes in certain of the water beetles. Its wings are feathery, its body black with a white stripe across the breast. The long legs are well suited to swimming. The whole creature is only 1 mm long. Some species of this family have atrophied wings and are unable to fly.

268

267

269

270

271

The third suborder of hymenoptera consists of the **Stinging Insects,** *Aculeata*. The females of this group have at their disposal a sting which is connected to poison-secreting glands. The *Aculeata* consist mainly of social, colony-forming insects, divided into three larger superfamilies: the ants, wasps and bees. About 200 genera of ants are recognised, and there are about 500 species, subspecies and forms determined to date. The nest of the large **Red Wood Ant, Horse Ant** or **Hil Ant,** *Formica rufa* [270], can contain up to on million inhabitants. It is found in partl shaded spots in mixed and coniferous forests The actual nest itself is underground o situated in the hollowed-out stump of a tree The hill consists of piled-up dry twigs an pine-needles and can reach a height of as muc as 1.5 m. In central Europe, the wood ant present in extremely large numbers; its area

272

273

listribution stretches as far south as the Mediterranean.

The workers [271] are between 6 and 8 mm long and have a brownish-red colouring.

The largest ant of central and northern Europe is the **Giant Ant,** *Camponotus herculeanus*. It is capable of causing extensive damage, due to its habit of selecting for its nest living trees as well as dead ones and gnawing its chambers concentrically, following the annual rings of the tree [272]. The female [273] is 16 to 18 mm in length; the workers, 14 mm; they are all black.

Ants inhabit huge, complicated social states or formicaries, in which each individual carries out the business of its whole life according to the particular function for which it is adapted. The founder, race-mother and head of each formicary, is the fertilised female—the queen. She is born with wings, and the sole function

274

of her existence is to provide for the continuation of the species. Our illustration [274] shows the winged female of the wood ant before fertilisation. The males too are winged; they are more numerous than the winged females. Their function begins with swarming and ends with the fertilisation of the young queen. After swarming, both males and females discard their wings. The largest caste in the ant state consists of sterile females—the workers, who are born wingless. Depending on the species they measure between 1 and 30 mm. Their function is to do all the work of the colony—collecting, storing and preparing the food, building and maintaining the nest and defending it against invaders. Above all, however, they tend, care for and bring up succeeding generations. They can always be observed hurrying about with the larvae and pupae—which are sometimes wrongly called "ants' eggs". Depending on the weather and the requirements of the young, they are bringing them to the surface to warm in the sun, or dragging them into the shade when the sun gets too hot, or back down again to a lower level; at the same time they keep them clean and ensure that they do not dry up. If the colony is situated near a river and if the nest is threatened by floodwater, the workers carry the larvae and pupae quickly to safety on higher ground or on the branches of a tree or plant. Illustration [275] shows ants of the genus **Leptothorax** clinging to the stem and leaves of a thistle, waiting for floodwater to sink again.

275

Some ants are predatory, some omnivorous, while certain species feed on seeds and others partially or exclusively on the excretions of other insects. There is a group which lives parasitically in the formicaries of other species of ants. Some tropical species cultivate fungi on which to feed. Connected with this practice is an interesting feature of the South American genus **Leaf-cutter Ants,** *Atta,* which does considerable damage to certain cultivated plants. Illustration [276] shows a worker of the species **Atta cephalotes,** which is 11 mm long. These dark brown ants march about in vast armies and cut pieces about 2 cm square out of the fresh leaves of young plants, which they carry off to their nests and use as the basic fertiliser for the fungus culture, which forms the principal ingredient of their diet. Illustration [277] shows a worker of the species *Atta cephalotes* from underneath.

A number of species of ants build their colonies inside buildings. These are generally small, but so numerous that they can have catastrophic consequences for the larder. Some tropical species have arrived in temperate countries with imported fruit or ornamental plants, have reproduced there and now live in houses and other buildings. A typical representative of these pests is the tiny **Pharaoh Ant,** *Monomorium pharaonis* [278]

276

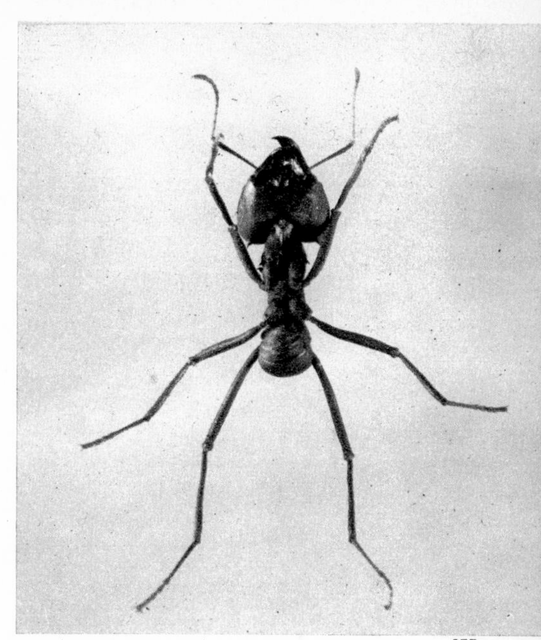
277

278

Some tropical species of ants have become nomadic in habit, whole populations of up to 100,000 individuals setting out on the march and attacking everything living which crosses their path. Illustration [279] shows a "soldier" of one of these predatory and hunting species, the Peruvian **Eciton rapax.** It is peculiarly adapted to its defensive role by the greatly enlarged toothed mandibles of its mouthparts. This soldier is 13 mm long and has a light yellowish-brown colouring.

The **Velvet Ants,** *Mutillidae,* are parasitic wasps but the wingless females resemble ants. They live in the nests of various

279

280

281

282

ɪther species of hymenop-
.era, on which they feed.
Γhe males are winged.
Mutilla marginata [281]
s about 13 mm long, has a
ed-brown breast with a
ough surface and a dark
ear part with light hairs.
Γhe **Dagger Wasps,**
Scoliidae, are distributed
hroughout the world in
nore than 1,000 species.
Γhe winged males and
emales quite often
neasure more than 6 cm
ınd are generally black
ınd yellow. They can be
ound on flowers, protect-
ng themselves with a
ɔowerful sting. The largest
European species is the
**Red-headed Dagger
Wasp,** *Scolia flavescens
sp. haemorrhoidalis* [280].
Γhe female, on the left,
neasures about 4 cm. She
ays her eggs in the larvae
f the stag beetles *Oryctes,
Lucanus,* etc. On the right
s the male.

The **Hornet,** *Vespa crabro,* is distribute throughout Europe and as far north as Laplan It measures up to 3.5 cm, has a black breast, yellow rear part with a dark brown patter and is a terrible predator, hunting down othe insects not only to feed itself, but also chev them and regurgitates them for its larvae. Th hornet is also very fond of nectar and fru juices and will even bite into the flesh of rip fruit. For human beings the sting of th hornet is not only extremely painful, but ca also be dangerous. These insects live in soci nests or colonies, which survive for on season. The nests consist of layers of comb with an outer covering, usually in hollow tree but also in buildings and nesting boxes. Th female has an extremely effective sting, wher as the male [282] has no sting at all.

The young, fertilised queen hornet [28

283

284

2

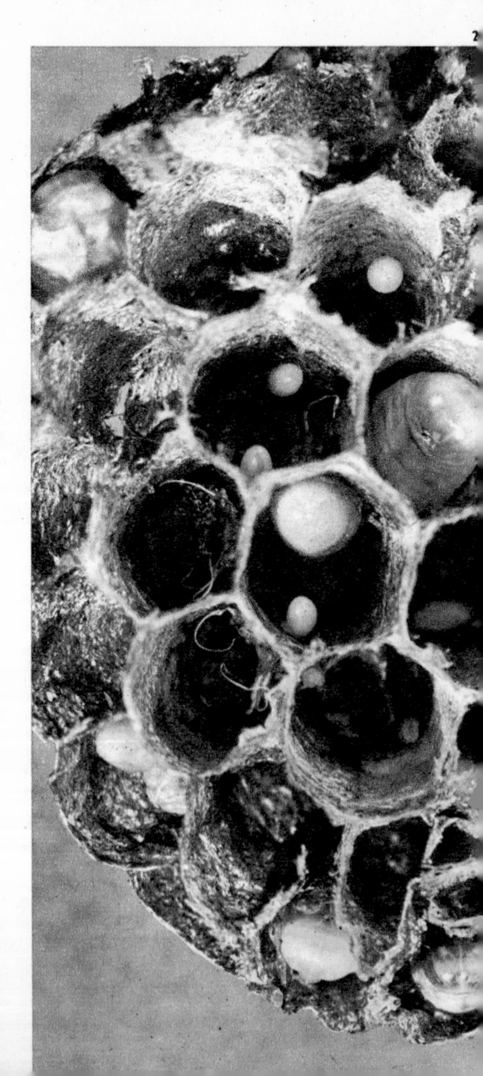

›asses the winter sleeping in hibernation, ›ften underground, and in the spring she ›ounds a new colony.

There are several species of wasp in Europe, ›ach of which builds a characteristic nest the ›onstruction material for which consists of ›vood chewed and mixed with saliva. Wasps' ›ests can be found in a variety of sheltered ›laces—in stables, summerhouses, rafters, ›ut more commonly underground. The **Saxony Wasp,** *Vespa saxonica* [286], has ›lack stripes running lengthways down its ›ead. It builds its balloon-like nest [284] for ›reference in covered places such as rafters. There are three layers of outer covering, with › narrow entrance. In illustration [285] is ›hown the comb with larvae in various stages ›f development, which continue with the pupal ›r motionless stage [287].

286

287

The **Field Wasp,** *Polistes imaculatus,* builds a comparatively small nest with no outer covering, which s attached by a strong talk to jutting cliffs [288]. These wasps live in small ocial groups and are completely harmless creatures. Some wasps are solitary. Their progeny develop in ells built in cracks in walls, in mud walls, or in ome other suitable hiding lace, or attached to plants, nd covered with mud rought by the parent. To hese breeding cells the arent brings the larva a utterfly, a beetle, or a awfly, which it has preiously paralysed with oison from its sting, and ins it down beside the gg. Then the cell is sealed nd the larva develops unlisturbed, amply proisioned with food. Once he fully developed wasp merges from the cocoon, t bites a way through the ell into the open air. There are various different enera of these solitary vasps throughout the vorld, in the tropics and ubtropics as well as in he temperate zones.

The largest European pecies is the **Potter Wasp,** *Eumenes unguiculatus.* The female measures p to 26 mm. Illustration 289] shows a male 24 mm ong. It is black, yellow nd reddish-brown. This otter wasp is an inhabiant of southern Europe, nd attaches its breeding ells close to one another o walls and stones. It overs them with a shallow, rotective layer of mud.

The superfamily of **Digger Wasps,** *Sphecoidea,* is composed of a

289

290

291

series of solitary predator wasps of all sizes, distr buted throughout th world. Many tropic species have a metalli glossy body. Some of the measure more than 5.5 c and are the largest hy menoptera in the worl outside the *Scoliidae*.

One of these is the **Locust hunter Wasp,** *Sphex ir gens,* from Argentin which is black with brown red wings. In illustratio [290] we see a male. Thes giants among insects fee on nectar, but provisic for their offspring in th following manner: the dig tube-like breeding cel in the earth with sever chambers, into which the bring insects which the have paralysed with the stings, and which the peg down beside their egg Some of the insects whic they seize in this way a bigger than they are them selves: the larger cock roaches, long-horne grasshoppers, locusts an even spiders. The large species of the *Sphex* gen do not hesitate to ent into mortal combat on a open piece of ground wi the large, poisonous bir spiders. A Europea species of digger wasp wi much the same way of li is the **Common San Wasp,** *Ammophila sab losa* [291, 292]. It is co oured black, only the wai and the front part of hin quarters being red. measures between 16 an 28 mm, prefers warmt and can therefore princ pally be found in sunn sandy, dry places. The dig ging of their breeding-ce tubes is accomplished b means of their powerf iting mouthparts. The iost common prey for ood for its larvae are the aterpillars of noctuid and eometrid moths. Since iese creatures are often eavier than the sand wasp self and it is therefore unble to fly with them, it ormally drags them with icredible strength along ie ground to the prepared reeding cell, and is not eterred by long distances ver difficult terrain. Once ie prey has finally been rought to its destination, ie parent sand wasp careilly seals the entrance to ie underground breeding ells with a stone which he selects from the neighourhood. These sand 'asps work very swiftly 'hen dragging away their rey, for it sometimes appens that another, arasitic species of wasp an arrive and lay its eggs n the paralysed body of ie prey, which of course estroys its value as longrm provisioning.

'he **Bee-wolf,** *Philanthus riangulum* [293], is 15 to 7 mm long, black with a vely yellow pattern on s head and breast. The bdomen is attached to the iorax without the usual iin waist, is yellow, and n each segment there is a iangular pattern, the tip f which is pointing backards. The bee-wolf uilds its breeding-cell ibes in sandy banks, in 'hich it feeds its progeny rincipally on the honeyee, but also on certain olitary species of bee. 'requently, the prey also erves the parent as food, i that it dismembers the ictim and drinks the oney from its crop.

292

293

294

The best-known member of the superfamily of **Bees,** *Apoidea,* is the **Honey-bee,** *Apis mellifera* [294], which is kept and bred in hives throughout the world. The history of the domestication of the honey-bee is extremely ancient, but the story appears to begin in western Asia. The original home of the genu *Apis* is southern Asia, where today there ar still three wild species known. The economi significance of the domestication of bees do not lie only in the advantages brought by the honey and wax, but also by reason of the

295

light from flower to flower, in which they collect pollen and nectar. They play a role which it is impossible to value too highly in the pollination of wild and cultivated plants. Bees have perhaps the most highly developed sense for the continuance of the species among all insects, living in complex communities which are marvellous examples of organisation and division of labour. Illustration [295] shows a breeding-comb with larvae. The workers emerge about 20 days after their eggs are laid, the drones after 24 days and the

queens after 16 days. All bees are winged.
In any one population of bees there is always only one reigning queen [296]. If a new one is about to emerge, she makes her presence known by a squeaking sound, which the old queen answers with a piping tone. The whole hive is literally buzzing with excitement. A large number of the bees gather around the old queen and finally fly away with her—the bees swarm. The queen does not possess any special organ for the pollen-collection like the workers. In illustration [297] she can be

296

297

XVII Chinese Tiger Beetle, *Cicindela chinensis*. China. 2 cm. long.

XVIIIa Right: Spanish Ground Beetle, *Carabus hispanus*. 35 mm. long.
Left: Golden Ground Beetle, *Carabus auronitens*. 28 mm. long.

XVIIIb One of the most beautiful Ground Beetles, *Coptolabrus coelestis*. China. 4 cm. long.

298

299

stinguished from the workers by her size. ee-keepers mark her out with a white spot the thorax.

hen the larvae pupate the workers seal off e cells with a wax cover [298]. It is possible tell by the size of the cell whether it contains worker or a drone. The drones [299] are out the same size and build as the queen, t have a shorter abdomen and no sting. hey are unable to feed themselves and are urished by the workers. It is the drones that out of the colony or hive with the queen on r "marriage flight". After it they either rish straight away or are refused admittance the hive by the workers.

300

301

The **Bumblebees** or **Humblebees** are distributed throughout the world and are social in the same way as the honey-bees. The populations of their colonies are, however, much smaller, with 50 to 100 members. In the colder countries the whole community perishes at the outset of winter with the exception of the young, fertilised queens which survive to provide new generations the following spring.

The **Earth Bumblebee,** *Bombus terrestris* [300], is distributed throughout Europe. It measures up to 28 mm long and builds its nests deep underground, sometimes as far down as 1 m. Illustration [301] shows this species in the nest. Under the outer covering there are individual barrel-like cells for pollen, honey and larvae.

The **Stone Bumblebee,** *Bombus lapidarius* [302], is likewise a very well-known European species. It is up to 25 mm long and coloured black with a red-tipped abdomen. It prefers to select rocky ground or piles of stones under which to build its nests.

Besides these and the honey-bee genus, which are highly social insects, there are also a whole series of families and genera of solitary bees. They inhabit both hemispheres in many thousands of well-known and surely also of yet undescribed species. Although by and large they prefer warmth, some species can be found as far north as the most northerly reaches of flowering plants.

302

303

304

305

Among the most well known is the **Mining Bee** family, *Andrenidae,* with about 3,000 species, which it is often difficult to distinguish from one another. The genus *Andrena* [303] digs holes in the earth and there builds a cell for its larvae, which it then tends and nourishes with pollen and honey. The bees of the genus **Parasitic Bees,** *Psithyrus,* on the other hand, resemble the bumblebees, in whose nests they live parasitically.

Psithymus rupestris [304, left] lays its eggs in the colonies of stone bumblebees, which then rear the larvae as their own. This "cuckoo bee" is distinguished from its host by the dull colouring of its wings, the more powerfully built body and the absence of "pockets" for the collection of pollen. It is about 26 mm long.

The second largest of the European solitary bees is the **Blue Carpenter-bee,** *Xylocopa violacea* [304, right]. It measures about 25 mm, is glossy black and has dark brown, opaque wings with a lilac-coloured sheen. It is a native of southern Europe and also lives in very warm spots in central Europe. It gnaws holes up to 25 cm long in dry, dead wood (such as dry tree-stumps and building beams) and inserts its larvae there.

The majority of solitary bees build their nests in tunnels which they make in the ground. The **Four-belted Mining Bee,** *Halictus quadricinctus,* tunnels a shaft some 10 cm long in the earth, at the bottom of which it constructs up to 25 interconnected comb-like cells [305], in which it rears its young. It is darkish in colouring, with four whitish diagonal stripes across its rear end.

The **Long-horned Bee,** *Eucera longicornis* [307], is one of the solitary bees which have large hair-like bristles on their tibias. The males are distinguished by very long, waved antennae and a long proboscis. The long-horned bees are up to 16 mm long. In summer they fly from blossom to blossom, sucking the nectar.

The **Leaf-cutter Bee,** genus *Megachile,* is distributed throughout the world, species being particularly numerous in the tropics. It builds its breeding cells preferably in hollow plant stalks or in the holes left by boring beetles in dry wood. A characteristic feature of their behaviour is that they cut little pieces out of leaves [306]—long ones, with which they paper the walls of their cells, and round ones, which serve as lids for them [308].

306

307

308

309

The first order of the eleventh superorder, the **Sheath-winged Insects,** *Coleopteroidea,* is that of the **True Beetles,** *Coleoptera,* one of the most numerous of all orders of insects. It contains about 250,000 species. We divide them into two suborders: 1. **Carnivorous Beetles,** *Adephaga* and 2. **Omnivorous Beetles,** *Polyphaga.*

One of the common features of all beetles is their development, in which they undergo complete metamorphosis. Another feature is the presence in all species of wing-covers. One family of carnivorous beetles is that of the **Tiger Beetles,** *Cicindelidae,* which are distriıted throughout the orld in about 1,500 ›ecies. These slender, redatory beetles with ng legs, large eyes and owerful biting mouth-arts are able to run ex-emely fast and take off ıto flight like a flash of ghtning. They often hunt ısects larger than them-lves and live in dry, arm places. The larvae ave a large, hard head nd a maggot-like appear-nce. They lurk in per-endicular shafts in the arth, which they make ıemselves, with open ıandibles waiting to eize passing insects hich they snatch down ıto their retreat.

'he **Forest Tiger Beetle,** *Cicindela silvatica* 309], is about 17 mm ɔng. It is dark brown with lighter pattern and lives n sandy coniferous ɔrests. The **Common 'iger Beetle,** *Cicindela ybrida* [310], is likewise bout 17 mm long, with opper-coloured wing-overs. It lives on the dges of forests, often on ıigh ground. The largest f all tiger beetles is found n the African savannah: t is the **Night Hunter,** *Mantichora herculeana* 311], dark brown and up o 7 cm long.

310

311

The **Masked Groun Beetle,** *Graphopter serrator* [312], is 17 m long and has a black-an white mask on its win covers. It lives on th edges of the North Africa deserts and has so adapte to its surrounding that bears a surprising resen blance to the tiger beetle which also prefer a sand habitat.

The **Garden Groun Beetle,** *Carabus hortens* [313], achieves a length up to 28 mm. The blac wing-covers have thre rows of little dimples wit a golden sheen. It is

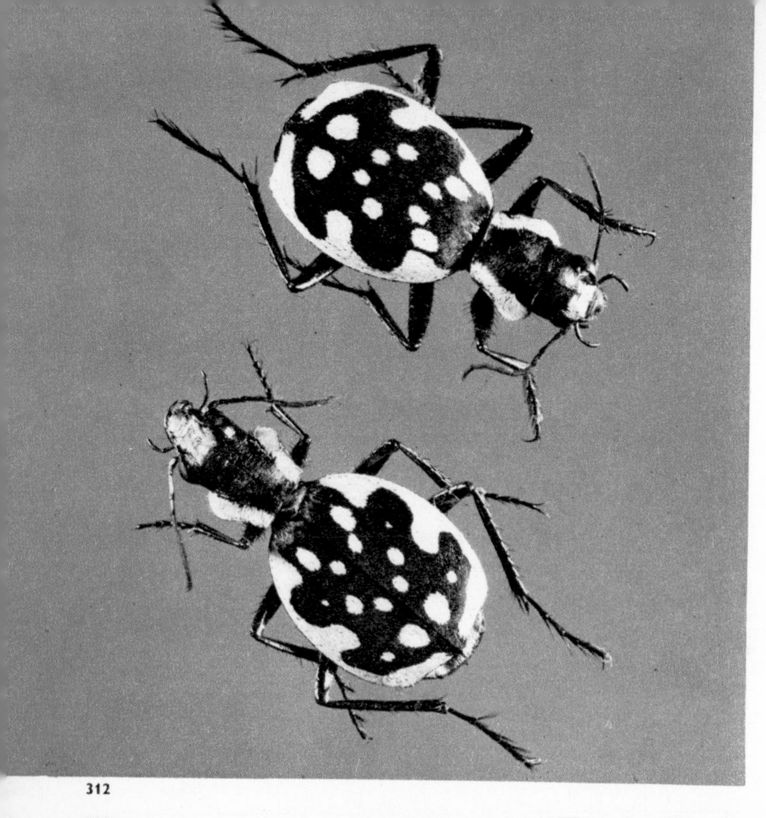

312

313

predator and scours forests and gardens for its prey. The form of its body is characteristic of all ground beetles, with a similar way of life.

The **Japanese Ground Beetle,** *Damaster blaptoides viridipentris* [314], is 35 mm long. The wing-covers and head are dark green, the strikingly elongated, narrow thorax is glossy copper.

The **Ghost Walker,** *Mormolyce phyllodes* [315], from Sumatra has a bizarre, atypical body form for a ground beetle. It measures about 6 cm.

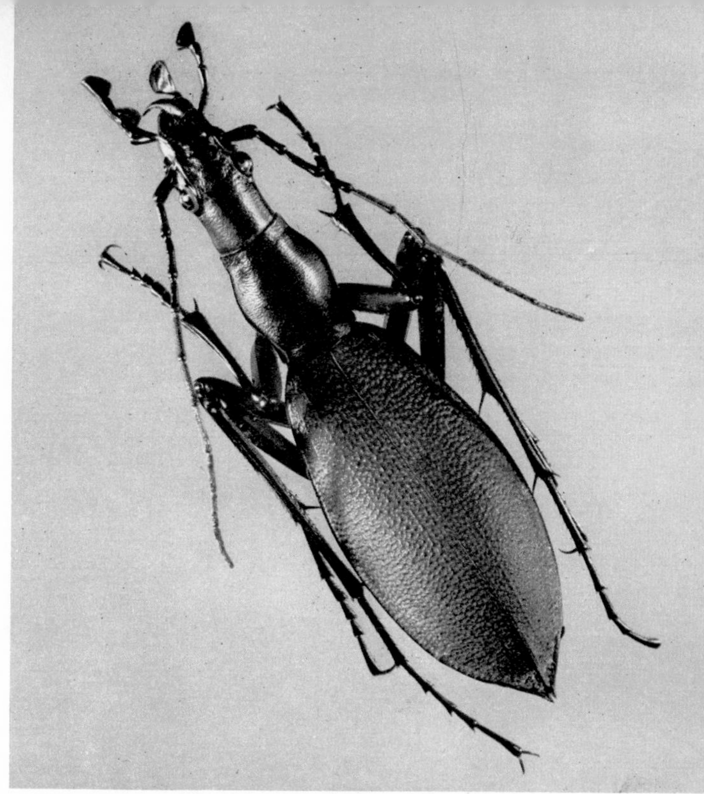

314

315

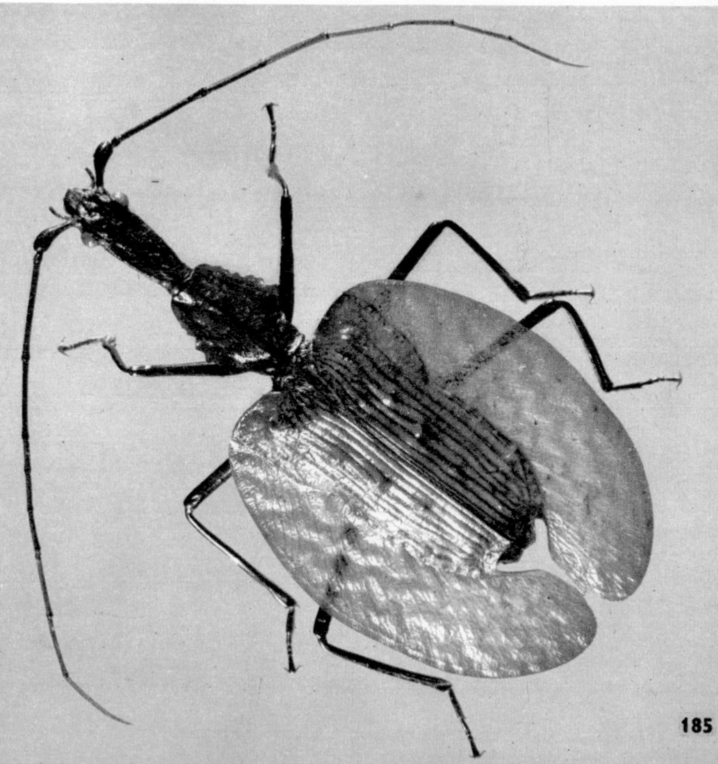

316

On illustration [316] three European species of ground beetles are shown. At the top is the 3 cm long **Pupa-predator,** *Calosoma sycophanta,* which has red-green wing-covers with a metallic gloss, and a steel-blue breast-shield. It lives principally in deciduous forests. Below on the right is the **Lesser Climbing Ground Beetle,** *Calosoma inquisitor,* which is up to 2.1 cm long and copper-coloured, but sometimes blue-green. This species likewise prefers deciduous forests. These two species of beetle play an important role in the biological war on plant pests, in that it is predatory chiefly on the larvae of harmful insects. On the left below is **Calosoma maderae ssp. auropunctatum:** this beetle is up to 3 cm long and coloured black with golden dimples on the wing-covers. It lives in fields in the warmer parts of central and eastern Europe. A remarkable black beetle about 1 cm long is

317

'latyrhopalopsis melyi [317], which in ;urma inhabits ants' nests. It is a member of ıe ground beetle family. It is fed by the ants, 'hich in return lick the sweet secretion from its body. Other genera of the family *Paussidae*, which inhabit ants' nests, are found in Australia and Africa. There are two genera in southern Europe.

318

319

The **Water Beetles,** *Dystiscidae,* are predatory beetles distributed throughout the world in more than 4,000 species. They are fully adapted to the aquatic way of life in which they pass the greater part of their existence. Their bodies are flattened and streamlined and the hind legs have become transformed into flippers. In some genera the fore legs of th males have broadened to form sucking, ad hesive pads with which it is able to maintain firm hold on the slippery body of the femal during copulation. All water beetles are pre datory and the larger species hunt the smalle vertebrates. They breathe through the tip c ıe rear end of the ody, which they stick out f the water, retaining a ıpply of air on their mem-ranous wings under the ing-covers. They are ood fliers and change ıeir hunting ground as ıd when necessary. They refer, however, to lurk in ıe plant growth in still aters. The best-known uropean species, the **reat Diving Beetle,** *ysticus marginalis,* is up 35 mm long and dark rown. The head, thorax ıd wing-covers have yel-w borders. Illustration 18] shows a female with riped, indented wing-overs and ordinary fore gs. The male [319], on ıe other hand, has suc-on pads on the front legs ıd smooth wing-covers. lustrations [320 and 321] ıow the larvae. These o are predatory and tack their prey with fang-ke biting mouthparts. hese have little openings the tips, from which ey inject their prey with gestive juices, in order liquefy the tissues. fter this, the larva allows me for the injection to ke effect, holding the ctim all the time firmly ith its mouthparts. Fin-ly, it sucks out the lique-ed food.

320

321

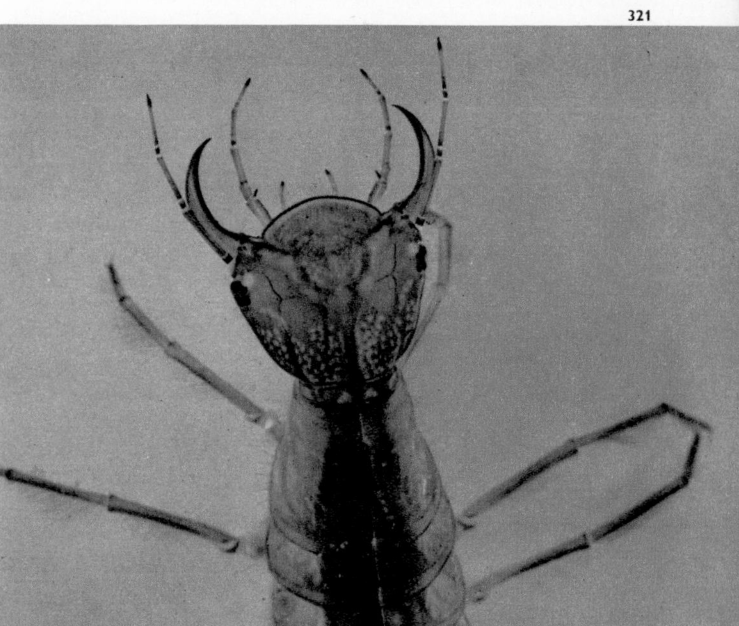

322

323

325

In the larger ponds and lakes of central Europe the species **Dytiscus latissimus** [322] was formerly very plentiful. It is up to 4.3 cm long, very stout in the middle of its body (up to 2.5 cm across) and dark brown with a doubled yellow border round the edge. On the right is a female; on the left a male.
A solitary species living in central Europe is **Cybister marginalis** [323]. It is between 3 and 3.5 cm long, and olive-green with a yellow border round the shield and wing-covers. In mountainous regions it can also be found in flowing water.
Graptoderes cinereus [324] is about 22 mm long and one of the most common of the smaller species of water beetle in central European waters.
Another comparatively common species in central Europe is the **Sulcated Diving Beetle,** *Acilius sulcatus* [325], which can be found even in the smallest pools and puddles. It is about 18 mm long and a flattened oval in shape. It bears a black pattern on the brown wing-covers.

326

327

328

XIXa Eyed Ladybird, *Anatis ocellata*, of coniferous forests of Central Europe.

XIXb Bee-wolf, *Trichodes apiarius*. 12 mm. long. The larvae feed on bee larvae within the hive.

XX Two Splendour Beetles from Madagascar. Left, *Polybothris quadricollis*; right, *Polybothris sumptuosa*. 33 mm. long.

The females have deeply channelled wing-covers with long hairs on them. On illustration [326] is a female and on [325 and 327] are males. The larvae resemble those of the great diving beetle, except that they are shorter, with a smaller head and stouter body [328].

The **Whirligig Beetle** family, *Gyrinidae*, is distributed throughout the whole world in about 400 species. The European **Common Whirligig Beetle,** *Gyrinus natator* [329], is about 7 mm long and the shape of a slightly-flattened egg. Whole swarms of these insects play on the mirror-like surface of still waters, turning like sparks in curves, arches and spirals round each other. They feed on insects which fall on to the surface of the water.

The vast majority of all beetles belong to the second suborder, the **Omnivorous Beetles,** *Polyphaga.* The difference between the two suborders lies in how the rear pair of legs are attached to the body and in the venation of the wings. The *Polyphaga* contains a numerous family of insects living in water: the **Water-scavenger Beetles,** *Hydrophilidae.*

Hydrophilus caraboides [330] is about 15 mm long. Its diet consists principally of animal matter.

329

330

332

Beetles of the species *Hydrophilus caraboides* will eat a dead fish down to the skeleton in a matter of minutes [331]. They can also be found in quite small pools.

The **Great Silver Water Beetle,** *Hydrous piceus* [332], is 45 mm long and the largest member of this family. It is black, with an olive sheen. The imago feeds on water plants, although the larvae are predatory. The female lays more than 50 eggs at a time in a kind of floating balloon. These water beetles prefer stagnant water as a habitat, in which they row

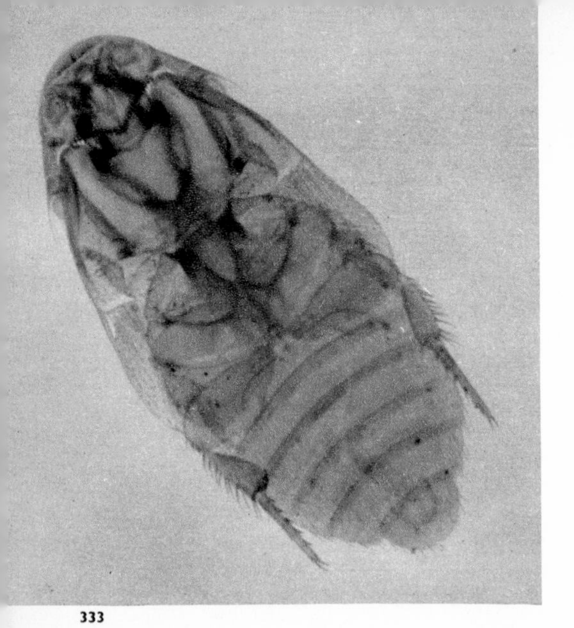

about with a jerky motion.
Only one species makes up the family *Platypsillidae,* and that is the tiny **Beaver Flea,** *Platypsyllus castoris* [333]. The remarkable feature about this creature is that it lives in the coat of the beaver. It measures about 3 mm, is yellow-brown and extremely flattened in shape; it is also wingless and blind. Its larvae develop in the pelt of the beaver. It has not yet been ascertained what forms its diet.
The **Rove Beetle,** family *Staphylinidae,* contains more than 20,000 species distributed throughout the whole world. The rove beetles have a long, extended body and shortened wing-covers. All are predators. The **Greater Rove Beetle,** *Staphylinus caesareus* [334], is up to 2.5 cm long and coloured black with red-brown wing-covers and tiny gold-sheened marks on the sides of the abdomen. It preys chiefly on the larvae of flies and is therefore

333

334

335

336

337

often found in the neighbourhood of dung and carrion.

The **Rove Beetle,** *Oxyporus rufus* [335], is up to 12 mm long. Its head and wing-covers are black. Part of the thorax and the front part of the abdomen are orange and the rear part black, as are the legs. The beetle and its larvae live in leaf-fungi. **Creophilus maxillosus** [336] is over 20 mm long and glossy black with shiny grey hairlines at the sides. It feeds on the larvae of flies and maggots in dung-heaps and carrion.

The sightless **Club-horned Beetle,** *Claviger testaceus* [337], lives in the nest of the **Yellow Meadow Ant,** *Lasius flavus.* It is yellowish-red and about 2.5 mm long. The ants feed and care for it, since it produces an excretion which they greedily lick off its body. This substance is exuded in tiny drops from pores on the beetle's wing-covers, which are set about with long, pale hairs.

338

339

'he family of **Carrion Beetles,** *Silphidae,* ontains about 2,000 species, the distribution f which is principally in the colder lands of the emperate zones. Both adults and larvae feed n carrion, fungi and vegetable detritus.
'he **Four-spotted Carrion Beetle,** *Xylo-repa quadripunctatus* [338], is 12 to 14 mm ong, yellow on top with black markings. 'he beetle and its larvae feed on oak pests.
'he largest member of the genus *Necrophorus* , the **Burying Beetle,** *Necrophorus germani-us* [339, male and female]. It is up to 3 cm ong, having black with brown-yellow sides. 'he burying beetles literally bury the carrion vhich they find, then lay their eggs in the valls of the burial chamber, so that the larvae eed on the decomposing matter. The most umerous central European burying beetle, he **Common Burying Beetle,** *Necrophorus espillo* [340], is black with two orange-ellow stripes on the short wing-covers and rows to a length of some 22 mm. The smaller eetle on illustration [340] is **Oecoptoma horacica,** which is about 16 mm long and lack with a brownish-red shield. It is found n carrion and ill-smelling fungi. The larvae f **Silpha obscura** [341] are about 17 mm

340

341

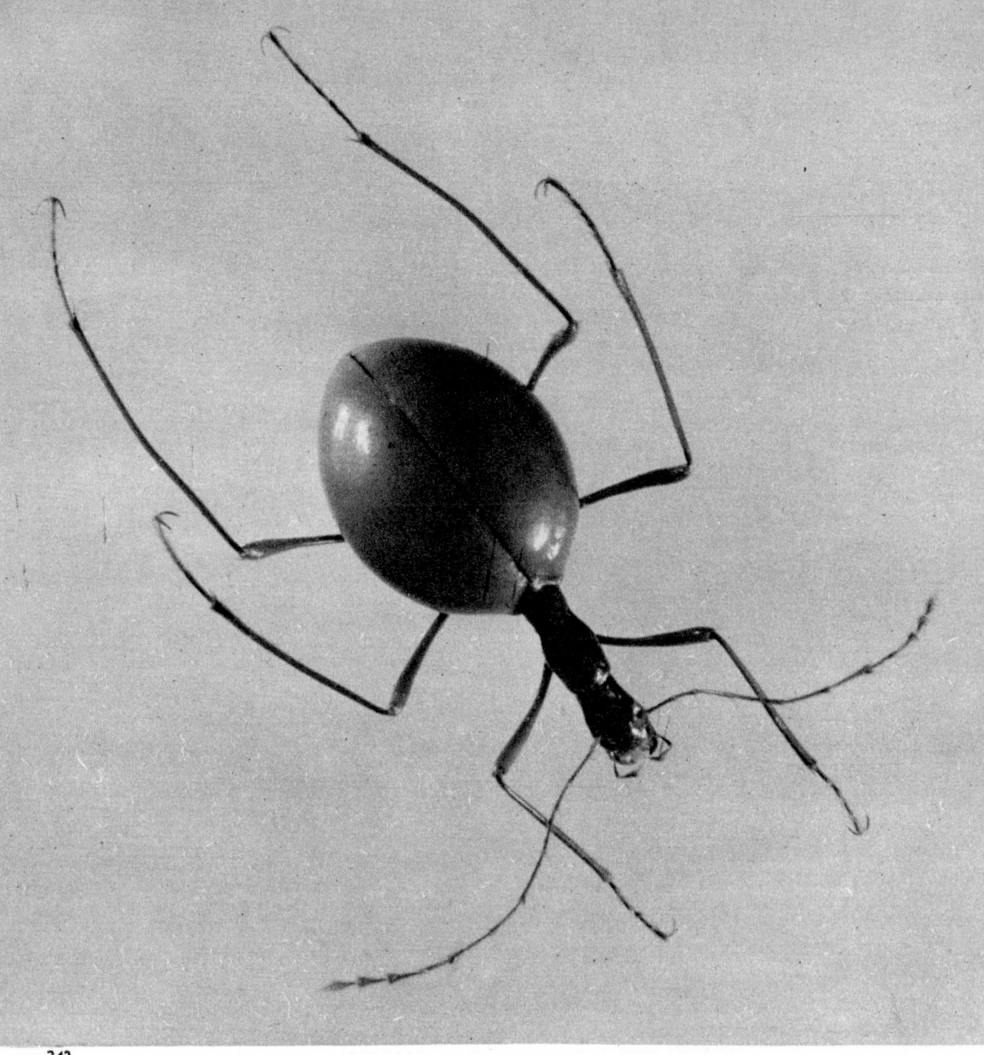

342

long and have a similar way of life to that of related species.

In the caves of southern Europe there live many sightless beetles of the family *Silphidae*. They feed on organic remains, often on the guano of bats. The species **Leptoderus hohenwarti** [342] is 7 to 8 mm long, yellowish-brown in colour, and has very long legs. Its larvae are not known.

The **Dwarf Beetle** family, *Ptiliidae = Tricho —pterygidae,* contains the smallest of all known insects. The tiny members of the tropical species **Nasonella fungi** measure only 0.25 mm. The largest dwarf beetles are only up to 2 mm long; their rear wings have long lashes. They are very active creatures, which feed on decaying organic matter. In th palaearctic region about 200 species are know **Acrotrichis brevipennis** [343] comes fro central Europe. **Acrotrichis sp.** [344] w captured under a heap of seaweed on th Dalmatian coast. Both these beetles are rath less than 1 mm in length.

The **Boat Beetle,** *Scaphidium quadrimac latum* [345], is up to 6 mm long. It is gloss black and bears two red marks on each win cover. It runs about extremely nimbly on i comparatively long legs. It is a member of th *Scaphididae* family, whose members live i fungi and under polypori.

One member of the **Prop Beetle** famil *Histeridae,* is **Hister fimentarius,** a sma

343

344

345

346

347

348

349

generally black creature, not quite 1 cm in length. It can be found in dung and corpses, but can also be predatory. If disturbed they stiffen and then simulate death for a long period.

The **Gloss Beetle** family, *Nitidulidae*, consists of some 2,500 of the smaller species of beetle. They are occasionally found in old pelts, but also on flowers, on exudations of sap on trees, under loose bark and in the tracks of bark beetles, *Scolytidae*. Some of them can also be counted as agricultural pests.

Omosita (Saprobia) colon [347] is light brown with darker markings. It measures 2 to 3 cm and feeds on animal and plant detritus. The maggot-like larvae of **Omitosa colon** [348] have three pairs of legs and are whitish and soft. The pupae of this species [349] are, like those of other beetles, quite motionless. The only exception is that when they are disturbed the pointed rear end of the abdomen twitches. All the parts of the body of the developing beetle can be clearly distinguished in the pupa.

The **Ladybird Beetles,** *Coccinellidae*, are very well known, and are great favourites with

0

351

352

ɔung and old alike. The members of this
mily have a hemispherical body, quite flat
ɪderneath and brightly coloured. They can
: found in every corner of the world in about
400 species. The names which they have
ɛen given in the everyday speech of people all
ver the world, as for example "sun calf",
God's little thief", "God's lamb", and so on,
dicate the popularity of these little creatures
nong mankind.
his is very well deserved, since the ladybirds
ɪd their larvae consume aphids and scale in-
cts with such voracity that some genera are
ecially cultured and set loose to free jeopar-
sed plants from these pests. Its great
rtility ensures that it is more than adequate
r such advantageous tasks.
he **Seven-spot Ladybird,** *Coccinella sep-
mpunctata* [351], is 5 to 8 mm long, displaying
ily coloured red wing-covers with seven
ack, symmetrically arranged spots. Its warty
rvae [350] are equally brightly coloured. The
ɪpae of the ladybird have a characteristic
rm and are fastened to a leaf by their tails.
lustration [352] shows the pupae of *Adalia
punctata*.

353

354

After they emerge the ladybirds are at first still soft and without markings. On illustration [353] we see a newly-emerged adult of the species *Adalia bipunctata*. It takes several hours for the chitinous covers to harden and its colour to achieve its full brilliance. *Adalia bipunctata* [354] is distributed throughout Europe, Asia and North America. This species often hibernates inside buildings.

Paramysia oblongoguttata [355] is brownish yellow with light yellow points, and inhabits chiefly coniferous forests. Th **Eyed Ladybird,** *Anat ocellata* [356, larva measures 8 to 9 mm and is yellowish-red. Its tw black points have white borders. This species als is an inhabitant of coni ferous forests.

355

356

357

The **Bacon Beetles,** *Dermestidae,* are smal oval little beetles, usually covered with fir hairs or with scales. They lay their eggs in ol furs and hides, feathers, bones and dung—i matter which can later serve as nourishmer for the larvae. In this they fulfil an importa role in the balance of nature in the countrysid although on the other hand they have becom harmful pests in human habitations, especial in larders and store-rooms. The brown, hai larva of the **Museum** or **Cabinet Beetl** *Anthrenus museorum* [357], is about 4 mm lon If provoked, it spreads out and raises the tu of hairs around its anus. The fully-grow beetle [358] is 2 to 3 mm long and coloure grey-black with dull-coloured ornament markings on the wing-covers. Museum cur tors and keepers of zoological collections a constantly on the watch for the larvae of th species, for they devour in a very short tir stuffed animals and whole insect collectio

358

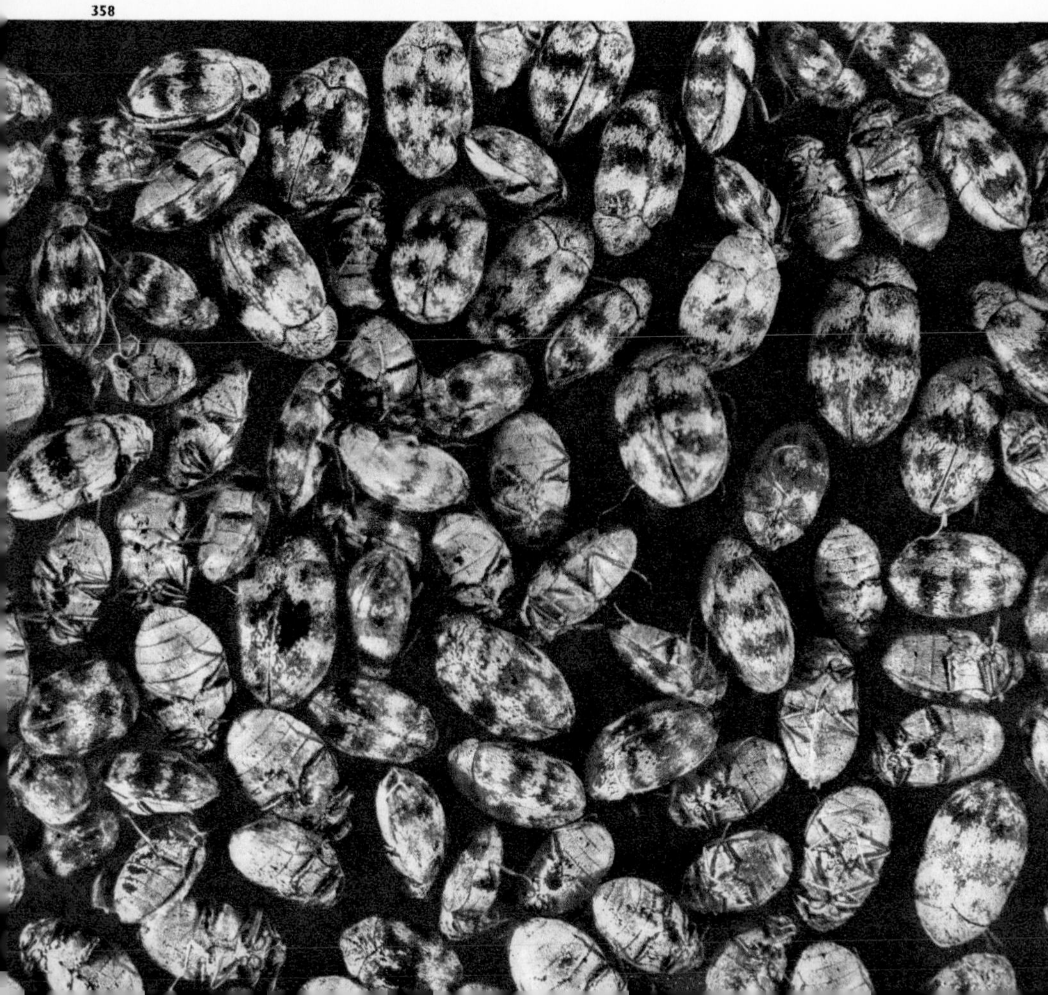

which have not been properly treated beforehand, leaving only a wretched little heap of remains. In the open, this beetle can often be found on blossoming hawthorn and spiraea, as a general rule in company with the **Carpet Beetle,** *Anthrenus scrophulariae* [359]. The carpet beetle is 3 to 4 mm in length, black and invested with groups of scales which on the shield are yellow-grey and on the wing-covers ginger-red. However, they are not as dangerous as their cousins of the same genus.

The **Common Bacon Beetle,** *Dermestes lardarius* [360, the larger species illustrated], grows to a length of some 9 mm and is dark with a broad, lighter band on the wing-covers. Its larvae are up to 15 mm long and feed in houses off the remains of bacon and other fatty foodstuffs.

The **Common Hide Beetle,** *Attagenus pellio* [360, the smaller species illustrated], is about 5 mm long and black with three white spots on

359

360

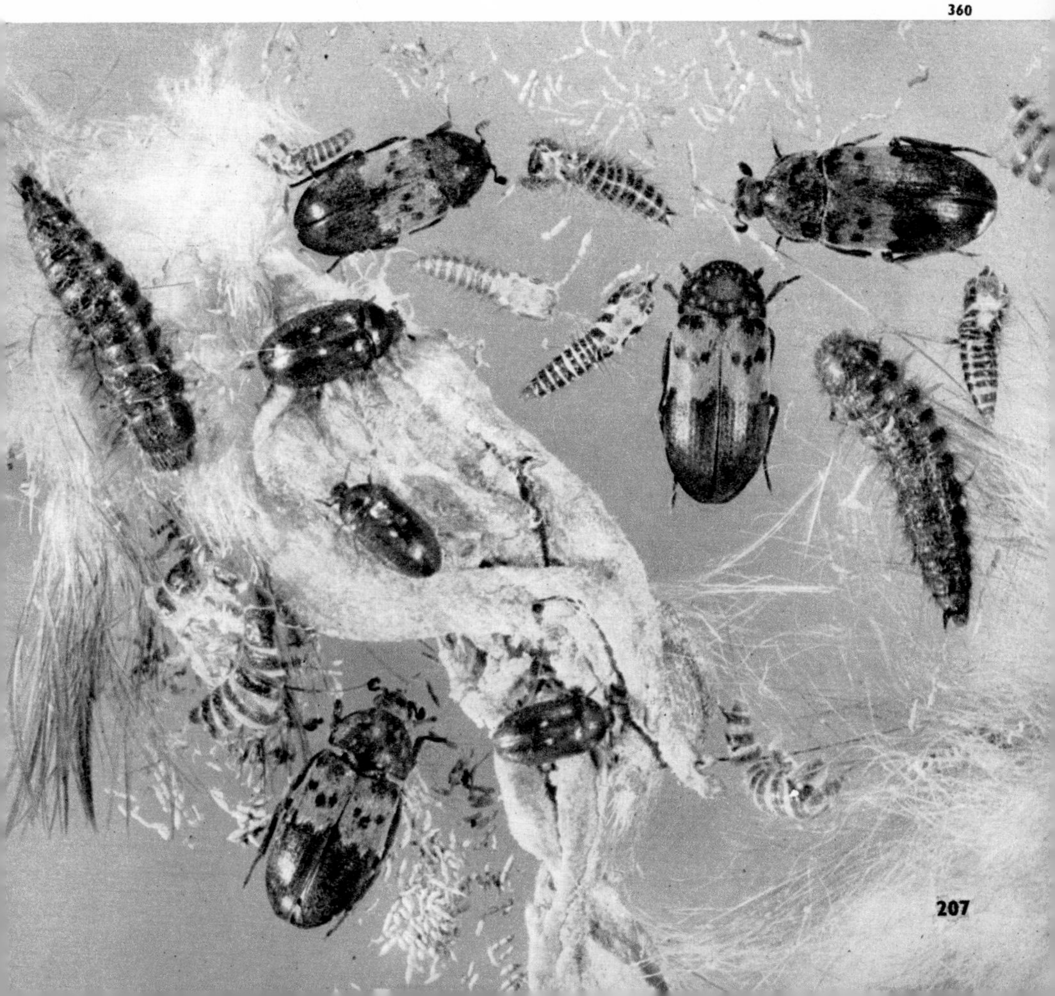

the shield and one in the middle of each wing-cover. Its larvae feed on hides and wool, bu also carpets, upholstery and furs.

The large family of **Soldier Beetles,** *Cantharidae,* contains among other species th very well-known **Common Soldier Beetle** *Cantharis fusca* [361], of central Europe. I has soft, black wing-covers, a reddish shielc with a black patch and black legs. This beetl is up to 15 mm long and can be found fron April to July on field, garden and forest plants on which it hunts other insects, especiall aphids. Its larvae are totally black, 2 cm lon and predators, living underground.

The **Glow-worm** family, *Lampyridae,* consists of about 2,000 species distributed throughout the world, especially in the tropics. Th males are winged, whereas the larva-lik females pass their lives on the ground and lov plants. Both sexes, but especially the males, a also the larvae and eggs, possess the ability t produce light. The cold, greenish light i emitted by yellowish-green luminescent plate under the side of certain abdominal segment

361

362

XXIa Longhorn Beetle, *Epepeotes togatus*. Solomon Islands. 4 cm. long.

XXIb Male Longhorn Beetle, *Macrodontia cervicornis,* from the primeval jungle of Peru. Up to 14 cm. long.

XXIIa Sack Beetle, *Clytra laeviscula*. About 1 cm. long. The larvae are predatory and infest ants' nests.

XXIIb Colorado Beetles, *Leptinotarsa decemlineata*. Up to 12 mm. long.

[363]. This consists of a group of cells which function in the manner of a gland and secrete the protein luciferin which is acted upon by an enzyme, luciferase, becoming luminous. The adult glow worms use their light signals to seek out their partners.

The **Redcurrant** or **Lesser Glow-worm,** *Phausis splendidula,* lives in damp woodlands in central Europe. On warm summer nights the males can be seen flying like fiery sparks through the woods, while the females flash their lights in the grass. The males [362, right, and 363] are brownish-grey and up to 1 cm long. The females [362, right] are wingless, wormlike and yellowish-green.

The second central European species is the **Greater Glow-worm,** *Lampyris noctiluca.* The male is up to 13 mm in length, the female 16 to 18 mm. The larva of the male [364] has dark grey spots on a greyish-red background, with a darker back. Glow-worm larvae are predators, their long, sharp, perforated maxillae to suck blood, especially that of small snails. The adult glow-worms either eat nothing at all after their emergence.

363

364

365

366

The **Gay-coloured Beetle** family, *Cleridae*, consists of some 3,000 species of predatory beetles which are distributed in every zone of the earth. Their common characteristic is their gay colouring. The adults and larvae of many species pursue mainly wood-eating insects; many of them also hunt bees in every stage of their development.

Some of the smaller species resemble ants. The **Ant Beetle,** *Thanasimus formicarius* [366], has a red-brown neck shield and two white bands across its wing-covers. Both the adult and larva live behind the bark of conifers, where they hunt bark beetles and their eggs. Illustration [365] shows the pupa.

The family of **Splendour Beetles,** *Buprestidae*, number upward of 15,000 species, which unfold their fairy-tale splendour mainly in the tropics. The smallest measure about 2 mm; the largest up to 7.5 cm. These beetles come out by day, loving sunshine and warmth above everything, and have a variety of blue, green, red and violet metallic colourings. They sip chiefly pollen and nectar. Their yellowish-white, generally legless larvae feed on rotten wood, but will also eat root vegetables.

A European species, the **Greater Jawed Splendour Beetle,** *Chalcophora mariana* [367], measures about 3 cm and is a glossy bronze colour. Its larvae [368] develop behind the bark of the stumps of old fir trees.

367

368

369

The **Two-spot Oak Splendour Beetle,** *Agrilus biguttatus* [369, 370], is about 12 mm long and dark blue to green with a white hair-line on each of the ribbed wing-covers. It develops in the thick bark of old oak stumps.

One of the largest members of this family, **Euchroma gigantea** [371, left] is a native of tropical America. It is about 7 cm long and decked out in the most splendid iridescent purple-reds and greens.

370

371

The Amerindians have since time immemorial been in the habit of making ornaments out of the wing-covers of this insect. It still enjoys the greatest popularity.

Sternocera boucardi [371, right] comes from the savannah of East Africa. It is up to 5 cm long, black with dark green wing-covers and yellowish-white hairline markings.

Catoxantha opulenta [372] comes from Sumatra. It is some 5.2 cm in length.

372

373

374

Another **Splendour Beetle** is *Capnodis cariosa* [373], which comes from the Dalmatian coast. It is 35 mm long. The breast shield and wing-covers are pitted with little holes which are stuffed with wax, giving them a dusty appearance. The larvae develop behind the bark of trees.

Julodis variolaris ssp. bucharica [374] is up to 33 mm long. Our illustration shows two beetles of the genus *Ephedra* on a shrub, in the roots of which its larvae develop. It lives in the hot sandy deserts of central Asia, coloured a glossy green and decorated with creamy-white areas of scales and hairs.

The **Click Beetles,** *Elateridae,* populate a large part of the earth's surface in something like 8,000 species. The smallest of them measure about 5 mm and the largest up to 8 cm. By means of a spring-like movement they are capable of leaping 8 to 10 cm into the air from a position lying on their backs, covering quite some distance forwards. In addition, numerous species are distinguished by beautiful colouring and a metallic sheen. Their larvae live underground, are very hard, coloured glossy yellowish-brown and feed on roots. These are the familiar "wire-worms". The adult beetles, too, feed on various parts of plants and are no less injurious.

One of the largest of the **Click Beetles** is *Tetralobus flabellicornis* [375], from west and southern

375

376

377

378

379

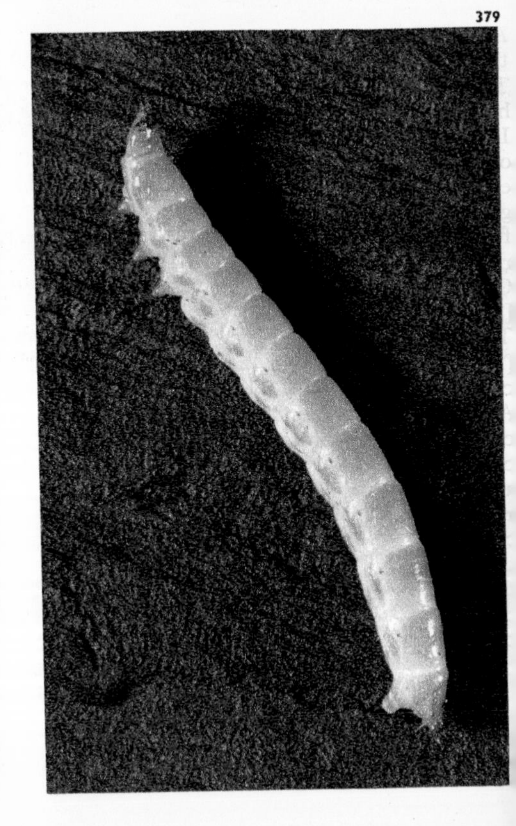

381

382

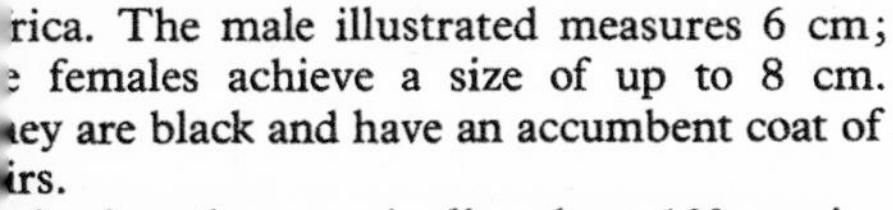

rica. The male illustrated measures 6 cm; : females achieve a size of up to 8 cm. ıey are black and have an accumbent coat of ırs.

the American tropics live about 100 species click beetles, the **Fireflies,** which are uipped with the ability to emit light like the ›w-worms, only significantly more powerıly. The light of four fireflies is sufficient to ıd by. The natives call them "cucuyo". ıe of them is **Pyrophorus noctilucus** 76]. It measures some 4 cm in length.

ıe **Click Beetle,** *Iphis madagascariensis* 77], is about 4 cm long, glossy black and vered in short creamy-white hairs.

already mentioned, the larvae and pupae of ck beetles are found either in the earth or in mpost. Illustration [378] shows the pupa of large central European species **Athous fus.** The larvae feed on those of their usins, *Chalcophora mariana.* The form illusted [379] is typical of the click beetle larvae, : "wireworms", which are feared agriltural, horticultural and forest pests, espeılly the members of the genus *Agriotes.*

383

38

385

The family of **Ground Beetles,** *Tenebrionidae* is distributed throughout the world in abou 17,000 species. They are small to middle-size and as a rule of a dark colour. Many of then come out in the dead of night in cellars, stable and sheds, where they feed on animal an vegetable detritus: others, on the other hand prefer the daytime, particularly in hot coun- tries with blazing sunlight. One of the bes known is the **Meal-worm Beetle,** *Tenebri molitor* [381], which is dark brown, glossy an up to 16 mm long. It inhabits mills an bakeries. Its larvae are the familiar "mea worms" [380], which are up to 3 cm long forming an indispensible foodstuff for the in- habitants of many cages and aviaries. Breedin them is a comparatively simple matter: th beetles multiply extremely quickly in warr surroundings, feeding on a variety of scrap and garbage. Illustration [382] shows a pupa Resembling this species, but only 5 mm long is **Tribolium destructor** [383], a world- wide pest in larders and storehouses.

386

On sandy, sunny paths we can find a grey, flat beetle, up to 8 mm long: the **Common Dust Beetle,** *Opatrum sabulosum.* The beetle and its larva are both very injurious to vines. Many ground beetles have atrophied wings and are unable to fly. Others, on the other hand, are distinguished by extremely unusual forms, which give rise to some extraordinary beings. Thus the Australian species **Helaeus perforatus** [385] looks like a shallow scale, and has a length of 25 mm. It is black with a striking brown patch of hair on the wing-covers. In the deserts of central Asia lives **Sternodes caspius** [386]. It measures 27 mm, is black and carries a clearly visible white design on the wing-covers.

387

The males of the **Fan Beetle** family, *Rhipiphondae,* have fan-shaped, spread-out antennae. The wing-covers of fan beetles are always set somewhat gaping apart, so that the wings can always be seen. Their larvae are parasitic on the *Hymenoptera*. The **Wasp Fan Beetle,** *Metoecus paradoxus* [387], grows to a length of up to 15 mm. The male has yellow-brown wing-covers, whereas those of the females are completely black. Their larvae develop in subterranean wasp nests.

The family of **Bristle Beetles,** *Mordellidae* number about 1,100 species, which are principally distributed in the tropics. Its members are small, agile beetles having a boat-shaped form and an arched upper side. The last seg-

388

ıent of the abdomen is extended into a point. ɑ danger they move away with a powerful ːap. A species which lives in warm places in 'urope is **Mordella aculeata** [388, 389], a ɫack, hairy beetle about 5 mm long. It is found on meadow flowers. The **Scarlet Fire Beetle,** *Pyrochroa coccinea,* is up to 18 mm long with soft, bright red wing-covers. Its larvae [390] develop in the wood of deciduous trees.

ɜ9

390

391

392

The family of **Oil** or **Blister Beetles,** *Meloidae,* is distributed throughout the world in roughly 2,300 species. The beetles of this family by-pass the usual scheme of development and undergo a complicated process called hypermetamorphosis. The larvae develop in three completely different stages. In these stages they are dependent on parasitism on other insect species. Most blister beetles emit an evil-smelling liquid from their knee-joints, which protects them from their enemies. On human skin it causes inflammation and blisters. It contains a poison (cantharidin), which was formerly much used in

393

he preparation of patent nedicines which, however, did more harm than ood, and indeed in some ases had fatal results. The **)il Beetle** or **May Vorm,** *Meloë violaceus* ıeasures between 10 and 7 mm, has a dark blue oft body and is flightless, aving no hind wings. The ıale [391] is smaller than ıe female [393], which an be recognised by its road abdomen. A female ıys up to 10,000 eggs. The il beetles creep about in ıe grass and feed on lants. Their larvae are alled *triungulins* in the rst stage [392].

'he **Oil Beetle,** *Meloë ariegatus* [394], is up to 8 mm long, copperoloured and green. It has ıuch the same way of life s the species described bove, in sandy spots in ıe warmer parts of urope.

'he **Spanish Fly,** *Lytta esicatoria* [395, the larger eetle], is a glossy goldreen, a good flier and rows to a length of up to 5 mm. It lives in the armer regions of Europe. ometimes masses of them warm like locusts and rip green leaves from sh, privet and lilac.

;erocoma mühlfeldi 95, the smaller beetle] is pproximately 12 mm long nd a shimmering green olour. The males have eculiar growths on their ont legs and antennae. his species lives in outhern Europe, occaonally being found in arm spots in central urope. It feeds on the ices of flowers; its larvae evelop in bees' nests.

394

395

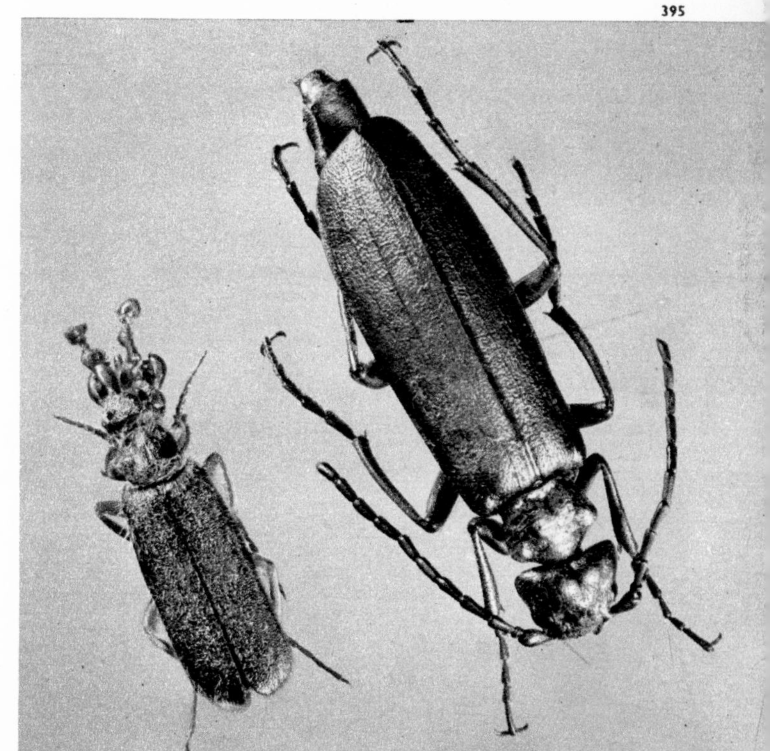

396

397

XXIII One of the most attractive European Leaf Beetles is *Cryptocephalus sericeus*, which is up to 7 mm. long. It is found in early summer principally on *Compositae* and St. John's wort.

XXIVa The Asparagus Beetle, *Crioceris asparagi*. 6 mm. long.

XXIVb The Twelve-Spotted Asparagus Beetle, *Crioceris duodecimpunctata*. 6 mm. long. Both species are inhabitants of Europe and northern Asia, and have been introduced into North America.

The family of **Longhorn Beetles,** *Cerambycidae,* contains over 20,000 species and is world-wide, being especially common in the tropics. Its members are distinguished by particularly long antennae, which are often longer than the whole body. They measure anything from a few mm up to 150. Many of them are injurious to forest and fruit trees. The extremely rare **Megopis scabricornis** [396, the smaller beetle] grows to a length of about 45 mm and is dull brown. Its larvae develop in old tree trunks, usually poplar, willow, aspen and chestnut. The longhorn **Callipogon barbatus** [396, the large beetle] is 85 mm long, has a black shield, red-brown wing-covers and a whitish covering of hair. Its powerful maxillae are equipped on the inside with thick, light-brown hairs. It lives in Mexico, Guatemala and Nicaragua.

The **Tanner** or **Sawyer Longhorn,** *Prionus coriarius* [397, the smaller beetle, male], measures 20 to 40 mm and is dark-brown. Its larvae develop in old beech or fir stumps.

The **Compost Longhorn** or **Carpenter Longhorn,** *Ergates faber* [397, the larger beetle, male], is, with a length of up to 55 mm, the largest of the central European longhorn beetles. It is a glossy dark brown and develops in the trunks of old conifers.

The species **Enoplocerus armillatus** [398, male] lives in Peru. It is 11.7 cm long, has a brown colouring and the sides of its shield are equipped with

398

399

400

red marks and sharp thorns. Another longhorn beetle from the American tropics is **Macrodontia dejeani** [399, male], about 90 mm long with comparatively short antennae, but strikingly long, saw-toothed maxillae. The effect of the patterns on the wing-covers and the brown colouring is to give this beetle a striking resemblance to the bark of certain trees.

Also from the tropics of South America is the largest of all longhorn beetles, **Titanus giganteus** [400]. It measures 15 cm excluding

401

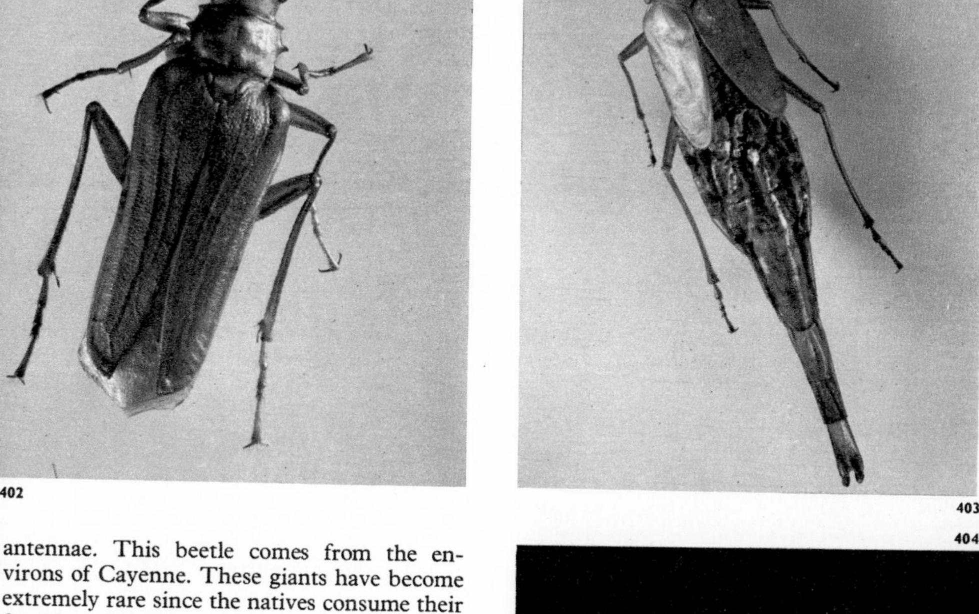

402

403

404

antennae. This beetle comes from the environs of Cayenne. These giants have become extremely rare since the natives consume their fat, 25 cm long larvae with great eagerness as a welcome delicacy. The same is also true of the natives of the Fiji Islands, who have in this way caused the extinction of the **Large Longhorn Beetle,** *Xixuthrus heyrovskyi* [401]. Only a few rare individuals are found here and there in collections. The male, photographed here, is 13 cm long and dark brown. It was caught on Viti-Levu, the largest of the Fiji Islands.

The antennae of some longhorn beetles are articulated in a peculiar way and developed into multi-branched sense organs which are among the marks of sexual dimorphism and enable the male to find a female even if she is thousands of metres away. On the antennae are located very fine, complicated organs of smell, which react to certain smells over great distances with a sensitivity which is incredible to human beings. The longhorn **Polyarthron komarovi** from Turkestan is ochre. The male [402] is 19 mm long, the female [403] 47 mm. Illustration [404] shows the head of a male.

405

The **Brazilian Longhorn,** *Hypocephalus armatus* [405, 406], has a remarkable appearance. Its thoracic shield—almost as long as the abdomen—covers the head completely. The form of its body is reminiscent of that of the mole cricket.
Some longhorn beetles have extremely shortened wing-covers and long, membranous rear wings, which they cannot fold together under the wing-covers, in which they resemble the Hymenoptera.
The **Greater Ichneumon Longhorn Beetle,**

406

Necydalis major [407], is reminiscent in appearance of an ichneumon. It grows to a length of about 35 mm; the head and shield are black and the wing-covers yellow-brown. Its larvae develop in poplars and willows, sometimes also in fruit trees. It is found chiefly in south-eastern Europe and is rare in central Europe. A longhorn beetle extremely common in the oak forests of central Europe is **Rhagium sycophanta** [408]. It is up to 2.5 mm long and is brownish-black. Its larvae live in the wood of dead, deciduous trees.

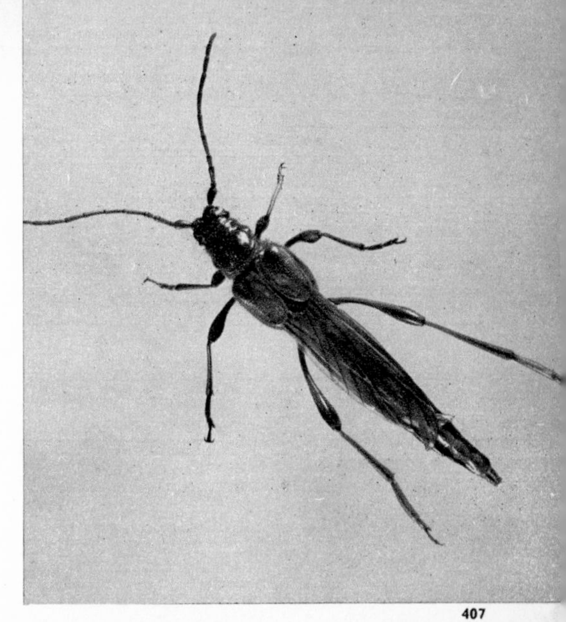

407

408

409

The **Great Oak Longhorn** or **Hero Longhorn,** *Cerambyx cordo* [409], has a length of up to 50 mm and is black with dark brown wing-covers coarsely wrinkled at the ends. Our illustration shows the male above and the female below. This longhorn lives in the warmer parts of Europe in oak trees. The larvae spend three or four years eating their way through the wood before they pupate. As well as old or dying oaks they can be found in ash, elm and walnut wood.

Some members of the tropical genus *Batocera* have antennae 25 cm long. One of the smaller species is **Batocera albofasciata var. saravakensis** [410], which is quite common and comes from Borneo. It is more than 4 cm long, light brown with yellow marks on the shield and white ones on the wing-covers.

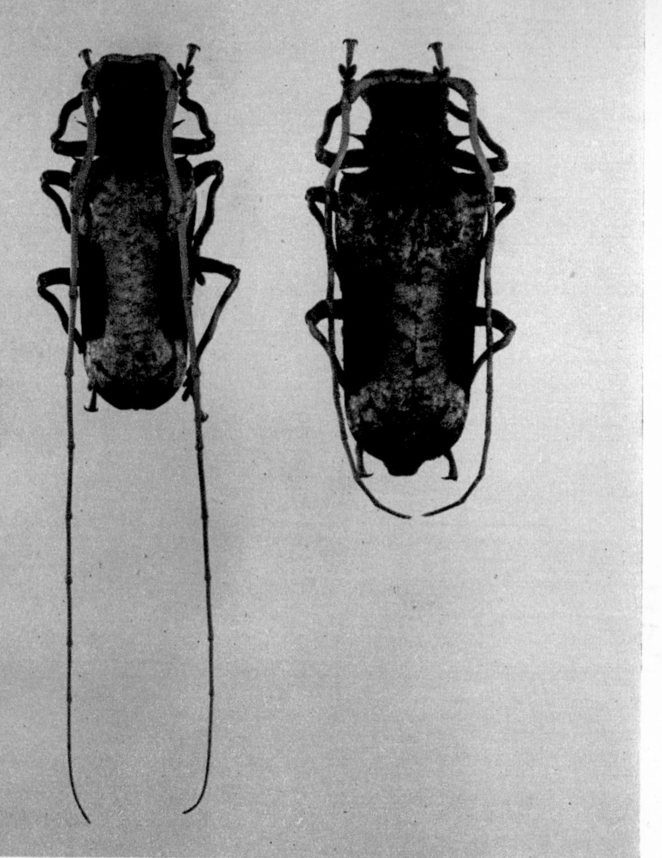

411

The female of the species **Petrognatha gigas** [411] from West Africa measures up to 7.5 cm excluding antennae; the male is smaller. They have a thick, short, velvety coat of hairs and their colouring generally ranges from dark to light brown.

The **Longhorn Beetle** *Epepeotes togatus* [412] from the Solomon Islands is 3 to 4 cm in length. A rare central European species is the **Alpine Longhorn,** *Rosalia alpina* [413]. It measures some 36 mm, is an attractive grey-blue colour and has on each wing-cover three large, velvet black marks. It lives in the mountains in old beech stumps and is protected by law.

412

413

The **Forest** or **Cylinder Longhorn,** *Spondylis buprestoides* [414], measures approximately 2 cm. It is a dull black, with short antennae and a cylindrical body. It is common in coniferous forests. Its larvae develop in the wood of freshly cut fir stumps.

414

415

416

417

The **Blue Disc Longhorn,** *Callidium violaceum* [415], has a red shield with a violet sheen. It is up to 13 mm long. Its pupae are found in the wood of conifers [416]. One of the most common central European longhorns is the **Weaver Longhorn,** *Lamia textor* [417], which is about 25 mm long and dull black. Its larvae [418] develop in the roots of willows, aspens and poplars. From the Tyrol and the mountains of southern and south-eastern Europe comes the wingless species the **Weeping Longhorn,** *Morimus funereus* [419]. It is up to 36 mm in length, grey, with two dull black markings on each wing-cover. It has its development in beech stumps, and can also be found in central Europe.

418

419

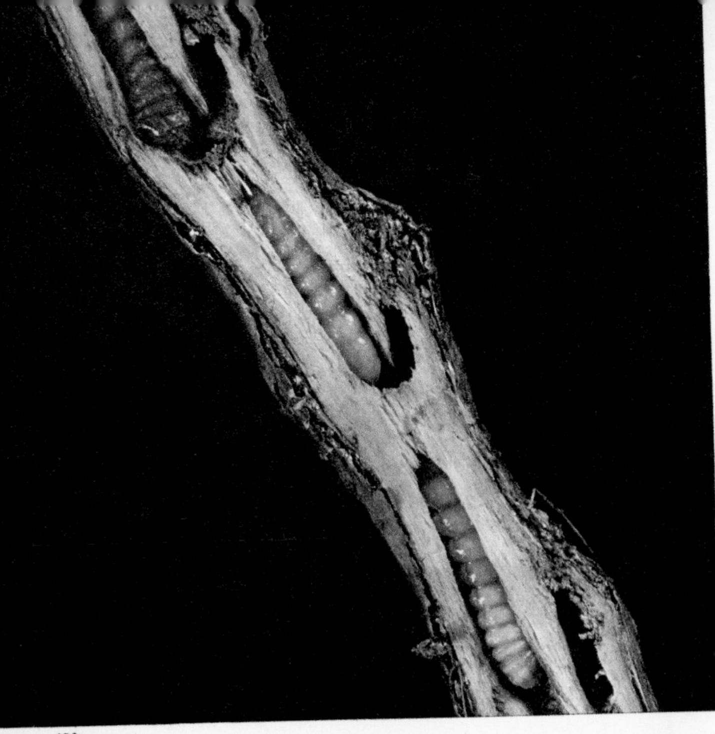

420

421

Knotty swellings on thin branches of young poplars and aspens are the chambers of the larvae [420] of the **Lesser Aspen Longhorn,** *Saperda populnea* [421, above left]. The beetle is 9 to 14 mm in length and dark grey with a close-packed yellowish pattern. The species **Saperda scalaris** [421, below left] is up to 18 mm long, coloured black and greenish-yellow. It develops in deciduous trees in forests and also in fruit trees. The **Greater Poplar Longhorn,** *Saperda carcharias* [421, right], achieves a length of up to 28 mm. It is light, yellowish-brown with black spots. Its larvae live beneath the bark and in the trunks of poplars, willows and aspens. The trunk is marked by rough swellings where the larvae are attacking and young trees quite commonly die as a result.

The **Carpenter's Longhorn,** *Acanthocinus aedilis* [422], grows to a length of about 2 cm and is grey-brown. The male has extremely long, thin, ringed antennae; the female has a projecting ovipositor. The larvae [423] have their development in the stumps of old conifers, or in the trunks of felled trees. Here, too, the pupae of this species [424, pupa of a female] can frequently be found.

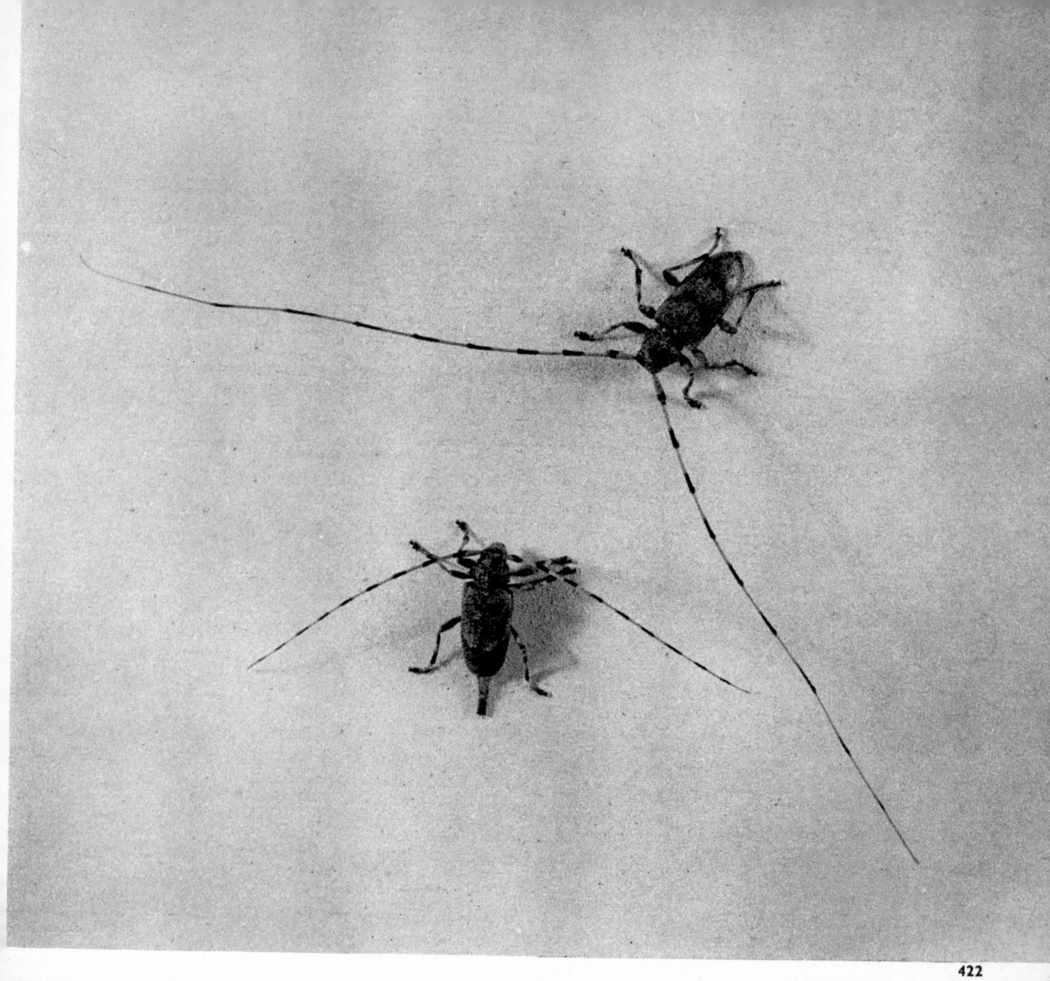

422

423

424

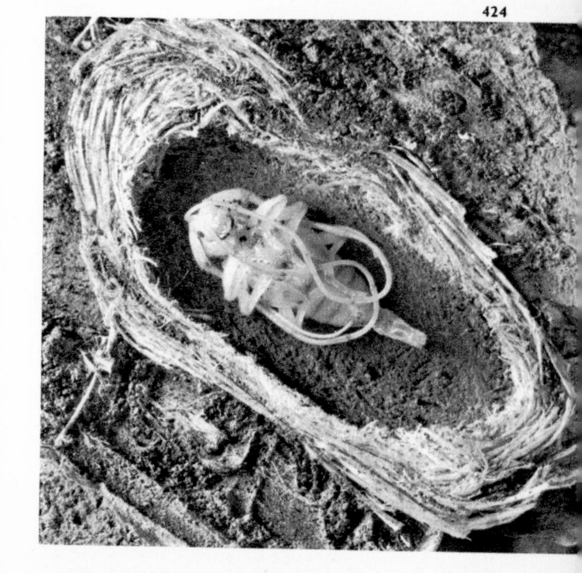

425

On the plains and steppes of the warmer parts of the palaearctic zone live the **Grass** or **Earth Longhorns,** *Dorcadion.* They have short antennae and cannot fly, since the rear wings are absent. Their larvae develop in the earth and eat the roots of grasses. **Dorcadion tibiale** [425] is 22 mm long, velvet-black wit cream-coloured stripes. It originates fro central Asia.

The species **Gnoma bisduvali** [426, 42 differs in appearance from the general run longhorn beetles. It is about 3 cm in length.

426

A bizarre appearance is presented by the **Harlequin Longhorn,** *Acrocinus longimanus* [428], from the tropics of South America. It is up to 8 cm long and dark grey with a brick-red pattern. During the day it stays hidden and only comes out to begin its secretive life at twilight. Its larvae develop in fig trees.

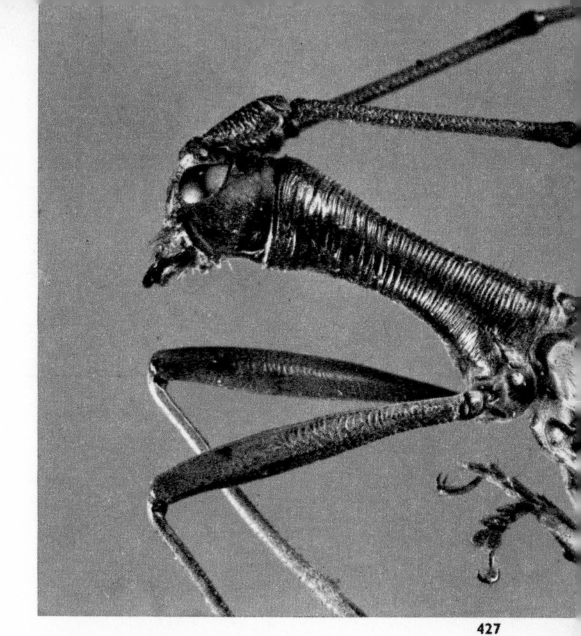

427

428

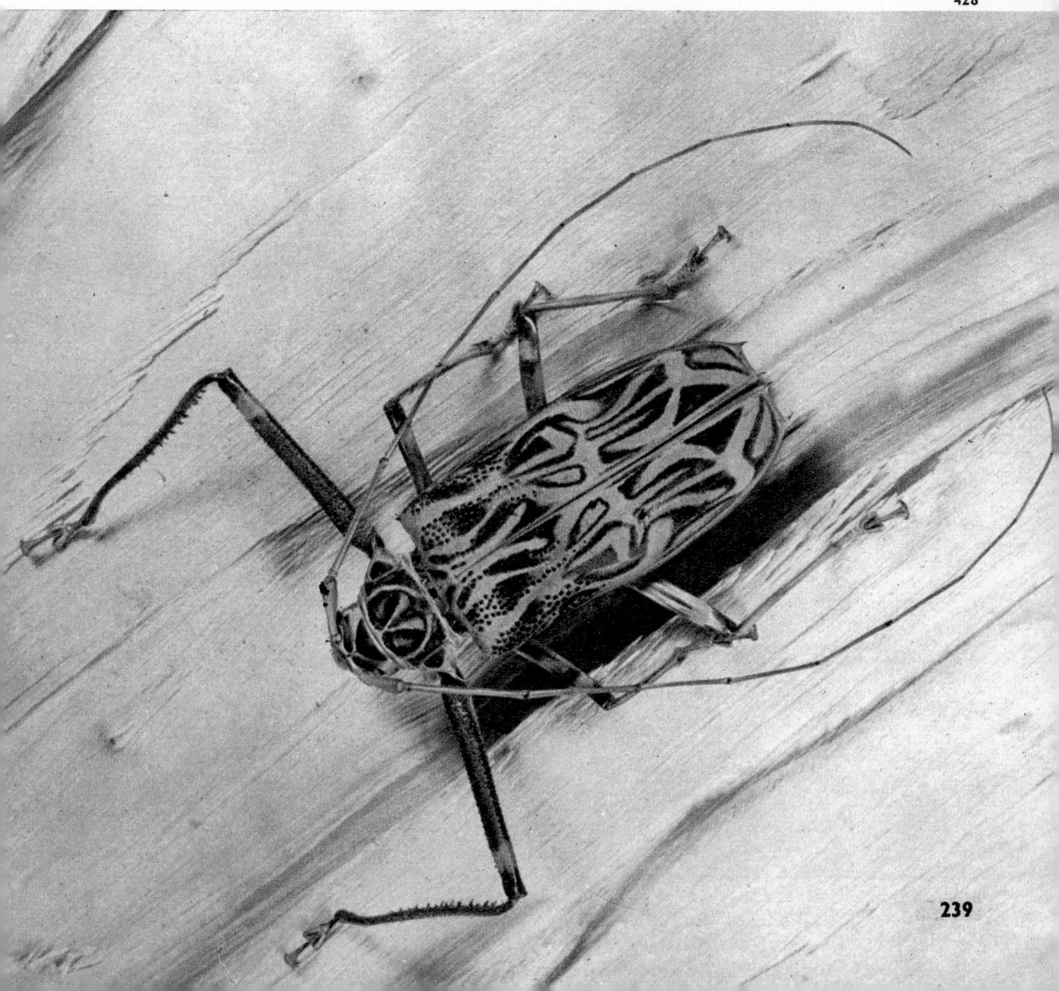

The **Leaf Beetle** family, *Chrysomelidae*, has world-wide distribution, in more than 25,000 species. They are small- to medium-sized beetles with highly arched backs and flamboyant colouring. Some species are greatly injurious to cultivated plants.

The **Blue Corn Beetle,** *Lema lichensis=cyanella* [429], is up to 4 mm long and dark blue with a metallic gloss on the wing-covers. Both the beetle and its larvae live on grasses. From time to time they start to increase in vast numbers, and as a result considerable damage may be done to grain. The pupa are encased in a dry, foamy substance and attached to grasses [430].

431

429

430

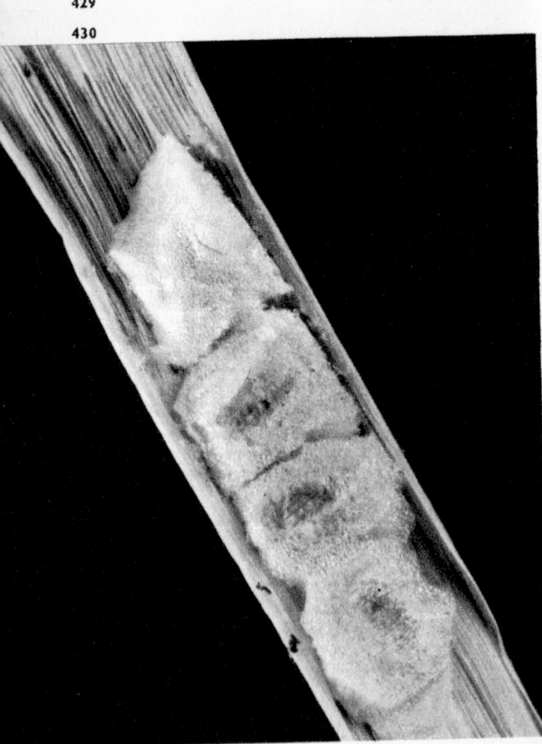

XXVa The Weevil, *Entimus imperialis*. Eastern Brazil. 3 cm. long.

XXVb The Tropical Leaf Beetle, *Sagra buqueti*. Java. The male is about 28 mm. long.

XXVI Stag Beetles, *Lucanus cervus,* from Central Europe. Pair.

achnaea sexpunctata [431] is approximately 10 mm long and light brown with lack markings. This beetle is common in outhern Europe, rare in central Europe; it is ound in oak forests.

The **Red Poplar-leaf Beetle,** *Melasoma opuli* [432], grows to a length of up to 12 mm, with a dark-green shield and red wing-covers. t is found on poplars.

Illustration [433] shows two of the largest leaf eetles: the larger is the Brazilian species **Eumolpus candens,** which is some 2 cm ong and a lively glossy green in colour. The maller is **Chrysochares asiaticus,** which is bout 15 mm long, comes from the plains of entral Asia. Its green shield and blue-green wing-covers have a metallic gloss. All these outhern leaf beetles are good fliers.

432

433

434

Once upon a time the **Potato Beetle** or **Colorado Beetle,** *Leptinotarsa decemlineata,* lived a more or less harmless existence, like a number of similar beetles, in North America on wild plants of the nightshade family. When the early settlers began to cultivate potatoes extensively, this beetle found a new, unlimited and satisfying supply of food. The result was that it increased very quickly. Soon it becam a serious menace, the greatest threat of a potato pests. In the year 1877 it was brough into Germany from America. From there i spread over the whole of Europe. Both th larvae and the adult beetle feed on the leaves c potato plants. Since they are extremely vora cious and need large quantities of food, thei presence in large numbers is a catastrophe. Chemical insecticides and other deterrents are not only costly but also limited in their effectiveness, since they destroy other insects which are enemies of the Colorado beetle. The best results are still achieved by picking the larvae and beetles individually off infested plants. The larvae [434] are red with black dots; the beetle itself [435] is 6 to 10 mm long and has a yellow body with five black stripes down the length of each wing-cover. The pupae [437] are orange-coloured and develop underground. The beetles emerge in the autumn, but the adult insect returns underground to hibernate. If they have no opportunity during their mass-reproduction to invade a new potato field or other growth of nightshade plants the adults will consume the tubers.

The **Spiny** or **Hedgehog Beetles,** *Hispinae,* mainly inhabit the tropics. Their bodies broaden out towards the back. The shield, which forms a neck-like ring, and wing-covers are set with spines, pits or small bumps. Their larvae are leaf-miners, burrowing a way along inside the leaves, which are left transparent. The **Hedgehog Beetle,** *Hispella atra* [436], is very common on dry slopes in central Europe. It is between 3 and 4 mm long and coloured blue-black.

435

436

437

438

439

The **Tortoise Beetles,** *Cassidinae,* have shield and wing-covers broadened into a disc-shape. The larvae have on their bodies tufts of thorny growths. On top of their bodies they carry a lump of excrement, which they use to conceal themselves. The **Tortoise Beetle,** *Cassida vibex* [438, 439], measures about 7 mm and is coloured bright green. It lives on the leaves of chickweed and tansy. There can be found its green larvae [440], as also the pupae [441] which are green.

Illustration [442] shows a number of tropical species of tortoise beetles:

centre: **Mesomphalia gibbosa** from Brazil 19 mm long; black and grey-brown.

above it: **Mesomphalia aenea** from Brazil 14 mm long; black with brownish-yellow marks.

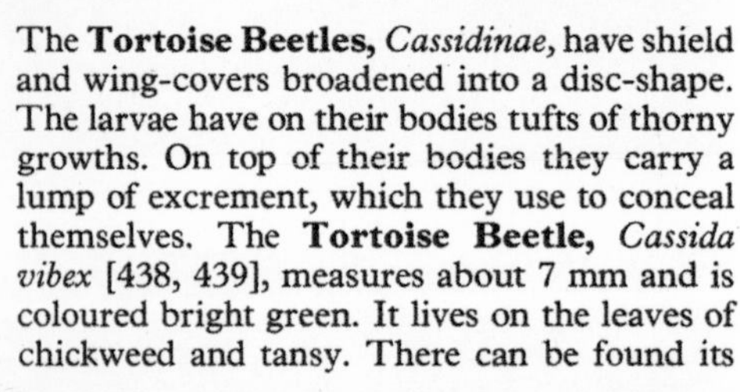

440

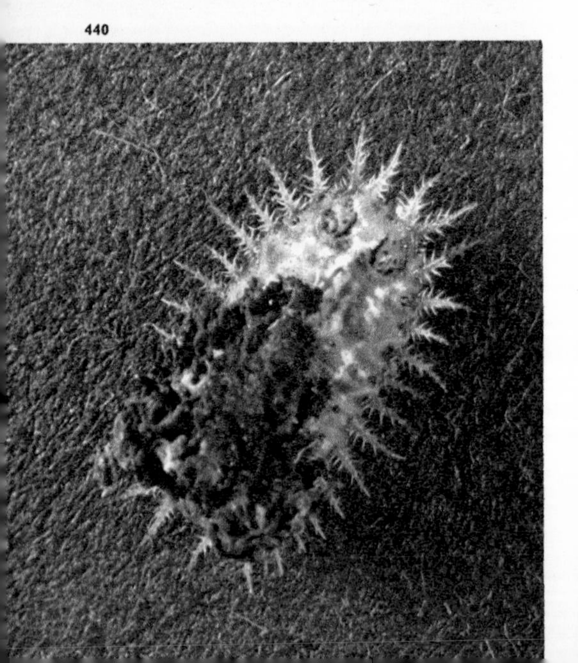

441

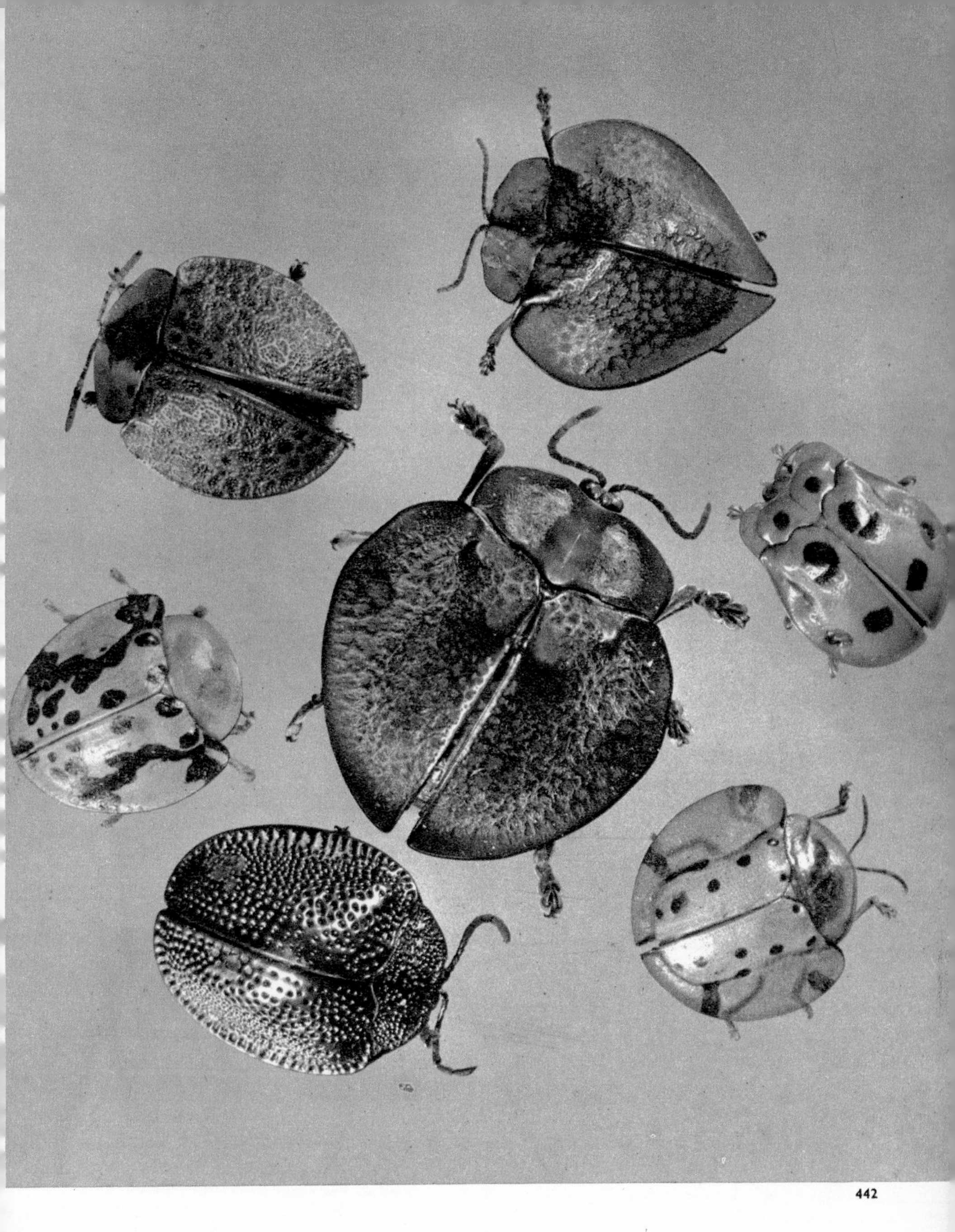

442

top left: **Poecilaspis nervosa** from Brazil; 12 mm long; black and blood-red.
bottom left: **Desmonota variolosa** from Brazil; 14 mm long; metallic green with glossy purple edges to the wing-covers.
bottom right: **Aspidomorpha inquinata** from Java; 12 mm long; yellow with black spots.
centre left: **Aspidomorpha miliaris** from the East Indies; 12 mm long; light brown with black markings.
centre right: **Prioptera decempunctata** from Malaya and Java; 11 mm long; light brown with red markings.

443

444

445

The members of the family of **Broad Weevils** or **Mole Beetles,** *Anthribidae,* are distributed throughout the world in about 2,000 species. One distinguishing feature of this family is the way in which the head is extended into a short, broad, snout-like process. Some of them are quite harmless, but others are extremely injurious to stores, especially to coffee. Certain species are parasitical on scale insects. In the palaearctic region only a small section of this family can be found.

The **White Spotted Mole Beetle,** *Anthribus* [*Platystomus*] *albinus* [443], is an inhabitant of central Europe. It is up to 1 cm long and dark brown with white markings. Its larvae develop behind the bark of dying trees.

The **Weevil** family, *Curculionidae,* consists of more than 40,000 species. They are distinguished by a snout-like extension of the head. Their wing-covers, and indeed the whole of their chitinous armour, are unusually hard. The average length must be round about 1 cm; the smallest measure 2 to 3 mm. Very few species are longer than 7 cm.

446

common weevil in central Europe is **Liphoeus tessulatus** [444]; it is about 8 mm long, rey-brown and feeds on low plants. The **hick-mouthed Weevil,** *Otiorhynchus laeigatus* [445], also from central Europe, is up o 7 mm long and black in colour. This genus ontains a large number of highly injurious ests to trees and crops.

The weevil **Scythropus mustella** [446] is about 8 mm long and bears yellow-brown markings. It lives on conifers in central Europe.

The weevil **Cyphocleonus tigrinus** [447] measures up to 11 mm; its wing-covers are covered with white hairs. It lives on cane flowers in dry, grassy places.

447

448

449

Some tropical weevils bea bizarre growths on thei wing-covers, have ex tremely long legs and greatly lengthened prc boscis. New Guinea boast such a weevil, of the genu *Eupholus* [448, 449]. is black and measures 2 mm. On each wing-cove there is a black thorn-lik growth. Its highly arche shield is marked wit wrinkles.

The weevils of the genu *Calandra* are greatly feare pests. The cosmopolita **Corn Weevil,** *Calandr granaria,* is a familiar an feared enemy in granaries and is related to gener much larger than itse which live in the tropics as for example the palı pest **Cyrtotrachelus du** [450, above]. Excludin the legs it measures 6 cn It is black and reddish brown and lives in Assan A more distant relative i the **Palm-borer,** *Rhyn chophorus palmarum* [450 left]. Its larvae are maggot like, white and about th thickness of a finger; the damage and even kill coc date and oil palms.

The **Brazilian Weevi** *Rhina barbirostris* [450 right], measures 42 mm It is black and has a re brush on its proboscis.

Weevils of the genus *Cur culio* have a very long arched proboscis for bor ing. As a rule they develo in the fruits of deciduou trees and shrubs. Togethe with related genera it i distributed throughout th temperate and warm zones The weevil illustrate [451] measures 22 mm, i black and white and native of Madagascar.

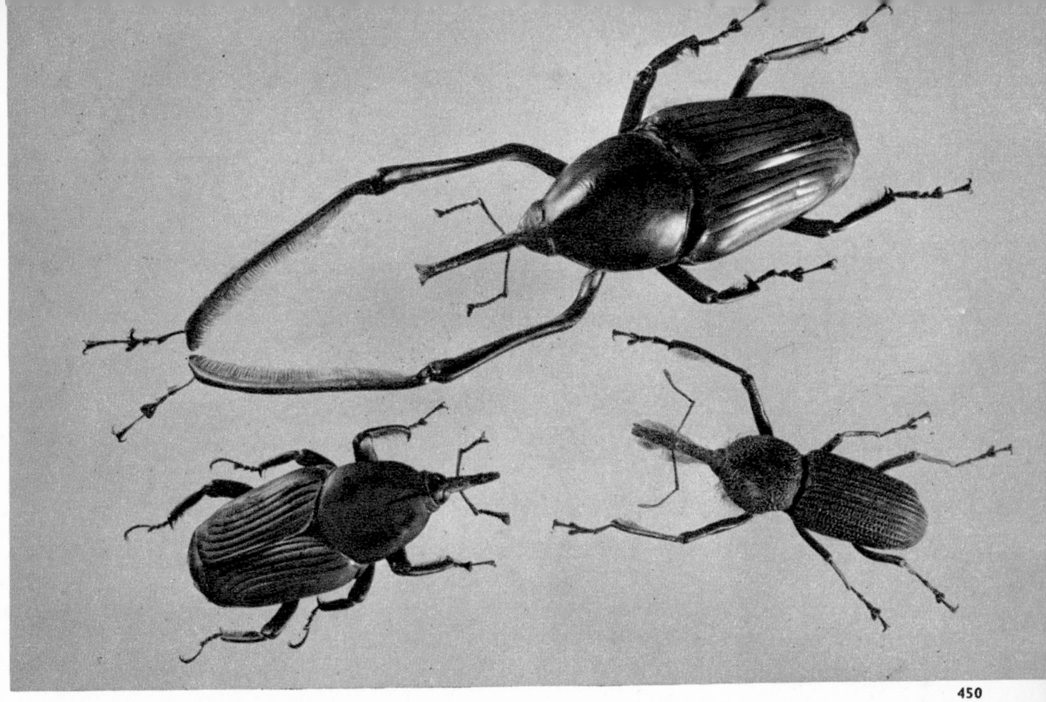

450

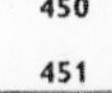

451

452

453

454

455

456

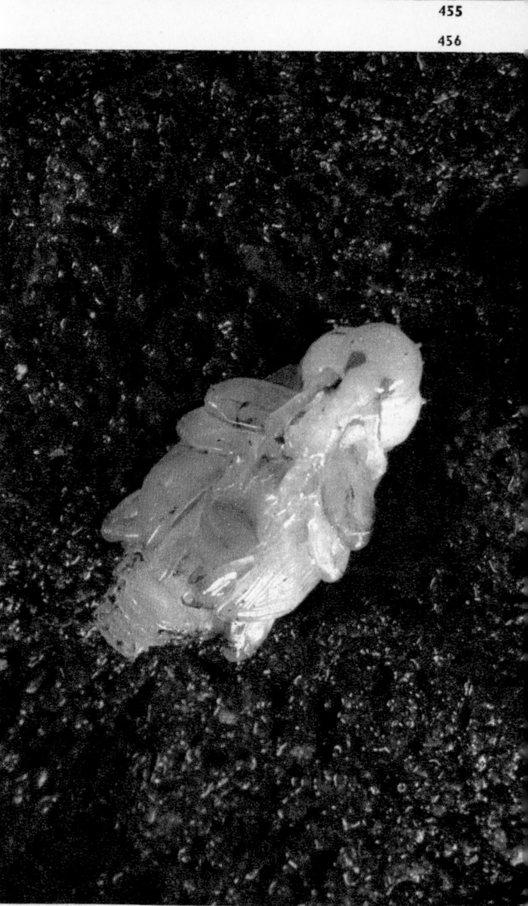

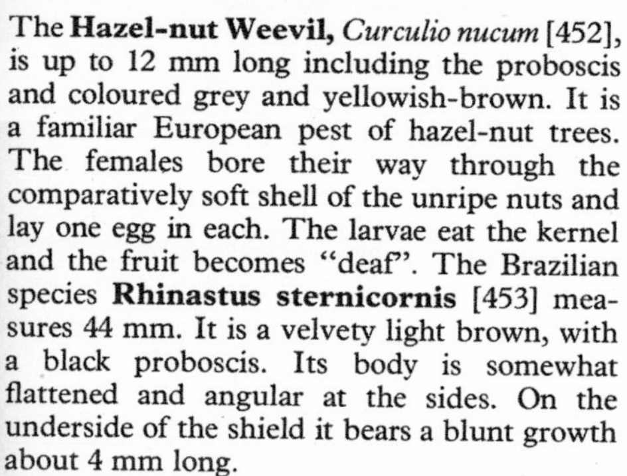

The **Hazel-nut Weevil,** *Curculio nucum* [452], is up to 12 mm long including the proboscis and coloured grey and yellowish-brown. It is a familiar European pest of hazel-nut trees. The females bore their way through the comparatively soft shell of the unripe nuts and lay one egg in each. The larvae eat the kernel and the fruit becomes "deaf". The Brazilian species **Rhinastus sternicornis** [453] measures 44 mm. It is a velvety light brown, with a black proboscis. Its body is somewhat flattened and angular at the sides. On the underside of the shield it bears a blunt growth about 4 mm long.

A common weevil of pine forests is the **Great Brown Weevil,** *Hylobius abietis.* It is up to 15 mm long, chestnut-brown in colour with a rusty-yellow coat of hairs. The adult weevil [455] gnaws the bark of extremely young pines and firs. Its larvae [454] live in the roots and in fir, pine and larch stumps. They are yellowish-white with a brown head. The majority of larvae hibernate and do not pupate [456] until the following spring.

457

459

458

460

461

The **Oak-leaf Roller,** *Attelabus nitens* [458], is up to 6 mm in length and black with a red neck-shield and wing-covers. The female cuts through the lamina of oak leaves and forms part of it into a cylindrical roll [457], in which she lays her egg. The larva feeds on the withering leaf.

The **Birch-leaf Roller,** *Rhynchites betulae* [459], does much the same thing with birch leaves, in which it lays two to four eggs. However, it does not confine itself to the birch, but attacks a whole range of other trees and shrubs, including fruit trees. In the tropics there are several related species of a very bizarre appearance. The thorny body of the **Attelabin** [460] from Madagascar is about 1 cm long and reddish-brown on the top; the legs are rusty-yellow.

Another of these weevils from Madagascar is **Tribus attelabini** [461], 25 mm long, with a glossy black head and shield and red wing-covers. Its head occupies half the total length of the body.

462

The family of **Long Beetles,** *Brenthidae,* inhabits principally the tropics, in about 2,000 species. They have elongated bodies and heads of almost the same length. They are generally social insects, and live behind the bark of trees. Their larvae develop in the wood.

The Madagascan species **Rhiticephalus brevicornis** [462] is 63 mm long. **Zetophloeus pugionatus** [463], likewise, comes from Madagascar.

463

The family of **Bark Beetles,** *Scolytidae, Ipidae,* contains tiny beetles generally of a dark colouring and a maggot-like body. They live behind bark and in the wood of forest trees, and in the event of mass-reproduction cause enormous damage.

The **Birch Bark Beetle,** *Scolytus ratzeburgi* [464], is 5 to 6 mm long and coloured black. Its larvae bore passages through the young wood of birch trunks. Affected trees can be recognised by the vertical rows of air holes.

Illustration [465] shows the larva and foodstuff of the **Greater Fruit-tree Bark Beetle,** *Scolytus mali,* a black, glossy beetle which grows to a length of 3 to 4 mm. It attacks every kind of fruit tree, as well as a number of other shrubs and trees.

464

465

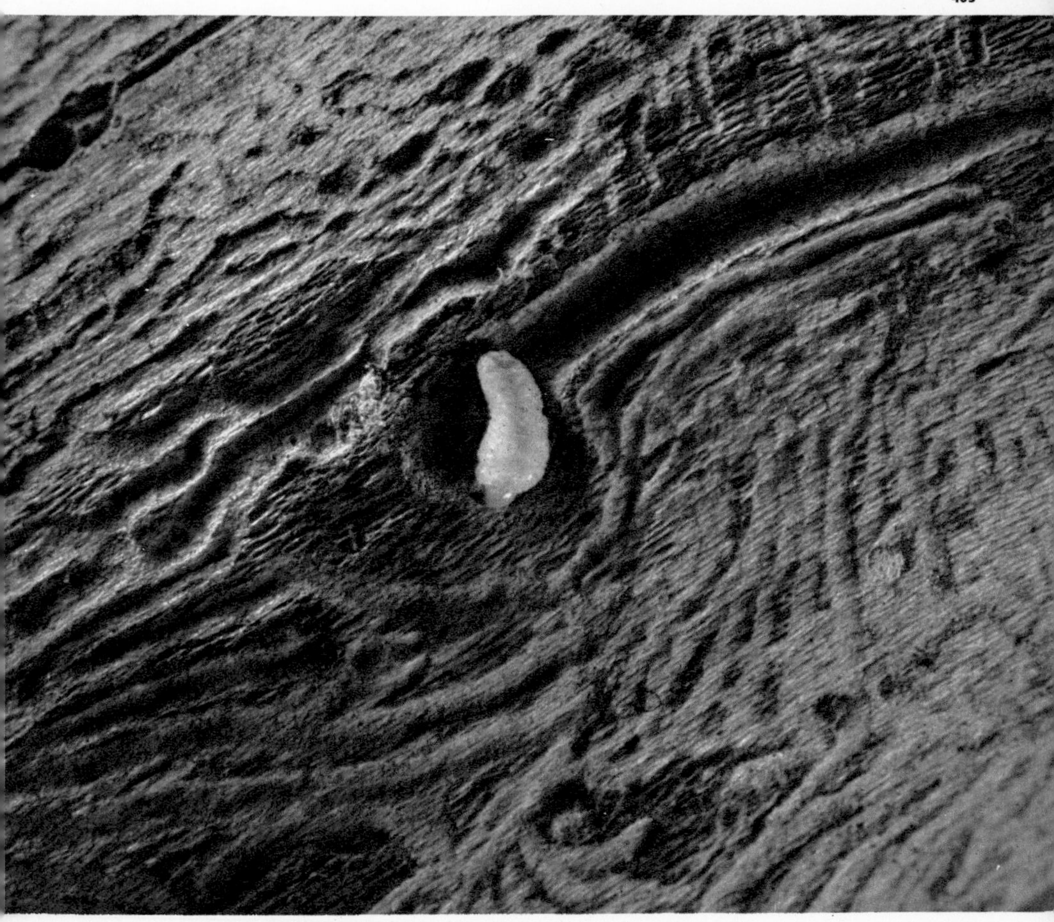

The **Wood-destroying Pine Bast Beetle,** *Hylurgus ligniperda* [466], is dark brown and grows to a length of almost 6 mm. Its larvae feed, burrow and gnaw on the roots and behind the bark of the trunks [467] of pine trees.

466

467

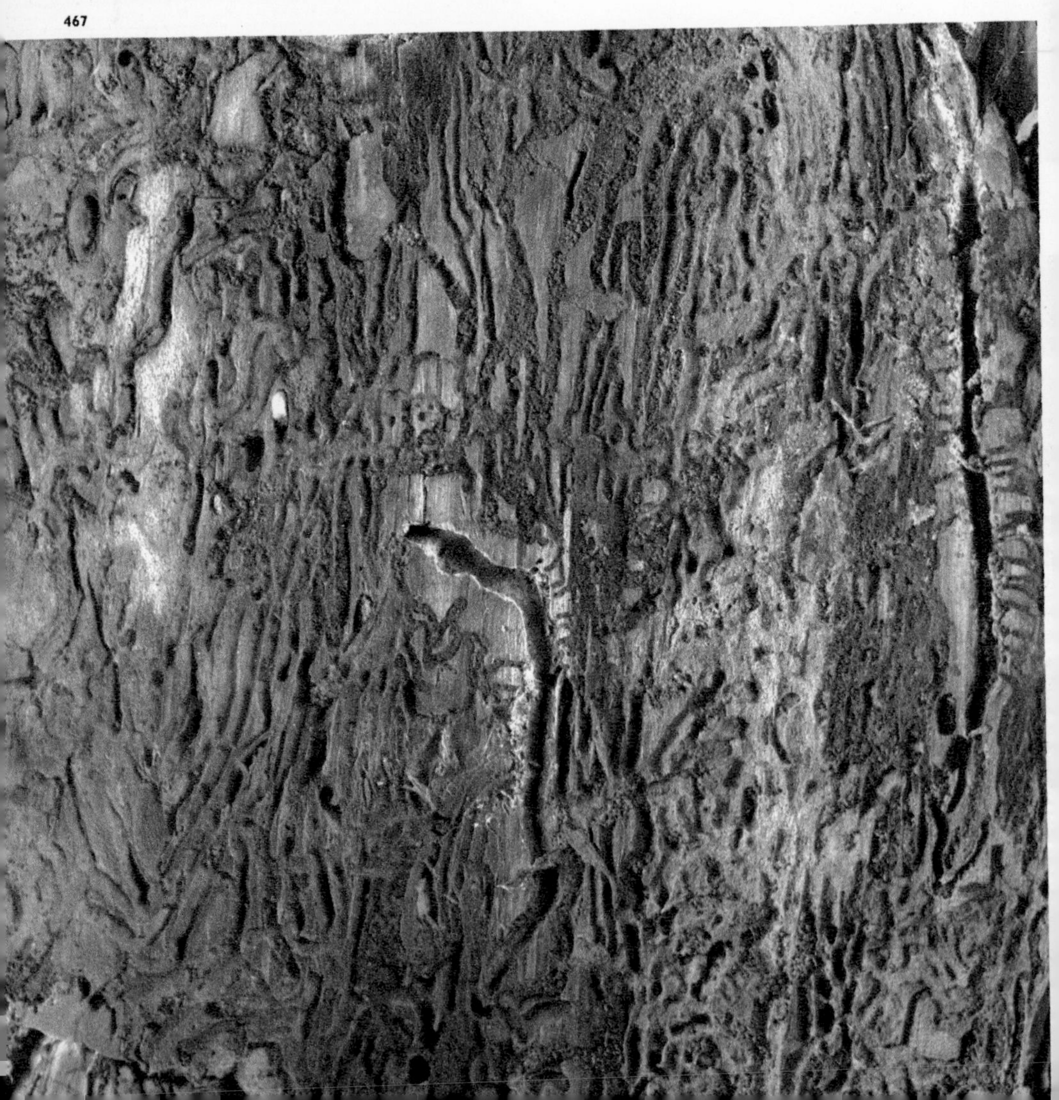

XXVIIa Tropical Stag Beetle, *Neolamprima adolphinae,* from the forests of New Guinea. 46 mm. long.

XXVIIb Male Dung Beetle, *Phanaeus imperator*. South America.

XXVIIIa The Brazilian Rhinoceros Beetle, *Enema pan*. Male. 55 mm. long.

XXVIIIb Hercules Beetles, *Dynastes hercules,* from the jungles of Central America. Pair. The male measures up to 17 cm. in length.

68

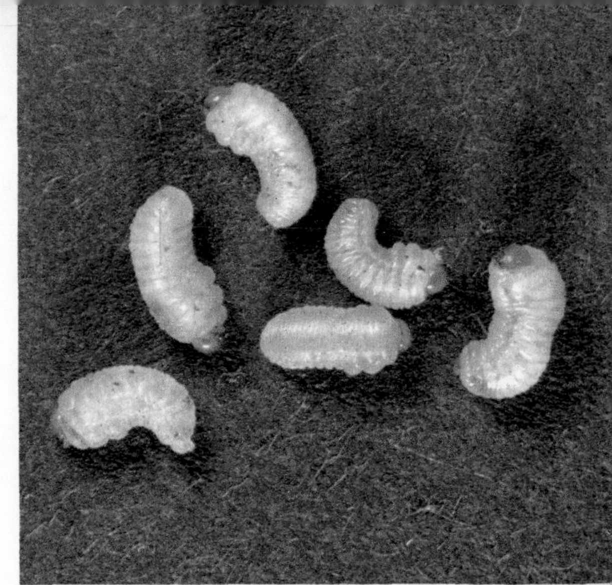

469

470

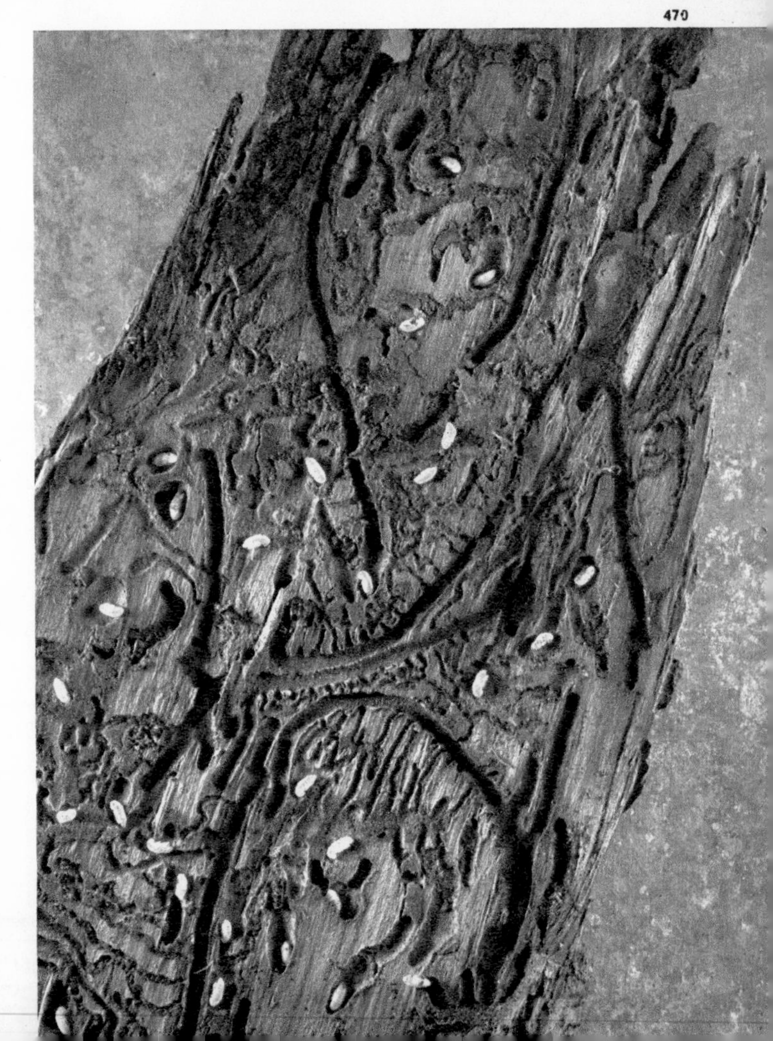

The most feared member of this family, causing tremendous damage, is the **Book-printer,** *Ips typographicus* [468], 4 to 5 mm long, dark brown and and covered with little yellow hairs. Its distribution ranges in Europe from the southern slopes of the Alps to the Arctic Circle and extends with the pine forests as far east as Vladivostok. As a rule it attacks fully grown and old trees. In feeding it forms characteristic patterns in the wood. In the middle, the fertilised female bores a straight, vertical entrance passage, from which a series of smaller larval burrows branch off—generally 10 to 25 on each side. The newly emerged larvae [469] are legless. As well as the wood itself, they also eat a fungal culture, in so far as it grows within reach of the entrance to the burrows, and they largely destroy the layer of bast on the trees they attack.

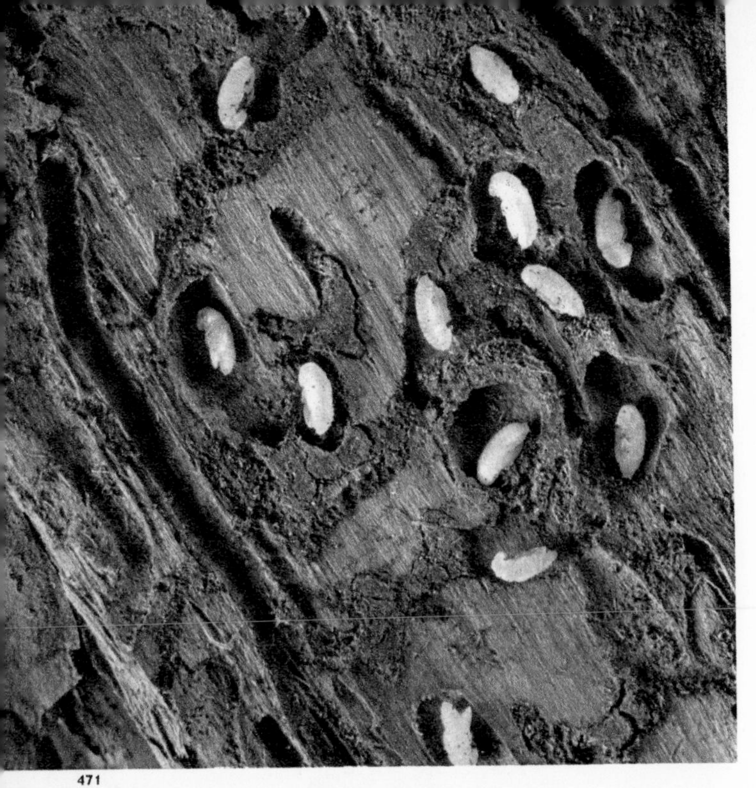

471

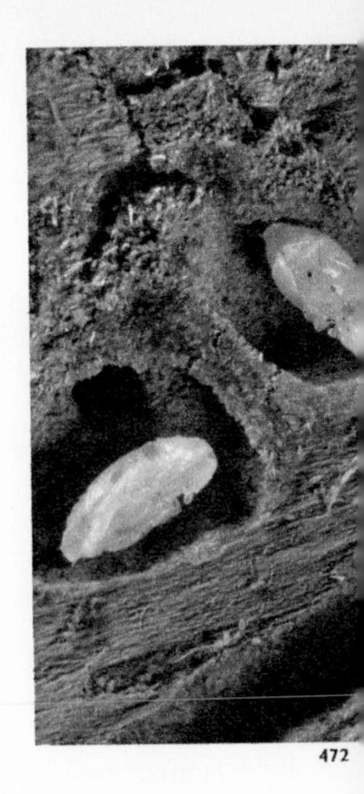

472

473

Illustration [470] shows the pattern formed by the larval burrows of the **Greater Larch Bark Beetle,** *Ips cembrae*.
The main burrow of the **Greater Larch Bark Beetle** is star-shaped, generally with three arms. The larval galleries, which are almost straight, lie close to one another. Illustrations [471, 472] show pupae in the pupal chambers. The imago of this species [473] resembles the book-printer in appearance.

'he superfamily **Leaf-horn Beetles,** *Lamel-cornia,* contains the family of **Stag Beetles,** *ucanidae.* It comprises over 900 species of eetles with strikingly elongated, antler-like ıandibles, which in the male sometimes equal ıe entire length of the rest of the body, lthough they seem less to be used as weapons ı battle than to serve to impress the females. 'he **Stag Beetle,** *Lucanus cervus* [475, pair], ; an inhabitant of ancient oak forests in 'urope and Asia. The male, including "ant-:rs" [474], sometimes achieves a length of 0 mm. It is dark brown, with a chestnut-oloured back. The larvae develop in the ompost of old oak trees and require a full ve years before they reach a length of about 0 cm. Then a further year passes before the dult beetle emerges from the pupa. The stag eetle is protected by law in civilised countries.

474

475

476

The **Giraffe Stag Beetle,** *Cladognathus giraffa* [476, left], is 90 mm long and coloured dark brown. It lives in Java, India and the Himalayan area.

Hexartbrius deyrollei [476, right] is stronger, but only 80 mm long. It is a native of Thailand and Sumatra.

A small species of stag beetle is **Leptinopteru tibialis** [476, above] from Brazil. It measure only 25 mm. The shield and head are brown the wing-covers light brown. At first sight i resembles very closely its European relations

477

Metopodontus bison [477] measures 60 mm, is chocolate-brown with a yellow border and lives in New Guinea.

Chiasognathus granti [478] grows to a length of 70 mm. It is copper-brown and shot with a greenish-red metallic gloss.

478

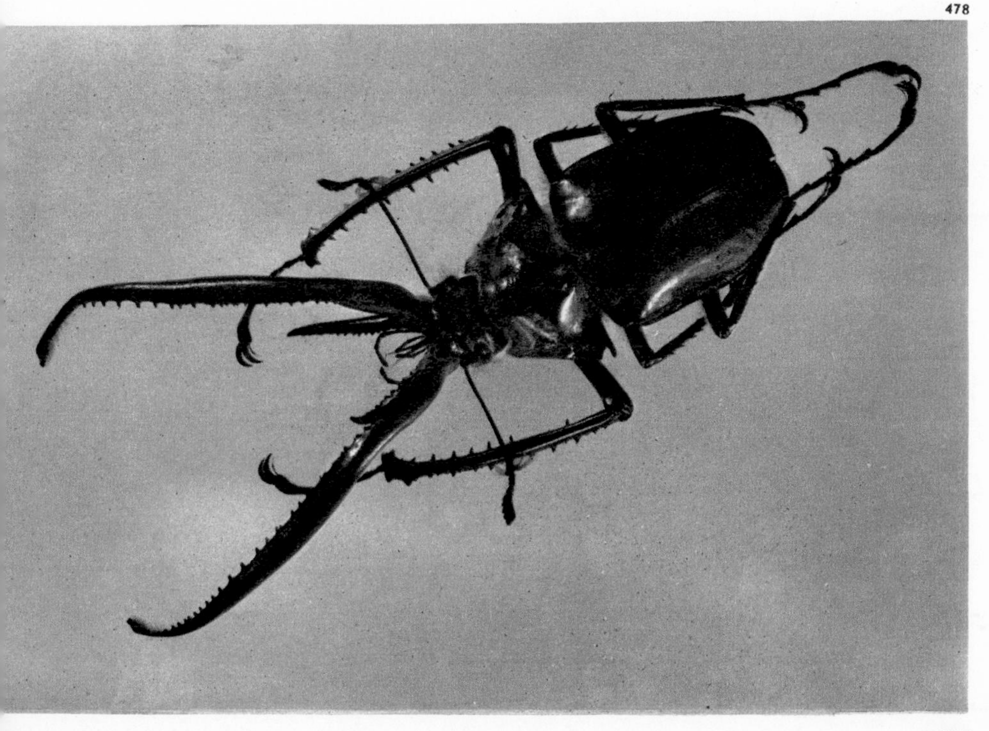

479

One of the most attractive stag beetles is the male **Neolamprima adolphinae** [479] from New Guinea. It is 45 mm long and golden-green with a fire-red head.

Almost all species of **Sugar Beetles,** *Passalidae,* live in the tropics. They lead a life of nocturnal habits and are remarkable in that both males and females feed their offspring. They obtain this foodstuff by mixing decayin wood and compost with their digestive juice Outside the tropics in North America lives th one species **Passalus cornutus** [480], a fla dark-brown beetle about 40 mm long. It ha a short process on the top of its head.

Proculus mniszechi [481], from Guatemal measures 60 mm, has a glossy black shield

480

4

lull black wing-covers and a thick coat of rust-
:oloured hairs along its sides.
The family *Scarabeidae* is distributed through-
›ut the world, in more than 20,000 species.
Among them is the famous **Sacred Scarab,**
vhose form has been preserved in thousands
›f symbolic renderings, which have survived
he passage of time, due to its place in the
nythology of the ancient Egyptians.
A pair of **Dung-pushers,** *Scarabaeus sacer*
483], are rolling balls of dung of a herbivorous
ınimal to a suitable place, where they will bury
hem in the ground.
Another species which provides for its off-
pring by storing balls of dung in underground
ıiding places is the European **Spring Dor**
Beetle, *Geotrupes vernalis* [482]. It measures
4 to 20 mm, is blue-green and has a metallic
heen.

482

483

484

485

486

The **Grape-cutter,** *Lethrus apterus* [484, male], measures up to 20 mm. It is black and forms an exception among related species in that it feeds on plants—generally on the winegrape.

The **Spanish Crescent-horned Beetle,** *Copris hispanus* [485, male], grows to a length of up to 30 mm, is black on top and has a rust-coloured coat of hair underneath. It lives in the Mediterranean countries and the adjoining parts of Asia.

The **Bull Beetle,** *Typhoeus typhoeus* [486, male], measures some 20 mm,

487

488

coloured black and lives in sandy forests in the Mediterranean area. It feeds on the dung of rabbits.

Onthophagus rangifer [487, male] is copper-red with a metallic sheen. It is found in central Africa.

The dung beetles of the American genus *Phanaeus* are gaily coloured, the males each bearing a horn on their head. The North American species **Phanaeus quadridens** [488, male] measures 20 mm. It is coloured violet. **Heliocopris gigas** [489, male], which is dark brown, comes from equatorial Africa.

489

490

The subfamily **Cockchafers,** *Melolonthinae,* has its distribution throughout the world, and contains over 7,000 species. Many of them cause a great deal of damage to agricultural and forest economies. The most well-known of these pests in Europe is the **Common Cockchafer** or **Maybug,** *Melolontha melolontha* [491, male]. It is up to 30 mm long, and familiar to young and old alike. Every three, or in some cases four, years they increase in vast numbers and completely strip deciduous forest trees and fruit trees bare of foliage. The larvae, which are called white grubs or rookworms [492] consume the roots of plants and cause much damage to field and garden cultivation. The cockchafers only wake up properly in the evening, when they set upon the tender, fresh greenery.

A species occasionally found in central Europe is **Melolontha pectoralis** [490, male]. It is a forest-dweller which moves about by day, and is most common in the forests of the Caucasus.

491

492

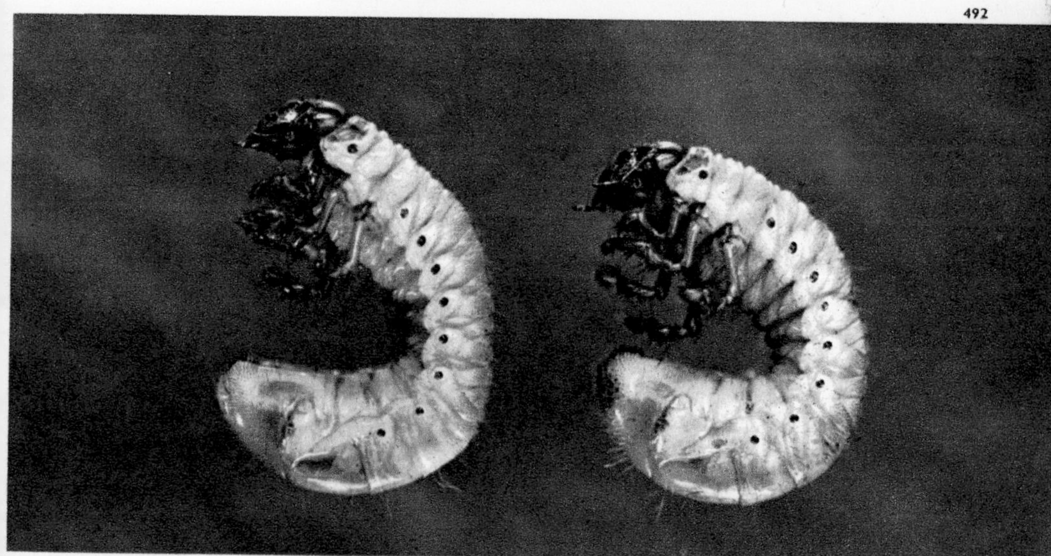

The **Fuller,** *Polyphylla fullo* [494, male], is up to 38 mm long. It is not only the largest of all European cockchafers, but also the most attractive of all the European members of the order of beetles. Its reddish-brown wing-covers are dusty and flecked with white. The males have exceptionally large antennae, which are broadened out into a leaf-like shape. It is an inhabitant of sandy coniferous forests, where its larvae feed on the roots of sand grasses. If one takes hold of it, it emits a chirping tone, which it makes by stroking the hind wings against the abdomen. It flies about in summer during twilight and in the night.

The subfamily of **Garden Chafers,** *Rutelinae,* includes the European **Common Corn Chafer,** *Anisoplia segetum* [493]. It measures 10 to 15 mm and has a black mark running down the middle of each wing-cover. The head

493

494

495

and shield are glossy green. Its larvae feed on corn roots and the adult beetles get their nourishment from the pollen of rye and wheat. The majority of related species of this beetle live in the tropics and subtropics. They are distinguished especially by their size and magnificent shiny, strikingly bright colours. In Peru and Ecuador lives **Chrysophora chrysochlora** [495, male]. Its egg-shaped body is green with a fiery red sheen; the wing-covers are pitted with dimples. The breast is glossy green and the legs a shimmering blue, violet and red. The rear legs of the males are extended into hook-like processes.

The beetles of the subfamily *Enchirinae* live in tropical Asia. There are only a few species and are all rare. The front pair of legs of the males is exceptionally long, and as a result they move along very slowly and with great difficulty. They can most commonly be found on the fermenting juices of palm-trees. One of them, **Propomacrus jansoni** [496, male], is about 60 mm long and brown in colour. It comes from southern China.

496

497

The males of the subfamily of **Rhinoceros Beetles,** *Dynastinae,* are characterised by a dark-coloured horn some 10 mm long and three knobs on the bulging neck-shield. They are the chief prize of every collector, not only on account of their unusual appearance but also because of their size. Among them is the longest known beetle in the world, the **Hercules Beetle,** *Dynastes hercules* [colour plate XXVIII]. It attains a length of up to

498

18 cm and is an inhabitant of tropical central America, along with other related species, such as the black and brown **Dynastes neptunus** [497, male] from Colombia. It grows to a length of some 13 cm and has a coat of golden-yellow hairs on its underside. **Megasoma elephas** [498, male] is a central American species, measuring about 13 cm. It is black with a thick coat of yellow-brown hair. The three knobs on the neck-shield can be seen.

499

500

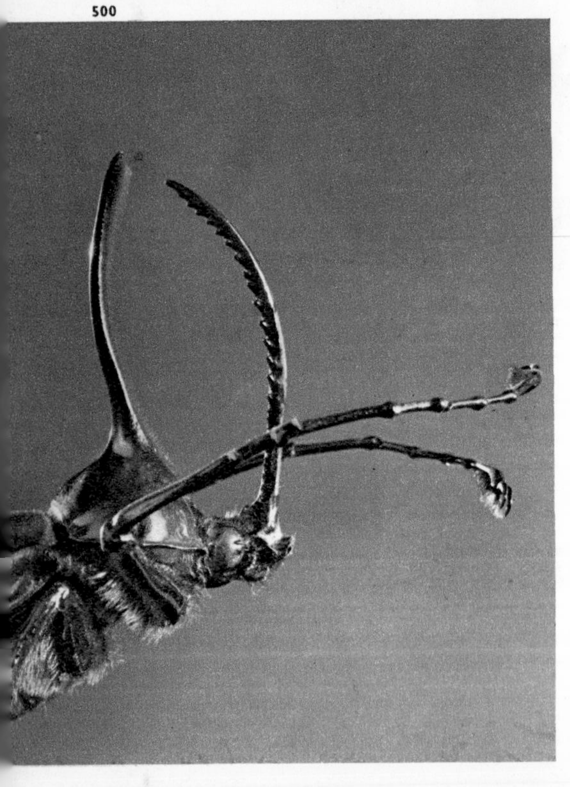

The bizarre forms assumed by the growths on the heads and shields of the male tropical rhinoceros beetles naturally call into question the purpose of these ornaments. Clearly they are in some measure analogous to the antlers of the stag beetles or indeed the true antlers of deer among the mammals. Since only the males are equipped with them, they would seem to play some part in the battle for females. However, observation of this battle leads one to suspect that these growths are more of a hindrance than a help to the warrior. In fact, only a few species are able to bring them into play as weapons of defence.

The **Mexican Rhinoceros Beetle,** *Golofa pizarro* [499, male], is approximately 50 mm long and coloured brown. The horn spreads out at the tip to form a highly arched, triangular surface, the underside of which is thickly covered with yellowish hairs.

Golofa porteri [500, male], from Colombia, is up to 80 mm long. It is reddish-brown, with darker colouring on the legs, horn and knobs. The crescent-shaped horn on the head, which measures about 40 mm, has two rows of sharp teeth on the inside.

XXIXa Banded Brush Beetle, *Trichius fasciatus*. About 13 mm. long. Lives in deciduous forests on high ground in Central Europe.

XXIXb The Tropical Flower Beetle, *Chelorrhina polyphemus*, from West Africa. The male is 6 cm. long.

XXXa The largest European Clearwing, the Poplar Hornet Clearwing, *Sesia apiformis*. 3 cm. long.

XXXb The Yellow-legged Clearwing, *Aegeria vespiformis*. 1.5 cm. long. The larvae live in the trunks of oak-trees. Central Europe.

The **Japanese Rhinoceros Beetle,** *Trypoxylus dichotomus* [501, male], attains a length of up to 75 mm and is dark brown. It carries on its head an extended horn ending in a doubled, antler-like fork.

In the East Indies and the Sunda Islands live black **Rhinoceros Beetles,** *Chalcosoma atlas* [502, male]. The largest are found on Sumatra and Java; they measure upwards of 12 cm. On the thorax they bear three prominent growths, and there is a fourth, curving upwards, on the head.

501

502

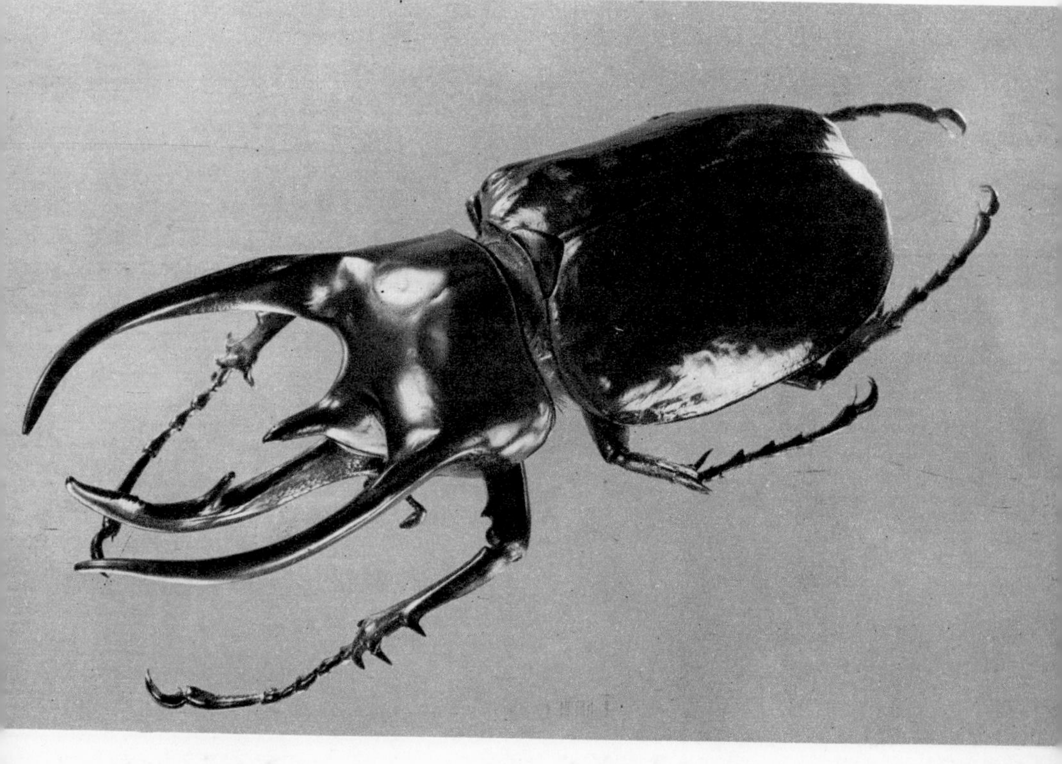

503

Eupatorus gracilicornis [503], from Vietnam, is approximately 65 mm long. The head and shield are black; the wing-covers light brown. On the shield there are four shorter, thorn-like growths; a longer one, curved sharply upwards, grows on the head.

The Brazilian species **Enema pan** [504, male] measures about 50 mm. It has a shiny dark brown colouring and the horn on its thorax is forked at the end.

The larvae of all rhinoceros beetles live in the dead trees, where they develop in the dead

504

505

wood, though they can also be found in rotting tree stumps, compost and dung heaps. The narrow worm-like larvae of the European **Common Rhinoceros Beetle,** *Oryctes nasicornis* [506], are usually found in the peat of greenhouses and early root crops, in tan bark and in heaps of sawdust which are kept for use in smoking rooms. The pupa of this species [505, from below] is brown-red and about 50 mm long.

506

507

508

The large subfamily of **Rose Beetles,** *Cetoniinae,* has a world-wide distribution. It includes beetles of strikingly beautiful colouring with a metallic sheen. In flight they do not spread out the wing-covers. They fly so well and for such long periods at a time that we can certainly look on them as the best fliers of all the beetles. On average they measure some 30 mm, though there are among them some giant species. Almost all feed, in the adult state, on tender flower-parts or suck sap exuded from trees.

The European **Common Rose Beetle,** *Cetonia aurata* [507], has a length of up to 20 mm. Its wing-covers are a metallic gold-green with short white stripes running crosswise. Underneath it is copper-red with a purple sheen. The larvae of rose beetles are broader and flatter than the rookworms of the cockchafer; they are yellowish-white with a brown head and brown hairs. They live in the decaying wood of dead trees and stumps. The larvae of the **Rose Beetle,** *Potosia cuprea* [508], which resemble those of the common rose beetle, grow up in ants' nests and feed on the vegetable rubbish available there. On illustration [509] we see the pupa of *Potosia cuprea,* slightly enlarged.

One of the most common Australian rose beetles is **Eupecilia australasiae** [510]. It grows to a length of about 20 mm, and its colouring gives it the appearance of a small, precious ceramic. On the glossy brown wing-covers it bears a yellow design. It feeds on the fragrant young leaves of flowers.

509

510

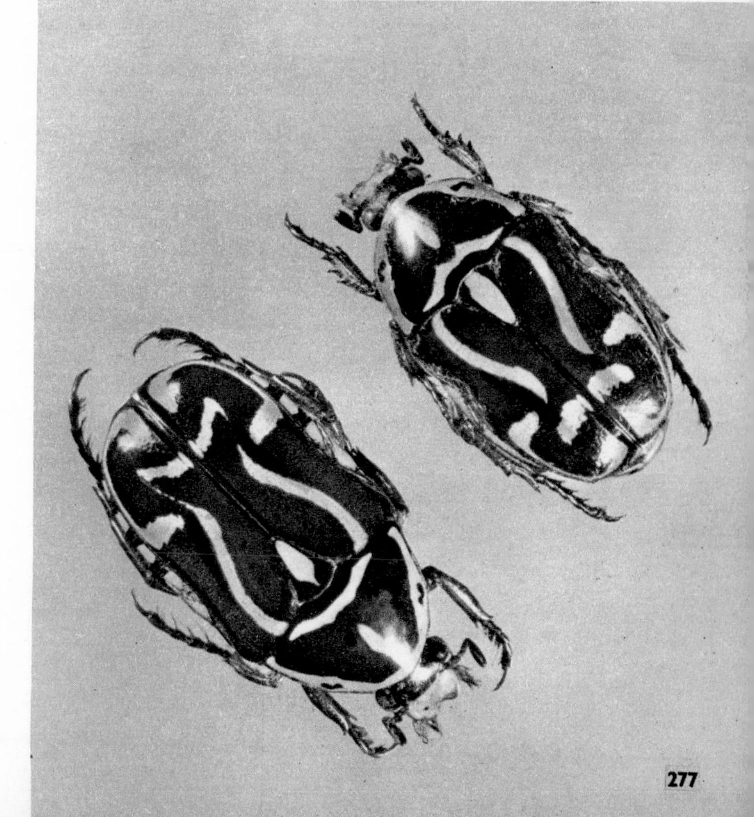

511

512

Among the rose beetles there are also some species the wing-covers of which are not glossy, but have a coat of hairs. Among these are the **Brush Beetles,** *Trichiinae,* as for example the **Banded Brush Beetle,** *Trichius fasciata* [511 and colour plate XXIX]. It grows to a length of up to 13 mm and flies with open wing-covers. It can frequently be found on blackberry blossoms and on wild roses.

Another species of the rose beetle family is **Epicometis hirta** [512]. It measures between 8 and 12 mm and has a thick coat of bristly hairs on the entire surface of its body. On its head the hairs are gold and yellow, on the breast and the wing-covers they are grey, and on the underneath of the body whitish. In the spring it can commonly be found on dandelion flowers. Its larvae are harmless, as they feed principally on rotting plant matter. This rose beetle is an inhabitant of central and southern Europe, but is also found in Asia Minor and adjacent countries.

A remarkable beetle, reminiscent in its appearance of the African goliath beetle, is **Dicranocephalus dabryi** [513] from northern China. It measures about 30 mm and is grey-brown with a velvety texture rather like suede. Its head bears a lyre-shaped ornamental structure.

513

514

In the primeval forests of Africa there are some marvellously beautiful insects of the rose beetle family, the males of which are distinguished by a striking head-dress.

Megalorrhina harrisii [514, male] is up to 50 mm long and coloured a dull dark green on the upper side. The shield is green with a yellow border. The wing-covers are spotted red and yellow. Its home is western equatorial Africa.

Mecynorrhina torquata [515, male] measures up to 70 mm including the horn. Its upper side is dull green and the head grey-white and equipped with a small, sharp horn. The lower part of the front pair of legs is set with thorns.

The giants in this group of beetle are the **Goliath Beetles.** Intact specimens are very rarely found in collections. The velvety-haired surface of their bodies is very sensitive to being touched and is therefore easily damaged by careless handling. Large specimens of males with well-developed horns are much sought after by collectors.

515

516

The goliath beetles fly through primeval forests at a great height through the tops of palms and other blossoming trees. They rarely alight on the ground. They lay their eggs in rotting wood, in which the larvae develop.

Goliathus meleagris [516, male] measures some 75 mm. It is grey-white and black in colouring, with a pearl-coloured sheen.

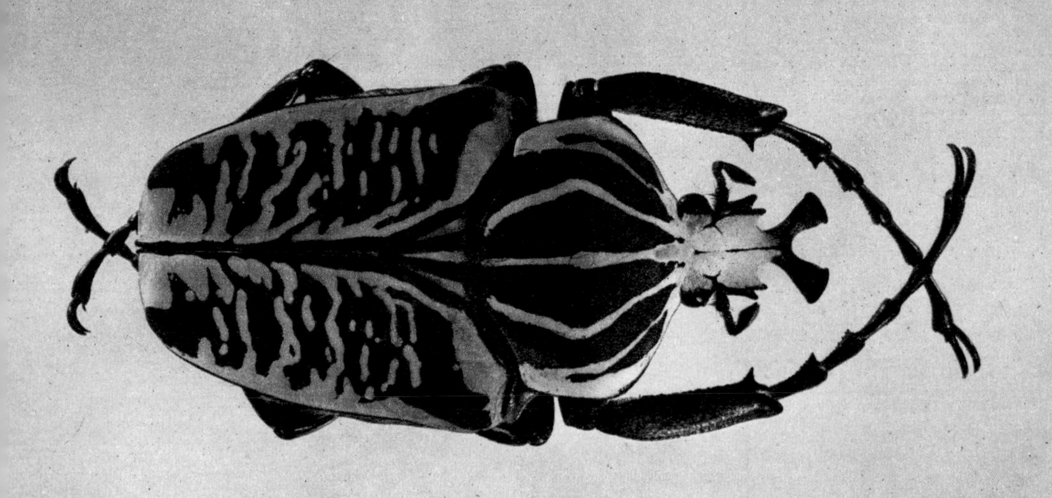

517

From eastern part of equatorial Africa comes **Goliathus kirkianus ssp. courtsi** [517, male]. It is up to 60 mm long and rusty brown to black.

One of the rarest goliath beetles is **Goliathu atlas** [518] from West Africa.
Goliathus regius [519] measures some 10.

518

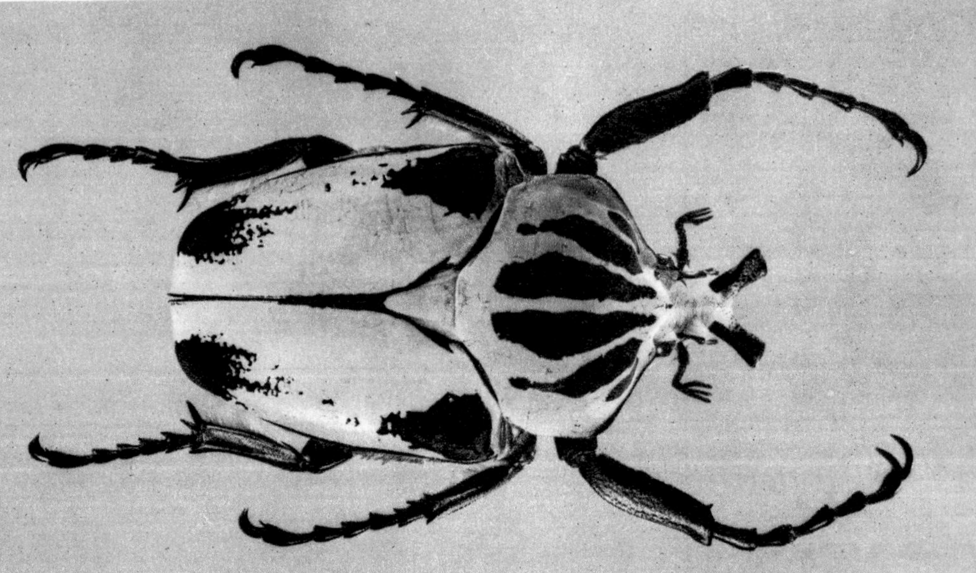

519

m and coloured dull white and velvet black. The largest goliath beetle is **Goliathus giganteus** [520, male]. It measures over 11 cm. Its dull velvet wing-covers are a dark reddish-brown, and the shield black with white stripes. It is found in equatorial Africa.

520

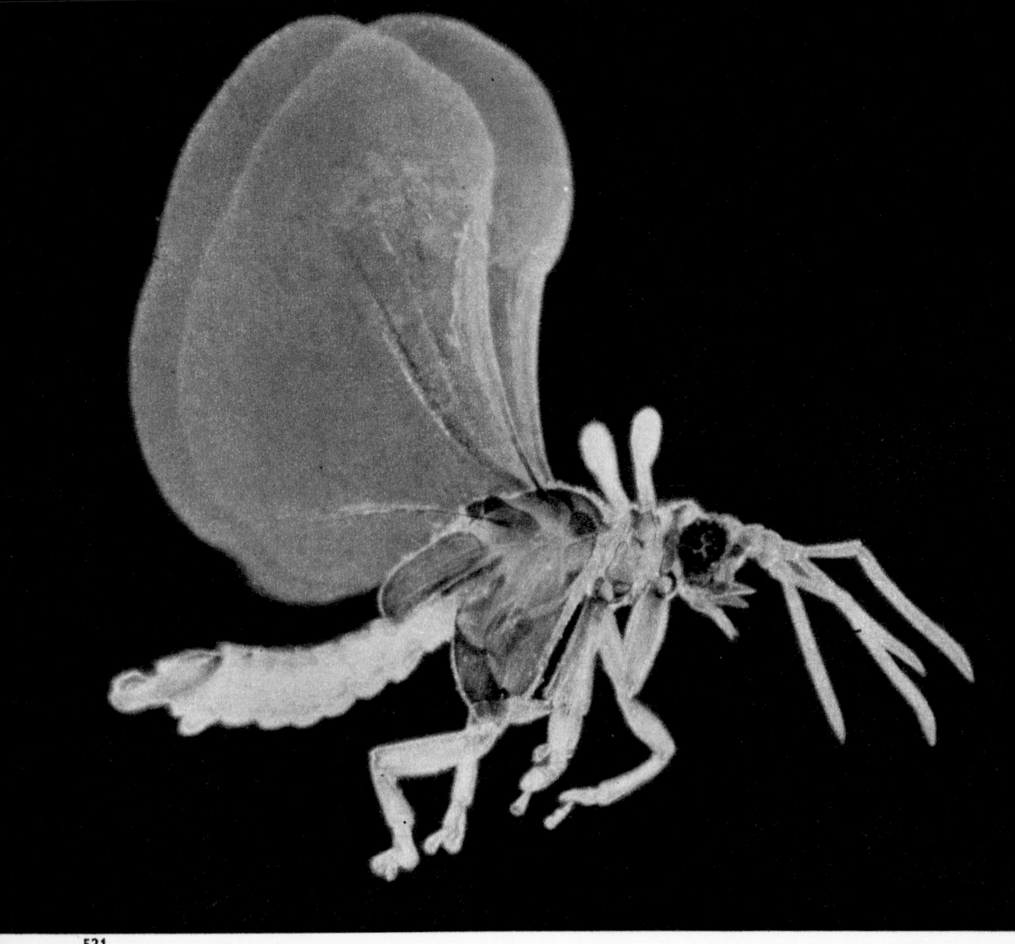

521

522

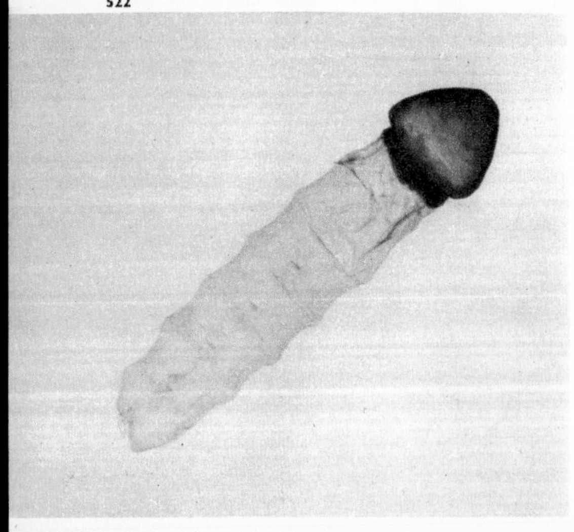

The second order which makes up the **Sheat winged** superorder, *Coleopteroidea,* is that the **Stylops,** *Strepsiptera.* It contains abc 300 species, which are very rarely seen, though they are distributed throughout t world. They only measure a few millimet and live parasitically in the bodies of otl insects, mainly in the *Hymenoptera*. The ma move about independently; their large, me branous rear wings enable them to fly qu quickly. The fore wings are degenerate a form small club-like processes, which app ently serve as balance organs (static organ Furthermore, they possess large, protrudi compound eyes and multibranched antenn The females on the other hand are sightle legless, wingless and incapable of independ movement. Their mouthparts are stron degenerate. Their bodies are soft and lar form. The head and thorax are merged int

523

524

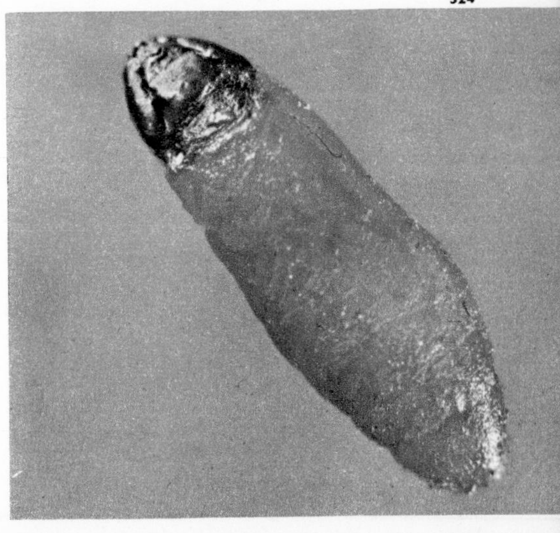

single chitinised section. The soft part of the female's body remains within the abdomen of the host, and only the flattened cephalothorax, on the underneath of which is situated the brood passage, protrudes from the host's body. On sunny spring days the males dart about. The male of **Elenchus carpathicus** [521], of the family *Stylopidae,* measures approximately 2 mm. This species lives parasitically in the cicada *Dicranotropis divergens*.

The female of the genus **Stylops** [522] is 6 mm long. It is parasitic on the **Earth** or **Sand Bee,** *Andrena rosae*.

Illustration [523] shows a male **Sand Bee,** *Andrena nigroaenea,* which has been attacked ("stylopised") by a female of the species **Stylops melittae.**

The female stylops of the species **Paragioxenos brachypterus** [524] is 6.5 mm long, has no feet and resembles a larva.

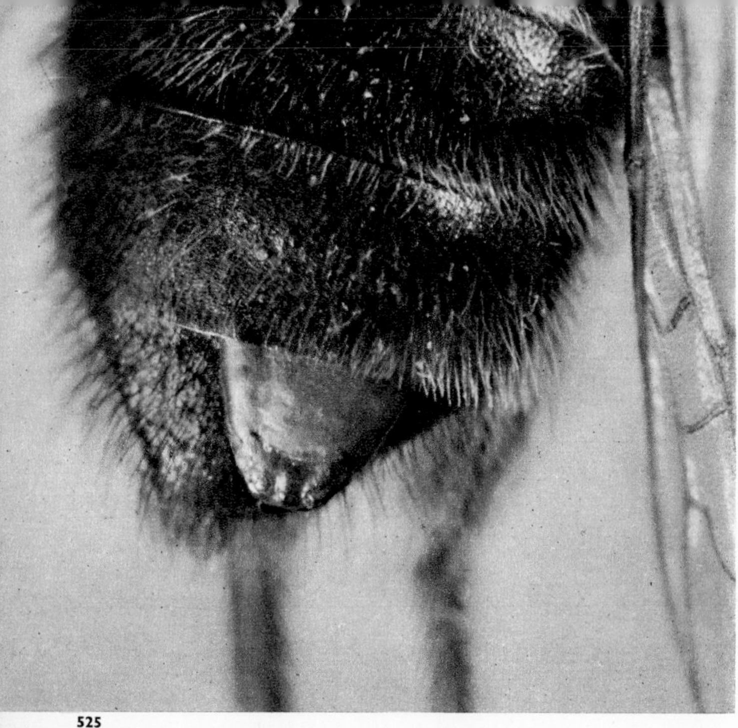

525

526

The cephalothorax of the female **Stylops melittae** [525; detail of 523] can be seen projecting from the rear end of its host's body. An interesting feature in the development of the stylops is that there are two different stages of larvae—an example of hypermetamorphosis. In the body of the fertilised female microscopically small larvae develop from thousands of eggs. These then escape into the open through the opening in the brood canal. This is the first stage, in which they are called *Triungulini*. They have between two and five eyes on their broad heads, long bristles at the ends of their bodies and little hooks or suction pads on their legs. They move extremely swiftly running about on the petals of flowers, where they wait for pollen-collectors and nectar-lovers, one of which can be a suitable host. It is, of course, a fact that the majority of them perish without being able to continue their development.
However, those larvae which are lucky enough to actually encounter a suitable host attach themselves firmly to its body with the hooks of suction pads on their legs, and are carried back to the host's nest. There they immediately bore their way into the abdomen of the soft, undeveloped host larva, shed their skin directly afterwards, and change into the motionless, parasitic larvae of the second stage. They live in the body fat of the host larva. As soon as they are grown up, they push the front part of their body out between the

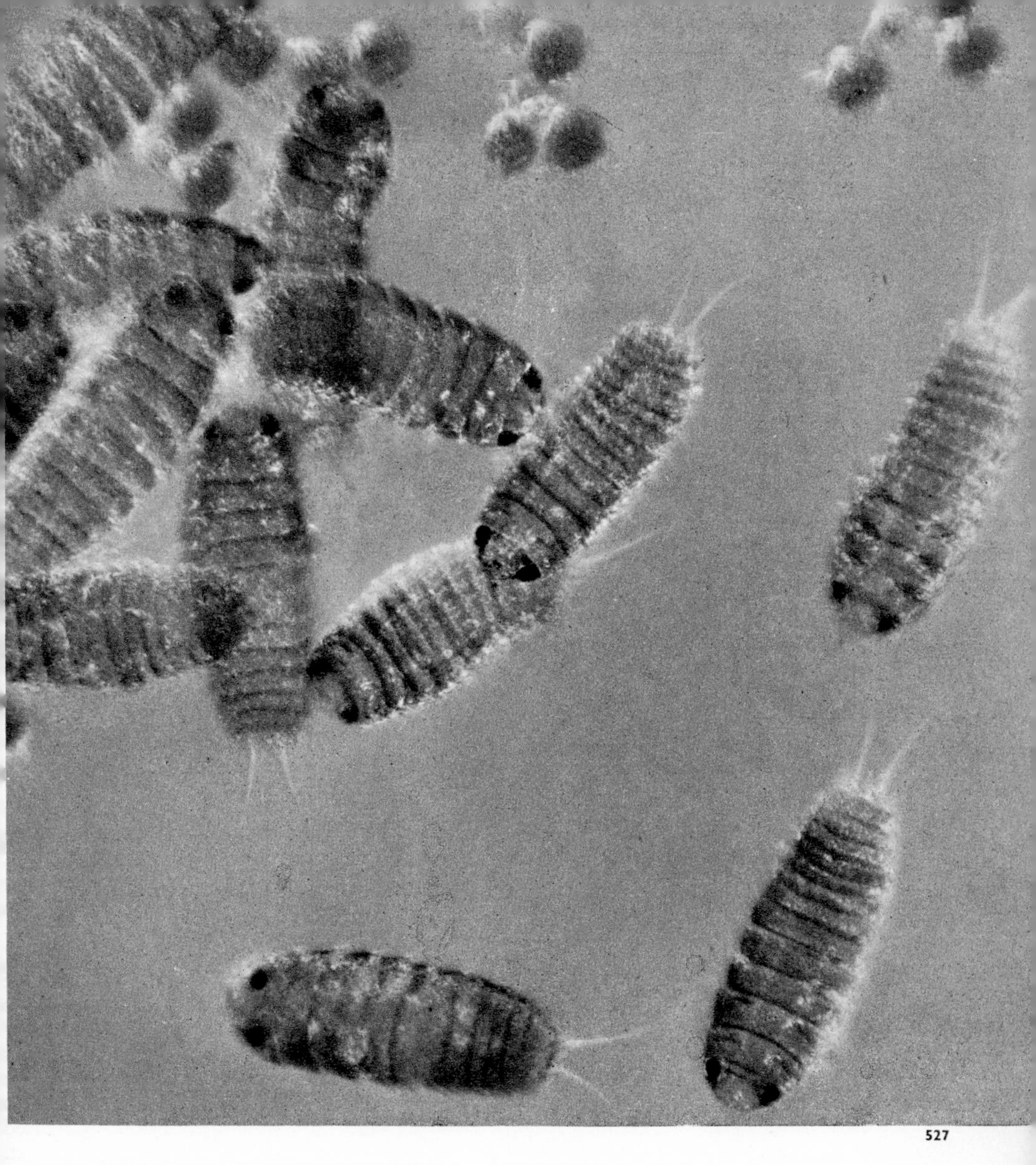

527

abdominal segments of the host and pupate. If there is a male in the pupa, it emerges from the pupa as an imago, leaves the host, and flies off to find a female. It takes no food, and its life only lasts a few hours. The female, on the other hand, only partly emerges from the pupa, in that it sticks the cephalothorax out, while the rest of the body remains in the pupal case, firmly fixed in the host, which has itself meanwhile become adult. Bees, wasps, cicadas, grasshoppers, bugs or any other attacked insect carry the parasites about with them everywhere. It has been observed that one effect of stylopisation for the host is the so-called parasitic castration.

On illustration [526] are Triungulini of **Stylops albofasciatae;** on illustration [527] those of members of the genus *Stylops* from the body of a stylopised **Sand Bee,** *Andrena bicolor*.

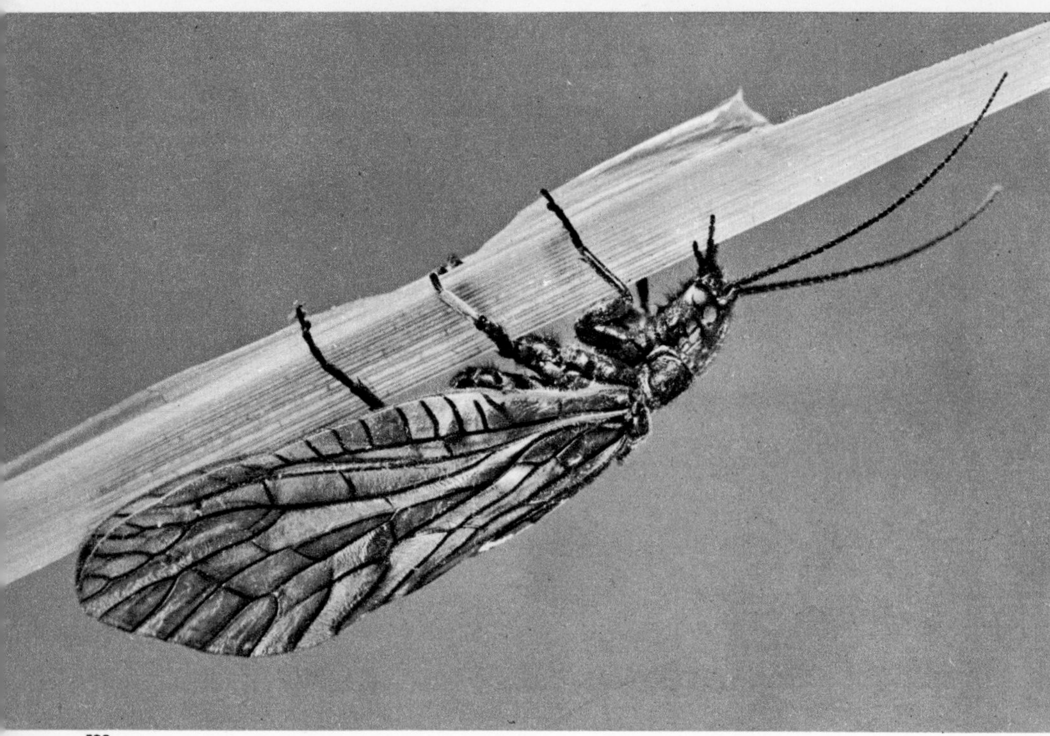

528

The twelfth superorder, the **Nerved-winged Insects,** *Neuropteroidea,* contains in all eight orders of insects of greatly varying appearance, whose common characteristic is a complete metamorphous cycle.
The first order is that of the **Alder Flies,** *Megaloptera,* which are insects with two pairs of membranous wings, which they carry folded into the shape of a steep roof when they are not flying. They spend the day at rest in the neighbourhood of water, and at twilight and in the evening they come out in swarms. We divide them into two families: *Sialidae* and *Corydalidae*.
A common alder which flies around still waters in Europe is **Sialis lutaria** [528]. It has a span of up to 26 mm, coloured dark brown, and its transparent wings have a smoky appearance.

529

XXXIa The Bloodwort Burnet, *Zygaena laeta,* a rare Central European species. Common in Southern Europe. 16 mm. long.

XXXIb The Iron Prominent, *Notodonta dromedarius.* About 3.5 cm. long. Central Europe.

XXXIIa Caterpillar of the Privet Hawk-moth, *Sphinx ligustri,* in the posture of a sphinx. Central Europe. 8 cm. long.

XXXIIb The Spurge Hawk-moth, *Celerio euphorbiae.* Attitude when suddenly disturbed. Central Europe. 5 cm. long.

The adult insects apparently do not take any solid food, but only drink.
The larvae of this order have large heads with powerful mouthparts adapted for biting, and on the sides of their bodies are rows of feather-like fan-appendages. They feed on small water insects and their larvae. The larvae of the **Alder Fly,** *Sialis fuliginosa* [529], the second central European species, which resemble the previous one in colour and form, live in mountain becks and other small, swift-moving streams.
Some of the remarkable tropical species are the **Giant Alder Flies.** The larvae are greatly prized by American anglers as bait for trout. One of the best known is **Corydalus cornutus** [530], with a length of about 10 cm. The biting mouthparts of the males are so well developed that they resemble long horns.
A member of the second order, the **Snake Flies,** *Raphididae,* is **Raphidia notata** [531, male]. It spans up to 28 mm.

530

531

532

533

The snake flies live on tree trunks, especially in oak forests, but sometimes in gardens on pear trees. The females have a long, flexible ovipositor which projects beyond the folded wings. All snake flies are good fliers. They are predators with voracious appetites, and as such play an important part as allies in the biological warfare on pests, in that they feed mainly on harmful insects like plant lice, scale insects, aphids and the larvae of bark beetles, among others. Unfortunately, many of these useful insects are destroyed together with the pests which form their prey during the spraying or dusting of wide areas with insecticides.

The females of **Raphidia flavipes** [532] have a span of about 27 mm. The snake flies lay their eggs singly in the crevices in the bark of trees. The larvae [534] have suction pads which they can project at the ends of their bodies, and use to hold on as they move about. They move about swiftly in the cramped space behind the bark, where they hunt the larvae of wood-eating insects. They pupate under moss or behind bark. The pupae [533] are at first motionless but shortly before the adult emerges change position.

534

535

536

537

The insects of the third order, the **Lacewings,** *Planipennia,* or **Nerve-wings,** *Neuroptera,* lead a predatory life, and are useful enemies of harmful insects.

The species **Boriomyia subnebulosa** [535] has a span of about 18 mm. It is grey-brown and lives in the forests and gardens of Europe and Asia.

Drepanopteryx phalaenoides [536] is up to 32 mm across and coloured rust-brown. Its broad wings are equipped with decorative designs and so shaped at the edges that they resemble butterflies. The rear wings have an opalescent gloss. These insects are encountered on fruit trees and near water. They are particularly common in the warmer parts of Europe.

The **Common Lacewing,** *Chrysopa carnea* [537], has a span of up to 28 mm; it is blue-green and possesses transparent, opalescent wings. Its eyes have a metallic sheen. Its winter generation is particularly well known, since it invades the living rooms of houses, often flying into electric lights. The summer generation has its habitat in forests, fields and gardens. Lacewings inhabit Europe, Africa, Asia and America.

The larvae of the "golden-eye" [538] are known as "plant-lice lions", since they play havoc among colonies of plant-lice. They catch these

538

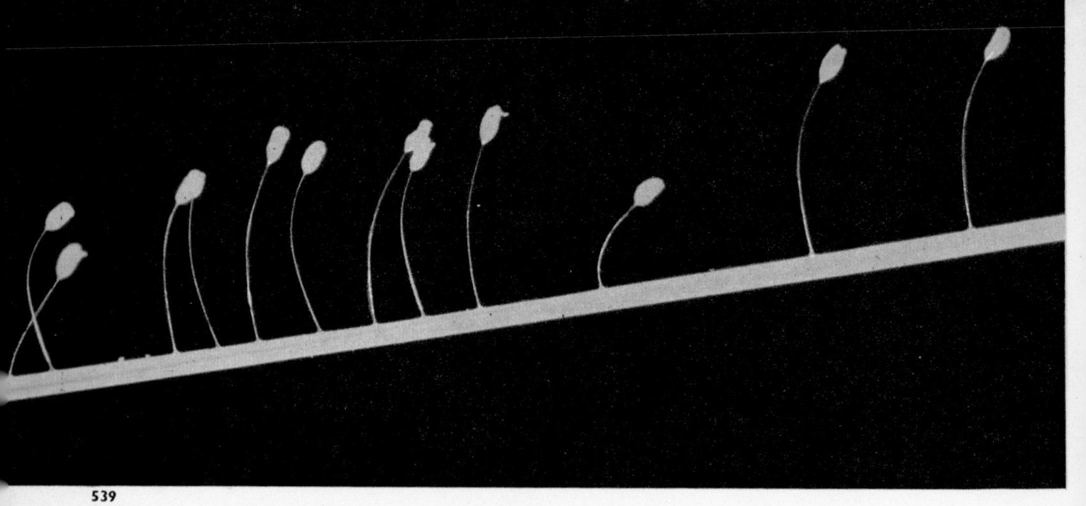

539

540

parasites and suck them dry with the aid of their long, hollow maxillae.

Female lacewings attach their eggs—up to 100 in number—with long pedicels or stems on to plants and various other objects [539].

The **Styrian Praying Lacewing,** *Mantispa styriaca* = *pagana* [540], has a span up to 35 mm. It is a yellow-brown predator with transparent wings, found in southern and occasionally central Europe. Its fore legs have developed in much the same way as those of the praying mantid into powerful organs for seizing the living prey. The larva which emerges from the reddish eggs hibernates for the whole winter without nourishment, and in the spring seeks out a female wolf spider (family *Lycosidae*). There it bores its way into the cocoon which the spider carries around on her abdomen, undergoes metamorphosis into the second stage of its development and eats the spider's eggs. After this it pupates. Before emerging into the adult form it leaves the cocoon in the form of a nymph, capable of independent movement. The tropical species of this and related genera grow to a length of up to 50 mm.

The **Thread Lacewings,** *Nemoptera,* inhabit the warm countries of Europe, Africa, Asia and Australia. Its rear pair of wings are greatly elongated and narrowed into a ribbon-like shape. Both the adults and the larvae are predators; the latter have a narrow, long neck and live in sand. The span of the front wings of **Nemoptera sinuata** [541] amounts to 55 mm. The light yellow-green wings bear a brown pattern. This species is extremely common in the Balkans.

541

543

The **Ant-lion** family *Myrmeleonidae* consists of insects with a slender, elongated body and four transparent wings with a network of venation. In some species they bear a variegated pattern of spots. They fly at evening and during the night, with an uncertain, giddy flight. Their larvae have a broad, short body and lead a predatory existence in the sand. In this they create little tunnels, at the bottom of which they prey on small insects which they seize with their long, powerful, hollow jaws, through which they suck juices of the victim.

Acanthaclisis occitanica [542, 543] has a span of up to 11.5 cm and is grey-brown with an irregular black stripe on its fore wings. Both the body and the head bear a coat of white hair. The larvae hunt their prey by burying themselves in the sand and lying in wait for passing insects with open jaws. This species inhabits southern Europe and North Africa.

An even larger related species lives in the Mediterranean area, the **Giant Ant-lion,** *Palpares libelluloides.* It has a span of about 15 cm with wings flecked dark brown.

544

The **Spotted Ant-lior** *Myrmeleon europaeu* [544], has a span up t 70 mm; its wings hav brown spots. It is found i sandy places in Europe often on the edges of coni ferous forests. One of it nearest relations is th **Unspotted Ant-lion** *Myrmeleon formicarius.*

The larva of the spotte ant-lion [545] is abou 12 mm long and grey brown. It builds a pit fo capturing its prey in th sand by moving backward round and round on th selected spot, with grea patience, loosening th sand with its abdomen an throwing it aside with it head. In this way a funne is made, at the bottom o which the ant-lion larv buries itself and awaits it prey. Only the wide-oper jaws protrude above the surface. When a passing insect slips over the edge of the funnel and tries to scramble back up again, the larva throws sand at it until it slips down within reach of its jaws. Having seized the prey firmly, the ant-lion injects it through its hollow jaws with a preservative juice which also serves to liquefy the victim's body, which it then sucks out. It throws the remains of its meal out of the funnel and settles down to wait for the next victim. The holes of the fully-grown larvae are up to 5 cm in depth and 8 cm across [546].

545

546

547

548

549

The larvae of the spotted ant-lion pass two winters in hibernation, and in the spring of the third year pupate. The silky cocoon of the pupa is covered with grains of sand or earth which are woven into it [547]. In illustration [548] we see an opened cocoon with a living pupa, and another from which the cocoon has been removed altogether.

The ant-lions are closely related to the **Butterfly-lion** family *Ascalaphidae*. These insects have very long antennae with a small, disc-shaped knob at the end. The head and thorax have a thick coat of hair. The wings are yellow with brown patterns. In contrast to the ant-lions, they fly well and quickly, keeping close to the ground. Furthermore, they are diurnal creatures and love the sun. Their larvae resemble those of the ant-lions very closely, but being broader and flatter. They hunt actively under stones and in the undergrowth, but do not build funnels, as do the ant-lions. This family is distributed in every warm country in the world. Some species are found in the northern parts of Europe and America. Both the species shown in illustration [549] are found on the rolling slopes of the central European plains. Both have a wing-span of about 50 mm. At the top is **Ascalaphus macaronius** and below **Ascalaphus libelluloides.**

550

'he fourth order of the *Ieuropteroidea* consists of ıe **Scorpion Flies,** *Iecoptera,* in which there re over 300 species of enerally small insects 'ith two pairs of mem-ranous wings. Some spe-ıes have atrophied wings r are completely wingless. 'he head has a beak-like rocess and equipped with iting mouthparts, long ıtennae and large eyes. 'he larvae have the ap-earance of caterpillars; ıost of them are predatory ke the adult insects, al-ıough others feed on ıimal and plant remains. 'he **Scorpion Fly,** *Pan-rpa germanica* [550], has span of about 25 mm. It as long, narrow wings 'ith dark-brown spots. 'he **Common Scorpion ly,** *Panorpa communis* 551, 552, male], has a span f some 30 mm; its wings re marked with black ɔots. It lives in damp laces near water.

Iale scorpion flies have angerous-looking chiti-ous segments at the ends f their abdomens which e curved upwards like ıe sting of the scorpion, though their only pur-ɔse is to steady the insect uring copulation. The :orpion flies are regarded useful insects, since they estroy plant lice and other nall pests. Occasionally ıey also feed on plant and uit juices.

551

552

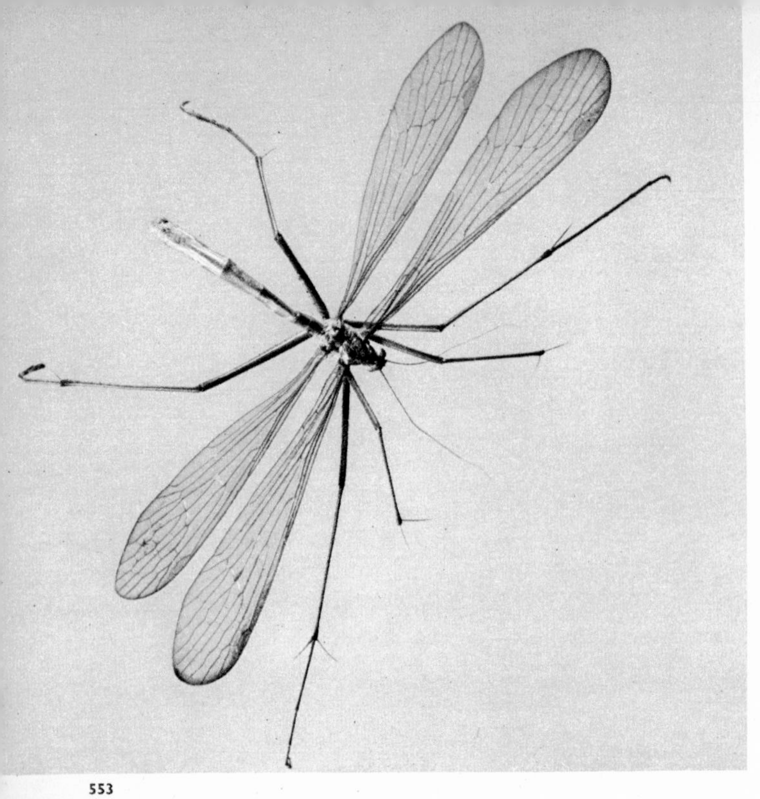

The **Gnat Scorpion-fly**, *Bittacus tipularius* [553], has a span of up to 40 mm. It resembles a crane-fly in appearance; it has two pairs of narrow, yellowish-brown wings and inhabits southern Europe and the warmer parts of central Europe. They come out at twilight, spending the day hanging by the front legs from the branches of shrubs growing near water with the other legs spread out like a spider's web. In this position they await small flying insects, which they seize with the tips of their prehensile legs. Besides southern and central Europe, all the warmer areas of both the Old and the New World can be regarded as the habitat of insects of the genus *Bittacus*.

553

554

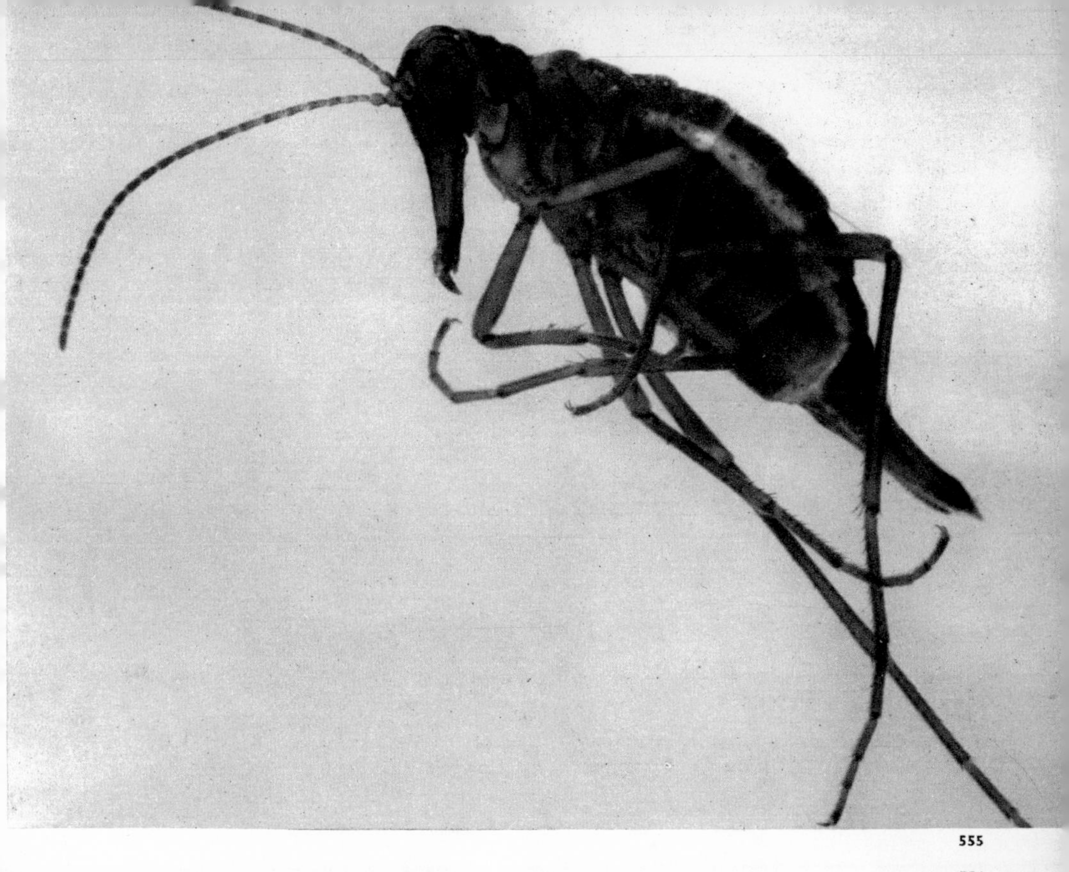

555

556

Insects of the **Winter Scorpion-fly** family *Boreidae* are small, measuring about 5 mm in length. They are confined to the northern hemisphere, but there are several species in North America, where they live hidden under trees and in moss on rocks. In winter they often appear on melting snow. They feed on moss, but are also thought to be predators, preying on springtails, especially on glacier fleas.

Boreus westwoodi [554, female] is a comparatively large species, measuring up to 6.5 mm. Its body is soft, dark-green and glossy, whereas the legs, ovipositor and the beak-like process on the head are yellowish. This species is a native of central Europe.

Boreus hyemalis has a hard body, is very dark brown and glossy. Small specimens measure only 2.5 mm, while the larger ones are up to 5 mm. They are found in central Europe. Illustration [555] shows a wingless female with ovipositor; illustration [556] is of a male with wing-stumps.

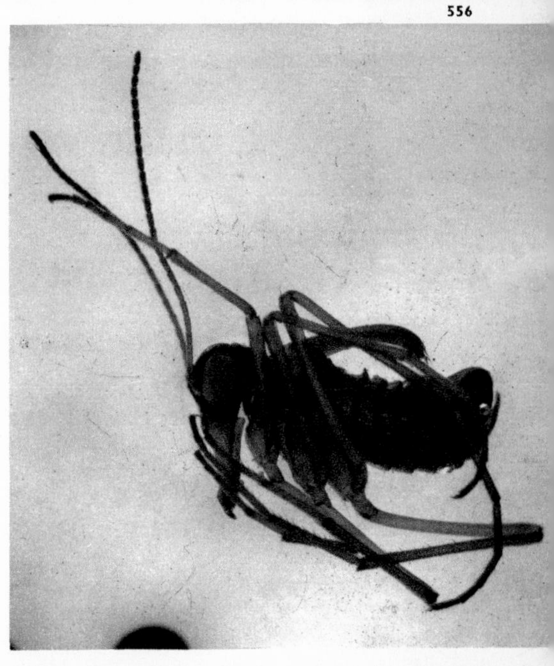

557

The fifth order of the *Neuropteroidea* consists of **Caddis Flies,** *Trichoptera.* In it are numbered about 3,000 species of insects of greatly differing sizes. Some measure only 4 mm, others up to 60 mm; they are distributed throughout the whole world. They are nocturnal and spend the day hidden on plants or under stones. The wings are folded to form a roof when they are at rest and the long antennae stretched out ahead of the insect. The mouthparts of the imago are atrophied, being used at most for licking up the juices of plants. The

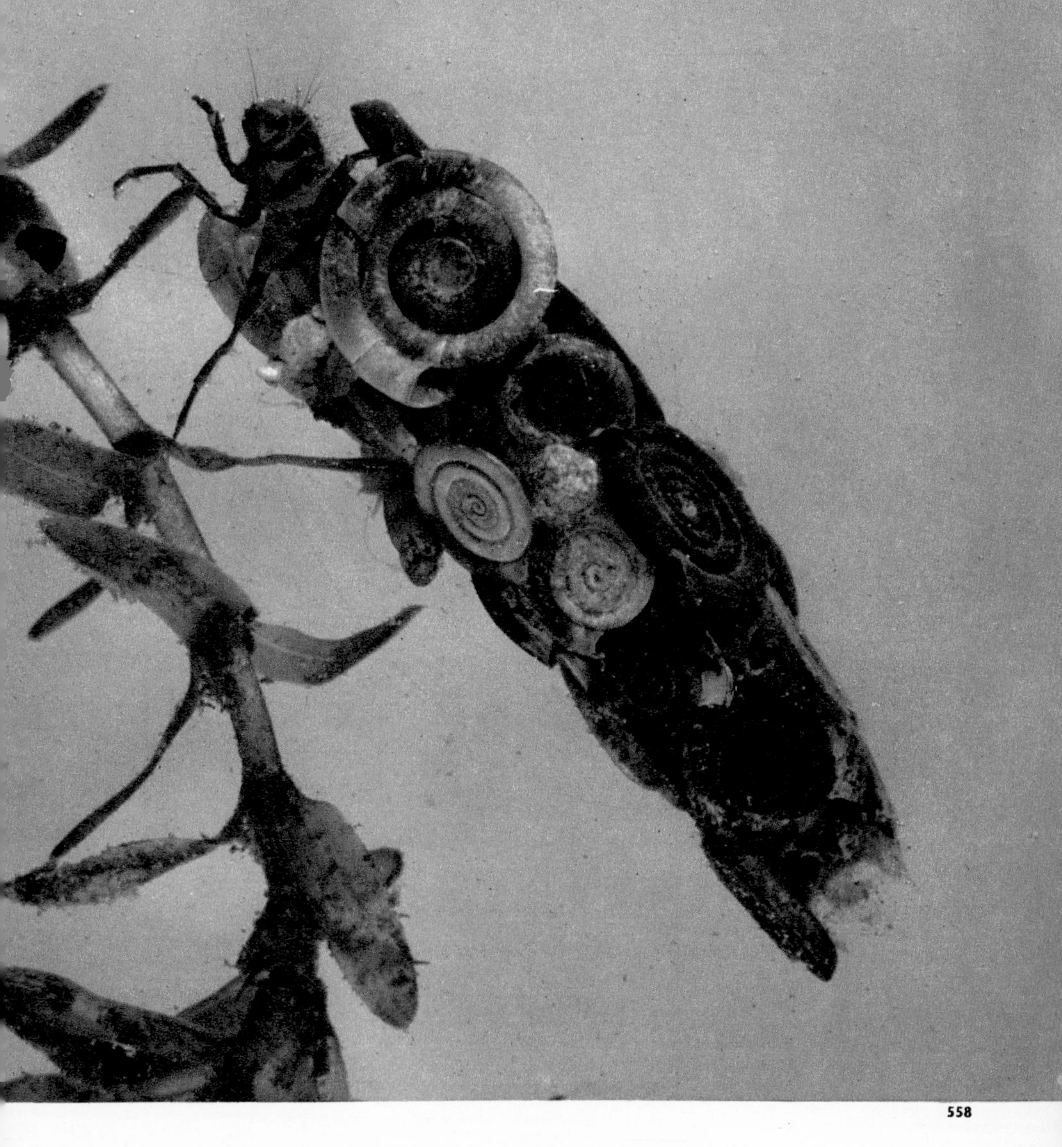

558

arvae have powerful biting mouthparts and ;enerally feed on plant matter. They live •rincipally in water, in "caddises", which onsist of material from various sources spun ogether.

)n illustration [557] we see a caddis fly of the genus *Limnophilus*.

The larvae of the species **Limnophilus flavicornis** [558] build little dwellings with the aid of small snails' shells, which go to make an extremely attractive caddis. They measure about 25 mm and live in stagnant water.

559

560

The larvae of the **Caddis Fly,** *Limnophilus rhombicus* [559], form cases for themselves out of living water-lentils and the flora on the bottoms of ponds. They measure 25 mm. The larvae of other species, on the other hand, live in running water and hide in their cases made of little stones. Some of them are predatory and spin nets or webs under water in front of their cases, in which they catch small aquatic arthropods. The larvae of the genus **Triaenodes** have ciliated legs adapted for swimming, with which they move swiftly through the water while still in their cases.

There are other species, by way of contrast, which crawl along the bottom or on plants; some remain quite motionless. The 25 mm long larvae of the genus **Stenophylax** [560] build their cases with small pebbles from the bottoms of streams.

Larvae of the genus **Odontocerum** [561] spin smooth caddises, which resemble a curved horn, and to which tiny grains of sand from the river bottom are attached for camouflage.

A larva of a caddis fly pupates within its caddis or case. After completing its development it bites through to the outside, crawls out of the water and changes into the imago. The adult caddis flies look very much like small butterflies, which have instead of scales a coat of small hairs on the wings, which can be quite thick in some species. All of them are an unremarkable grey or brown.

561

The sixth order of *Neuropteroidea* is that of **Butterflies** and **Moths,** *Lepidoptera*. There are over 100,000 known species distributed throughout the world, wherever insects can live. They undergo a complete metamorphic cycle and in the adult state possess four membranous venated wings, which are generally decorated with brightly coloured scales. The colouring is characteristic of individual species and, apart from small variations, is stable. Their brightness and gay colouring give the butterflies a place among the most beautiful creatures on earth. The mandibles are atrophied and the maxillae transformed into a tube for sucking. This is composed of two halves, forming a single tube. When not being used

562

563

this maxillary tube or proboscis is rolled into a spiral. The larvae of butterflies (caterpillars) have a very primitive structure. The vast majority of them feed on a plant diet, and most of them are confined to a particular species or genus of plant. Some species of butterflies belong among the most harmful pests to agricultural and forest economies.

The first suborder, *Homoneura* = *Jugata*, contains the most primitive of all small lepidopterans. They have a small lobe at the base of the middle edge of their fore wings which catches under the middle edge of the rear wings and in this way stabilises flight.

Some families in this suborder are equipped in the adult or larval state with chewing mouth-parts, such as the members of the family of **Primitive Moths,** *Micropterygidae,* which eat grains of pollen and spores of low plants.

Micropteryx thun-bergella [562] measures 8.5 mm and has golden-yellow wings flecked with red. It lives in the warmer parts of Europe. Other members of the genus *Micropteryx* are scattered extremely irregularly throughout the world.

A large number of these small moths live in South Africa, Australia and New Zealand. The family *Eriocraniidae* is found in Europe, Asia and North America. The European species **Eriocrania spermanella** [563] has a span of 10 to 12 mm. Its wings are golden-yellow, marked with a network of steel-blue venation; the fringes of the wings are yellowish-grey. The caterpillars burrow in birch-leaves—that is, they eat passages (leaf-mines) between the upper and lower surfaces of the leaves.

Another family which is classified with the primitive moths is that of the **Root-borers** or **Swifts,** *Hepialidae,* some species of which have a span of more than 25 cm. The European, Asiatic and American species are generally brown or grey, whereas the African and Australian representatives tend to be brightly coloured with silver markings. The caterpillars of this family bore their way into wood and roots, and for this reason they are regarded as pests.

The **Ghost Moth,** *Hepialus humuli,* which is a European species, has a span of about 60 mm. The female [564] is yellow-brown with red marks, the male [565] white with a pink rim on the leading edge of the wing. The caterpillar feeds on the roots of carrots, sorrel, meadow-sweet and hops and can reproduce massively to become a pest.

564

565

566

567

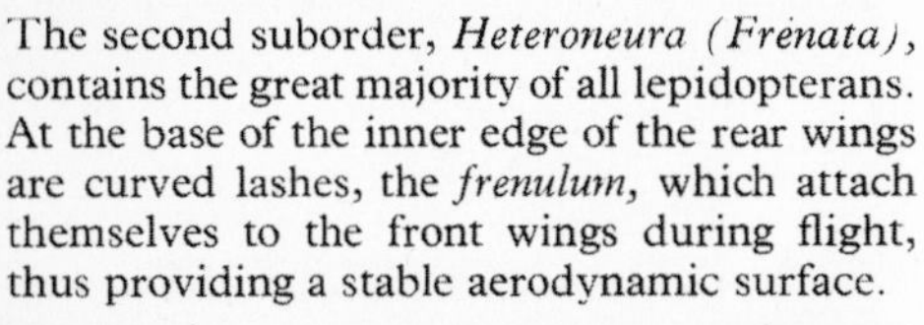

The second suborder, *Heteroneura (Frenata)*, contains the great majority of all lepidopterans. At the base of the inner edge of the rear wings are curved lashes, the *frenulum*, which attach themselves to the front wings during flight, thus providing a stable aerodynamic surface.

One member of the *Tischeriidae* family is **Tischeria ekebladella** [567]. It has a span of 8 to 9 mm; the front wings are of an ochre colour and the rear wings grey. The caterpillar mines cross-shaped marks on sweet chestnut and oak trees [566].

568

569

570

The family *Incurvariidae* contains the species **Incurvaria muscalella** [568], which is 14 to 16 mm long and dark brown, with two triangular white marks on each of the fore wings.

The caterpillars hide themselves between two shields which they bite out of leaves [569]. Evidence of their activity can be found in fallen leaves on the ground [570].

571

572

The **Wood-borer** famil *Cossidae* is distribute throughout the world i about 800 species. Thes moths are generall medium-sized and mea sure at the most up t 10 cm; some species onl achieve 20 mm. A gian species is known, how ever: it is **Xyleutes bois duvali,** which lives i Australia and is coloure grey. The female is large than the male and has span of over 25 cm. Th wood-borers have heav bodies, generally coloure grey or brown with ir regular wavy lines o marks. They feed on sa exuded from trees; thei mouthparts are consider ably atrophied, wherea those of their caterpillar are well developed; the live in wood, boring thei way into trees, shrubs roots and the woody stem of certain herbs. Th development of the large species takes a year.

The **Leopard Moth** *Zeuzera pyrina* [571, male] has a span of up to 66 mm The males are smaller tha the females and hav broadened antennae. Th translucent grey-whit wings have blue markings The caterpillars of thi moth need two years fo their development. The bore holes in fruit trees deciduous forest trees anc ornamental trees, especial ly in the trunks of young weak trees. They are whitish with black spots and have a dark head anc dark abdomen. This Euro pean species has also been imported into North America.

The **Goat Moth,** *Cossus cossus* [572, male at rest] is an inhabitant of Europe

573

It measures up to 85 mm in span. Its wings are brown and grey with a fine black wavy line. The caterpillar is reddish and bores shallow holes under the bark of various deciduous trees, especially poplar, willow, alder, apples and mountain-ash. They hibernate for two to four winters before they spin cocoons. The fully-grown caterpillar [573] measures up to 80 mm, is the thickness of a finger, has a black head, reddish-brown colouring and betrays its presence for a considerable distance by the strong, sour smell which it emits, as also does the chrysalis. The woody remains, and indeed everything with which the caterpillar or its cocoon have come into contact, retain this smell for decades afterwards.

574

575

One member of the family of **Long-horned Leaf-miner Moths,** *Lyonetiidae,* is the **Fruit Leaf-miner Moth,** *Lyonetia clerkella* [575]. It has a span of 8 mm and is silver-grey with yellow-brown wingtips. The caterpillars bore tunnels through deciduous leaves. Illustration [574] shows a birch leaf which has been mined in this way. This species is an inhabitant of Europe and North Africa.

The family of **True Moths,** *Tineidae,* is distributed throughout the world and contains many harmful pests which wreak

576

577

578

havoc in human dwellings and in storehouses. **Nemapogon granellus** [576] has a span of about 10 mm. It is dark brown, with a sprinkling of white markings. The larvae live in stored grain, dried fruit and dried fungi.

A very injurious pest is the **Clothes Moth,** *Elatobia* = *Tineola biseiella* [577]. It has a span of up to 9 mm. Its caterpillars consume woollen textiles and furs.

In the same way, the caterpillars of the **Common Tapestry Moth,** *Trichophaga tapetzella* [578], feed on textiles. The imago measures almost 20 mm.

579

The family of **Bag-worm Moths,** *Psychidae,* contains moths whose caterpillars spin bag-like dwellings for themselves rather like those of caddis flies. The females are wingless, often legless, and larva-like. In some species they remain in their bags even after attaining full development. Most of the members of this family are small, although certain tropical species achieve a noticeable size, and spin bags for themselves up to 18 cm long. Each species has a characteristic bag, both in shape and in the material used in its construction. The bag-worm moths are distributed throughout the world in about 400 species.

On illustration [579] we see above left a female bag-worm moth, **Fumaria casta,** and below it its bag. The wingless female, the bag on the right, and the male belong to the species **Fumaria crassiorella.** It has a span of 15 mm and is bronze-yellow and brown. Both these European species live on steppes and forested plains. Their caterpillars feed on grasses.

580

1

lustration [580] shows **Oreopsyche plumi-ra.** In the centre is a female with her bag, d at the sides two males. Their wings have scales and are almost transparent; the body black and hairy. They measure about 15 mm span. This species is distributed here and ere on sunny, forested hillsides in central urope.

he caterpillar of **Lepidopsyche unicolor** 81] is extremely common in fields and forest earings in Europe, and sporadically in north-n Asia. The imago is dark brown and has a an of some 15 mm. Illustration [582] shows chrysalis of this species.

582

583

584

The caterpillars of the **Sack-moth** family *Coleophoridae* live in little bags which are attached to various parts of plants. As a rule they are connected with a particular species of plant and start to burrow through the leaves on the first day of their lives. The imagines have very narrow wings, generally light in colour, with long fringes and a span of some 10 to 25 mm. There are a large number of species distributed throughout the world, though they are most common in the temperate zones of the northern hemisphere.

Coleophora onosmella [583] has a span of 18 mm and is light grey and brown. The caterpillars live mainly on **Viper's Bugloss,** *Echium vulgare,* and **Ox-tongue,** *Anchusa officinalis.*

The family of **Ermine Moths,** *Yponomeutidae,* consists of moths whose caterpillars live in large numbers in a thick, grey-white material with which they often cover entire trees and shrubs.

Yponomeuta euonymellus [584] measures about 25 mm in span. Its wings are white with black flecks. The caterpillars live principally on the bird-cherry.

Yponomeuta padellus [585] is smaller, with a span of some 20 mm and strongly resembles the previous species.

The ermine moths have a world-wide distribution and when present in large swarms they can completely strip cultivated plants, and for this reason are regarded as considerable pests.

585

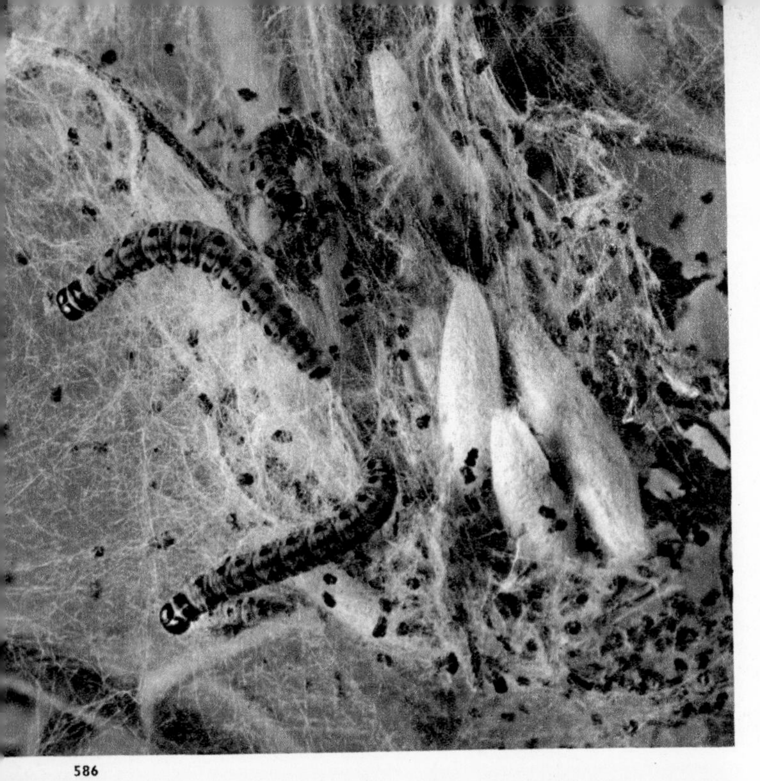

586

587

The caterpillars and the cocoons of the pupae of **Yponomeuta padellus** can be clearly discerned in a part of the web among the leaves [586].

The numerous family of **Palp Moths,** *Gelechiidae,* has among its number the magnificent-looking little moth **Aethmia bipunctata** [587]. It measures some 25 mm in span and is black and white. It is found on the plains of the warmer parts of Europe and North Africa. Its caterpillars feed on **Viper's Bugloss,** *Echium vulgare.*

The imagines of the **Plume-moth** family *Pterophoridae* have wings divided into feather-like lobes. They fly at dusk.

The **Plume Moth,** *Alucita pentadactyla* [588], has a span of 25 mm and is snow-white. Its caterpillars feed on the leaves of bindweed. It is a Euro-Siberian species.

The family of **Many-plume Moths,** *Orneodidae,* comprises not quite 100 species, whose wings are each divided into six feather-like plumes. They are most common in Africa and Asia, though rare in Europe. Their caterpillars live on flowers or form swellings in plant stems.

Orneodes grammodactyla has a span of some 7 mm, is grey-brown and an inhabitant of southern, and occasionally central, Europe and of Asia Minor. The caterpillars are found on **Scabious,** *Scabiosa cohroleuca* [589].

588

589

590

591

The inconspicuous little moth **Diurnea (= Chimabacche) fagella** is a European species of the family *Oecophoridae.* The male [591] has a span of up to 28 mm, being lighter in colour than the female [590], which has shortened wings and spans at the most 20 mm. The rough appearance of the surface of the wings is caused by groups of the little bristles which stand upright. The caterpillars live between two leaves, which they spin together and partly eat away. They emit a chirping noise by rubbing the third pair of legs against the leaves. They live on deciduous trees in forests and gardens. The imagines appear as early as March.

Another member of the same family is **Hoffmannophila pseudospretella** [592], a brown moth with black markings on the broadened, egg-shaped front wings. It has a wingspan of up to 23 mm. Its off-white caterpillars eat various kinds of rubbish in store-rooms and houses. This species is familiar in Europe and

592

the neighbouring parts of Asia.

The caterpillars of moths of the **Tortrix** family *Tortricidae* pass their lives in concealment, in hiding-places which they make out of rolled-up leaves. After first tunnelling through them, they later eat them. Some tortrices propagate in great numbers and cause massive damage to deciduous forests and orchards.

One of these injurious species is the **Oak Tortrix,** *Tortrix viridana,* whose caterpillars attack mainly oak trees. Illustration [593]

593

594

595

596

shows a fully grown caterpillar enveloped in an oak leaf.
On illustration [594] we see the chrysalis of the **Oak Tortrix.** The imago [595] has a wingspan between 18 and 23 mm; its front wings are light green, the rear wings glossy grey. When it wants to relax, it always attaches itself to the underside of a leaf. These handsome moths appear in large numbers towards the end of May, when they attack oak forests of Europe. Their green caterpillars can strip whole trees bare of foliage.
Another member of the tortrix family is **Acleria (= Acalla) schalleriana** [596], a snow-white little moth with projecting scales on the front wings. In span it measures 18 mm. The caterpillars live principally on snowball and aspen trees and shrubs.
The **Apple Tortrix,** *Ernarmonia (= Laspeyresia) pomonella* [597], has a span of 14 to 18 mm, is an inconspicuous grey-brown with a glossy copper design on the inner edge of the front wing. Its caterpillars bore their way into apples, pears and apricots. On illustration [598] we see the effect of this on young apples.

597

598

599

600

The **Pine-bud Tortrix,** *Eretria (= Rhyacionia) buoliana* [599], has a span of 18 to 23 mm. The fore wings are rust-red with wavy silver lines running across them. The caterpillars cause very considerable damage in pine forests by eating the buds of young trees—generally those between five and twenty years old. These then develop a stunted growth. Illustration [600] shows an affected pine shoot.
The larvae of the **Olive-brown Pea Tortrix,** *Ernarmonia nigricana* [601], feed on green peas within the pod. They hibernate underground, wrapped in a cocoon, pupate in the spring and in May the imagines emerge. They are dark brown, small, with a yellow pattern.
This tortrix is a native of Europe and Asia, although it has also been carried to North America.

601

602

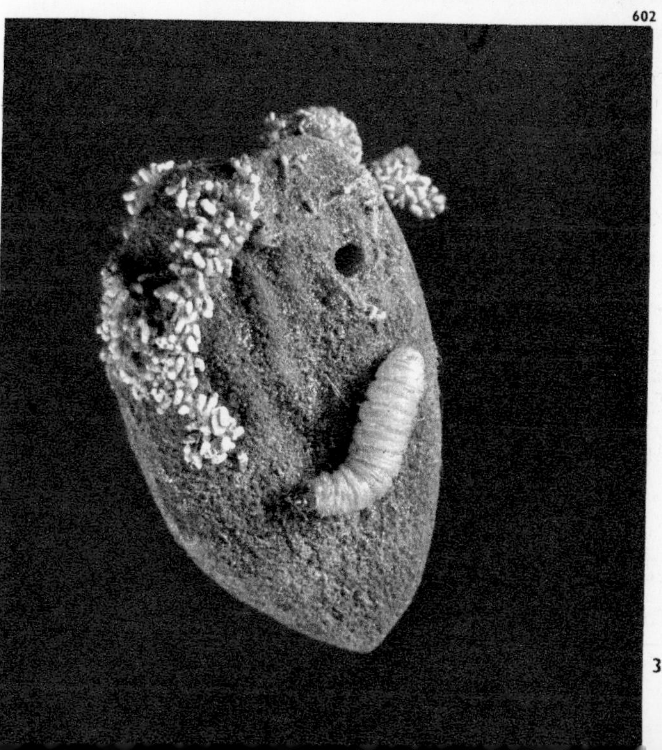

603

604

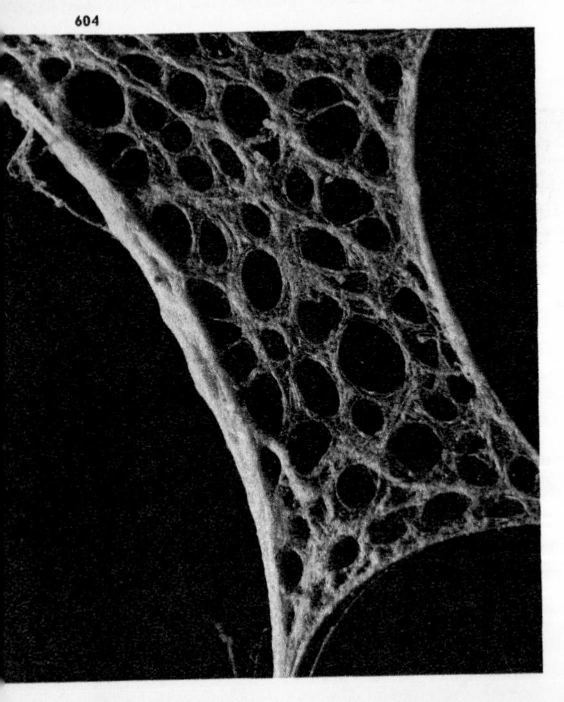

The **Zünsler** family *Pyralidae* contains the **Indian Meal Moth** [603]. It has a span of about 15 mm. The front halves of its wings are ochre-coloured, the rear red-brown. The caterpillars spin a fine web round their food [604, enlarged].

The caterpillars of the **Indian meal moth,** *Plodia interpunctella,* feed voraciously in food stores and larders. Illustration [602, on page 329] shows an almond kernel which has been attacked by this species.

The **Mediterranean Flour Moth,** *Ephestia kühniella* [605], has a span of 25 mm. Its wings are grey, with an irregular wavy line running across them. Its larvae develop in flour, binding it together in lumps with a web [606, on page 332]. These moths are often used in laboratories for the purpose of research into heredity.

605

606

607

The **Grass Zünsler,** *Crambus (= Cataptria) permutatellus* [607], has a span of about 23 mm. Its front wings are brown and yellow, having a dark brown and white pattern with a mother-of-pearl sheen; the rear wings are light brown and grey. The caterpillars spin little silken dwellings among the roots of grasses or in moss. Some species are injurious to rice and sugar cane.

Another member of this family is **Aphomia sociella** [608, male]. It has a span of some 30 mm, red and grey front wings with two black marks at the base and two wavy lines with blurred edges. Its yellow-grey caterpillars have reddish heads [609]; they live in nests of bees and wasps. Here they find a suitable store of foodstuff, especially if the hymenoptera have made their nests in bird boxes, where feathers, hair and other animal and vegetable remains have been collected to form a base. The caterpillars spin a communal nest together, where they hibernate and pupate, also communally, in the spring. This nest can serve as a breeding place for several generations. An infestation by swarms of these insects can have serious consequences for the bees or wasps: the moths spin a net round the comb, which they eat together with the larvae.

608

609

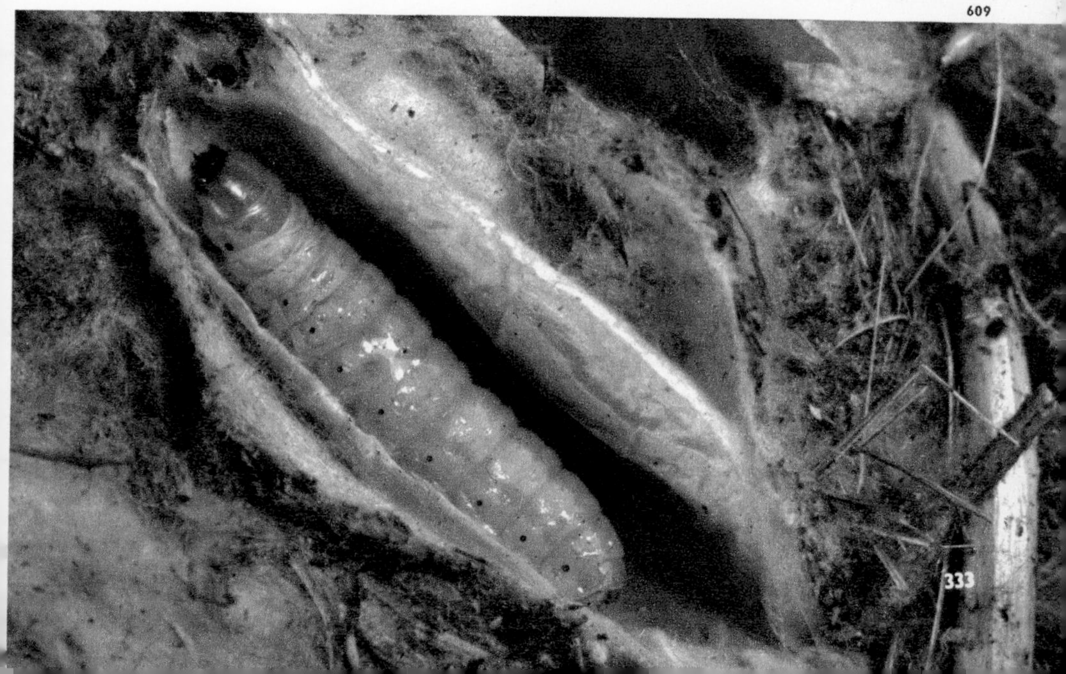

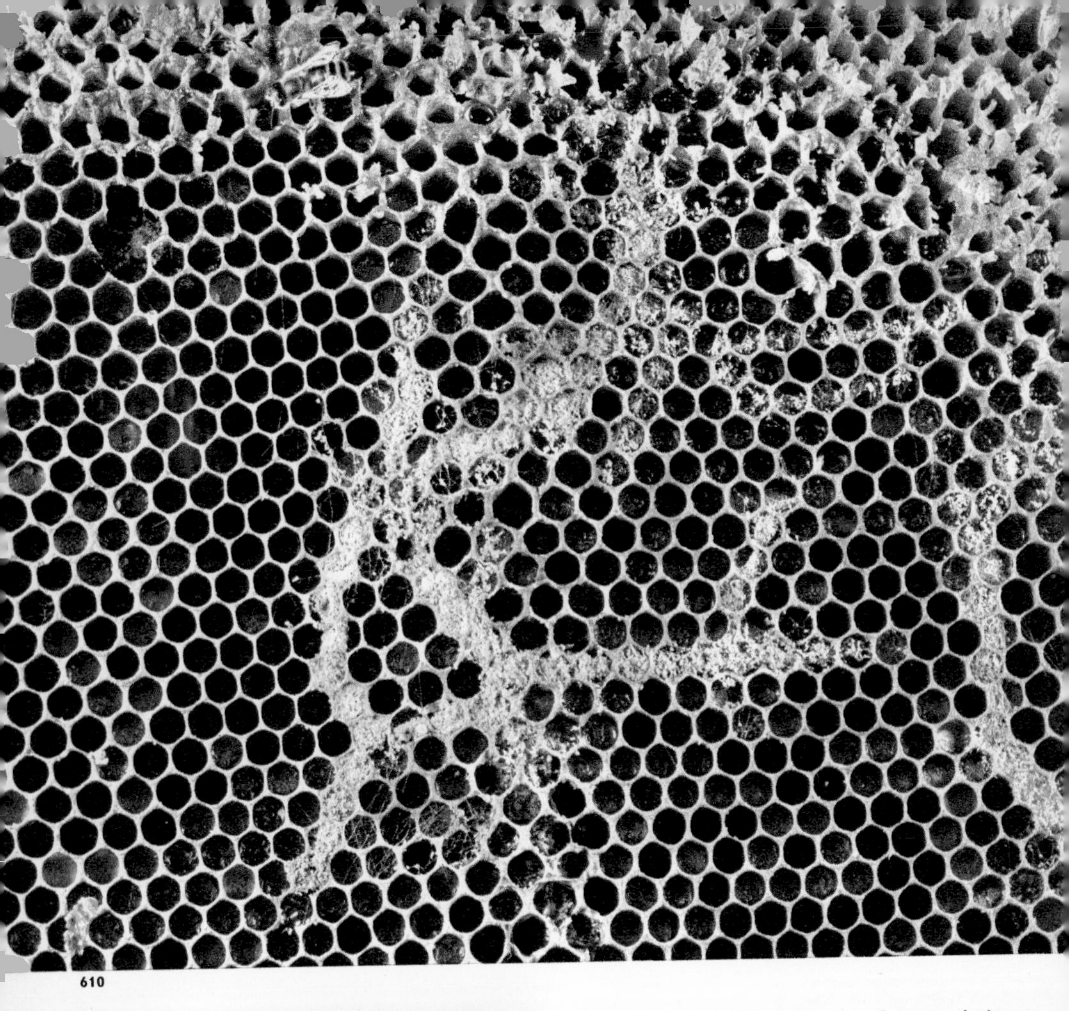

610

611

Illustration [610] shows the characteristic passageways in the comb of a beehive, with the larvae enveloped by the web. Here a female **Greater Wax Moth,** *Gelleria mellonella*, has penetrated the hive and laid its eggs directly on the comb. The larva [611] feeds on the wax of the comb. When present in larger numbers, they also feed on the offspring of the bees, and because of this they are much feared as pests. The adult moth [613] has a span of about 30 mm. Its fore wings are grey-brown with small dark spots; the rear wings are light grey with a dark border. In accordance with the location of

612

613

its life-cycle, it can best be found flying in the neighbourhood of beehives. The greater wax moth is distributed throughout the world.

The **Meal Zünsler** or **Flour Moth,** *Pyralis farinalis*, has a span of up to 25 mm. The ochre-coloured forewings bear four red and violet fields; the rear wings are light grey with olive-coloured marks. The caterpillars feed on flour and straw, flour products and vegetable and other remains. These insects have adapted to live in the presence of human civilisation, and when there are masses of them they can wreak a great deal of havoc, especially in warehouses.

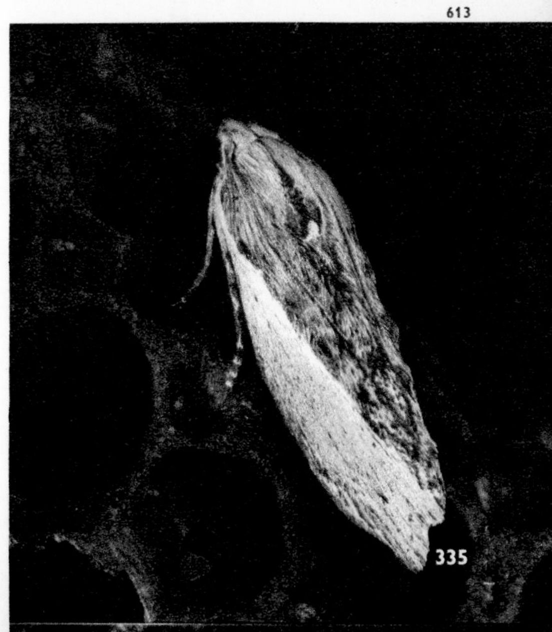

614

The moths of the family *Castniidae* inhabit the tropics of both the Old and New Worlds in about 160 species. They are very swift fliers and defend their quarters in the tops of the trees of the tropical rain forests with great ferocity. Their caterpillars tunnel in the tubers, bulbs, roots and stems of tropical plants.

Castnia cacica [614] has a span of 12.5 cm, is coloured brown with a green gloss and lives in the tropics of Central America.

The **Burnet** family *Zygaenidae*, numbers about 900 species of small and medium-sized moths. Like the *Castniidae*, they are classed with moths, which are chiefly nocturnal, although they are actually day-fliers. They exude a kind of oily sweat, which protects them from insectivores. Although they are found throughout the world, most species come from the plains and fields of the Old World. The caterpillar of the **Bloodwort Burnet,** *Zygaena laeta* [615], is greenish-yellow.

615

616

617

618

Illustration [616] shows the pupa of **Zygaena angelicae,** attached to a butterfly flower. The moth on the right of illustration [618] has a span of about 30 mm. It bears red spots on the dark green fore wings; the rear wings are also red.

The **Green Burnet** or **Forester,** *Procris statices* [618, above left], lives in damp meadows in Europe. It has a wingspan of about 25 mm and is a lively green.

The large burnets of the tropics and high ground in warm countries are very beautiful. **Erasmia sanguiflua** [617], has a span of some 10 cm. Its wings are purple and rust-brown, and blue with a violet sheen, and spotted black and white. It is common in northern India and Burma.

The burnets resemble the moths of the family *Syntomidae,* which is related to the tiger-moths. It contains about 2,000 species, most of which come from South America. One of the less numerous species is **Syntomis phegea** [618, below]. It has a span of up to 40 mm, is green and black and white and has two yellow rings round its body.

619

620

The family of **Prominents,** *Notodontidae,* is present throughout the world in some 2,500 species. Their fore wings are distinguished by tooth-like tufts of scales projecting from the middle of the inner edge. The caterpillars of the prominents often have extremely bizarre forms.

The **Puss Moth,** *Dicranura vinula* [620, female], has a span of up to 65 mm. Its wings are pale grey with dark grey markings. The male [621] is smaller and has wide, comb-like antennae. The fully grown caterpillar [619] is about 6 cm long and coloured green with a violet, white-bordered back and crimson border round its head. When disturbed, it spins a red thread from each of its tails. It is also able to spray its enemies with an evil-smelling liquid from a cleft across its back, between its head and the first thoracic segment. It feeds on willow and poplar leaves.

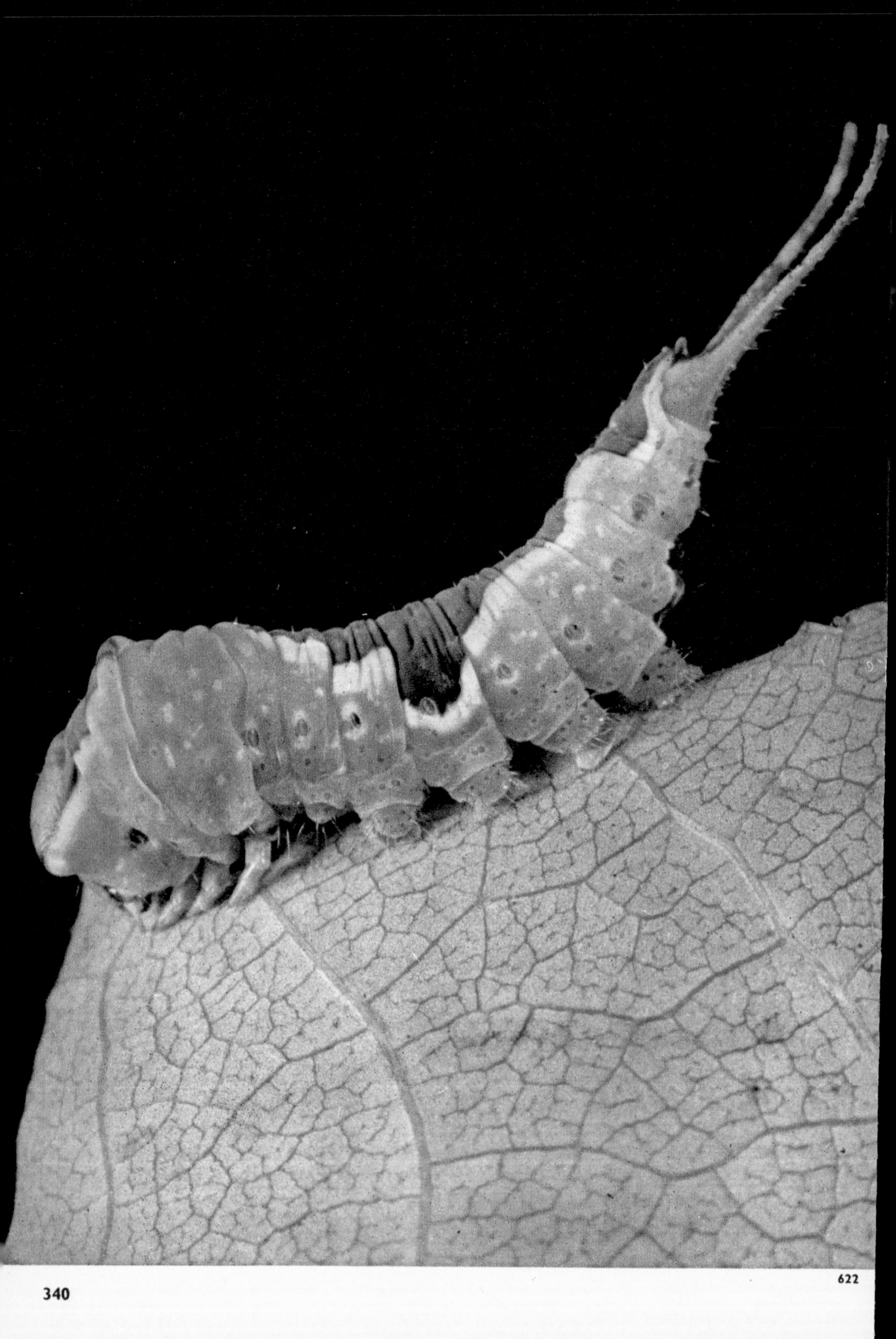

622

A somewhat small European species is the **Poplar Kitten Moth,** *Cerura bifida* [623]. Its average size is about 45 mm. Across the middle of the wings, which are grey, is a darker band, with black borders. There are also black markings. The caterpillar [622] grows to a length of about 40 mm and is green with a red and violet back. In just the same way as the previous species it has the ability to produce thread from the tail-like appendages of its abdomen. It eats aspen and poplar leaves.

The **White Prominent,** *Leucodonta bicoloria* [624], has a span of about 35 mm. Its fore wings are white with attractive orange and black markings. In central Europe it is found between April and July in light birch groves. The caterpillars feed on birch leaves.

623

624

The **Iron Prominent,** *Notodonta dromedarius* [625, male], has a span of some 45 mm. Its fore wings are brown, with rust-coloured markings. The caterpillar [colour plate XXXI] has a bizarre appearance; it consumes birch and alder leaves.

The **Buff-tip Moth,** *Phalera bucephala* [627], has a span up to 60 mm. The outside edge of its silver-grey and brown wings is marked by a large, ochre-coloured, moon-shaped area. The yellow and black striped caterpillars [626] feed

625

626

chiefly on lime, oak and willow leaves. They live in groups, and when they feel themselves threatened raise the rear ends of their bodies menacingly. The moths may be seen from May to the end of July and the caterpillars appear in July and August. The female usually attaches her eggs [628] to the underneath of a leaf, close together, with a secretion which she produces when laying. The eggs are about 1 mm in diameter, light green, porcelain-white on top and have a clearly visible green spot.

627

628

630

629

631

The moths of the **Looper** or **Spanner** family *Geometridae*, dominate the whole world, with about 15,000 species. The majority are small, measuring between 5 and 6 mm. Their wings are for the most part grey or brown and are therefore somewhat inconspicuous. A number of species, however, are extremely beautiful, having light green colouring with darker or yellowish markings. They are in the main nocturnal, flying during the day only when disturbed. They instinctively land in places which are best suited as camouflage for the colouring and patterning of their wings. The geometrids have at the ends of their bodies a well-developed tympanum or hearing organ. Their caterpillars have two pairs of abdominal feet fewer than the caterpillars of other lepidopterans. As a result they move along by stretching the front parts of their bodies out forwards as far as they can, anchor themselves by the thoracic feet, and then bring the rear end up right behind the front feet. This gives rise to the characteristic "looping", "measuring" or "spanning" effect of their movement which is responsible for their popular names in various localities.

The **Pale Brindled Beauty,** *Phigalia pedaria* [629, male], has a span of about 40 mm. The fore wings are grey-green, sprinkled with a dull pattern of darker cross lines. The female [630] is wingless, about 15 mm long with a sack-shaped body. The pale brindled beauty has frequently been seen in the deciduous forests of central Europe as early as January, even in cold years. The caterpillars feed on the wood of deciduous trees.

The **Poplar Beauty,** *Biston stratarius* [631], has two dark-brown bands across the fore wings, which are grey and look as if they have been dusted with black powder. It has a span of up to 50 mm and is extremely common in the deciduous forests of central Europe, where its caterpillars feed on deciduous trees.

The **Birch Beauty** or **Peppered Moth,** *Biston betularia,* measures some 40 mm across. It is of variegated and widely varying colouring, ranging from white through varying degrees of peppering with black spots to complete blackness. It is a good example of what is called industrial melanism – that is, it adapts itself in colouring to the smoke in the atmosphere in industrial areas and is of a darker colouring. It lives in the parks, gardens and tree-lined avenues of our cities. The colouring of the caterpillar [632] is variable: it is brown, grey, yellow or green and lives on deciduous trees.

632

The **March Moth,** *Alsophila* (= *Anisopteryx*) *aescularia* [633], has a span of some 35 mm. Its fore wings are brown and grey; in the middle of each of the rear wings is a black patch. The female is wingless and has a tuft of bristles on her abdomen. The moths emerge from the hibernated pupae between February and the end of April. The caterpillars feed on various deciduous woods. The range of distribution of the March moth covers the deciduous forests of all the warmer parts of Europe.

The **Orange Underwing,** *Brephos parthenias* [634], has a span of about 35 mm. Its fore wings are red-brown with yellow markings; the rear wings are an orange with a dark brown pattern. The caterpillars feed on birch leaves; the moth alights on a variety of decaying substances.

An extremely common species in the birch forests of Europe is the **Clover Heath Moth,** *Fidonia plumistraria* [635]. They measure roughly 40 mm in span. The fore wings are light golden-yellow with a dark brown pattern; the antennae have strikingly long lashes or pectinations. The moth itself never feeds; it lacks a proboscis or maxillary tube. The caterpillars appear in two generations; they feed on a species of clover (*Dorycnium*). This insect species comes from the western Mediterranean and North Africa.

634

635

636

637

The **Great Oak Beauty,** *Boarmia roboraria f. infuscata* [637], has a span up to 65 mm. Its wings range from light to dark grey-brown, with a darker pattern. The darker specimens have a light crescent-shaped area at the wing-tips. The caterpillars feed principally on oak. Like most geometrids, the oak beauties spread out their wings horizontally when at rest, in such a way that they are almost invisible against the bark of a tree. The caterpillars form a further excellent example of the power of adaptation in both form and colouring shown by these insects. They have the appearance of oak twigs, with unevenness and rough growth. Illustration [636] shows four of these caterpillars.

The caterpillars of the **Early Thorn,** *Selenia bilunaria f. illunaria,* are brown to reddish, with an ochre-yellow pattern. They live in deciduous woodlands and parks on shrubs and trees. Illustration [638] shows a caterpillar in the rigid rest position, imitating a twig with buds. Illustration [639] shows it in the course of its characteristic "looping" forward movement.

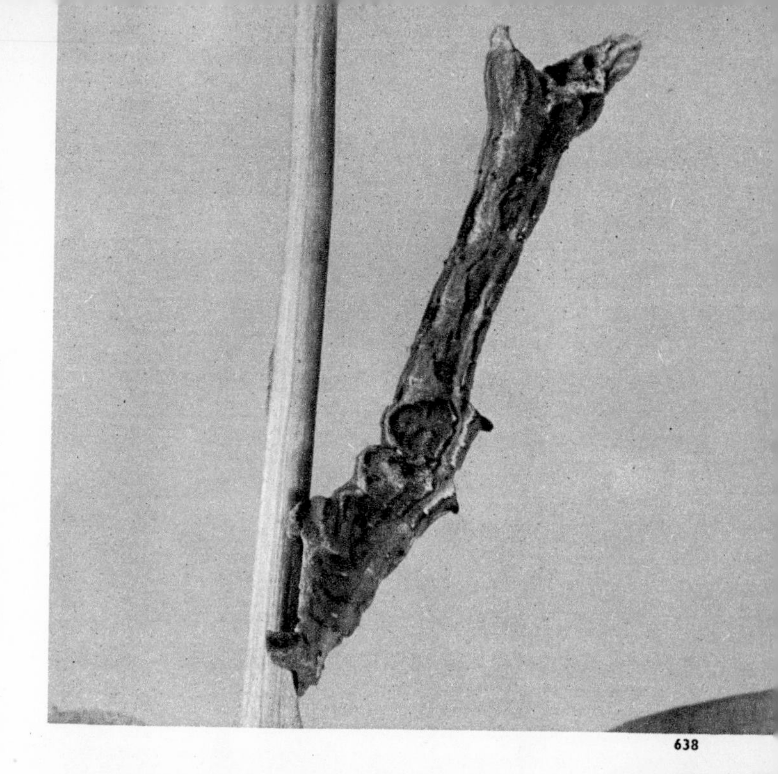

638

639

640

The moths of the genus *Selenia* fold their wings vertically together when they are at rest, like the butterflies, or else they hold them slanting open slightly. The outline of the wings and the colouring of the underneath give the impression of a dry leaf caught on the bark of the trunk. Illustration [640] shows the early thorn at rest. This species is an inhabitant of

641

central and northern Europe.

The **Magpie Moth,** *Abraxes grossulariata* [641], has a span measurement of about 40 mm and is coloured white, black and yellow; the caterpillar is yellow with black spots. It feeds on gooseberries, redcurrants, sloes, plums, cherries and the fruit of the spindle tree, a species of euonymus.

642

643

The **Bordered White Beauty,** *Bupalus piniaria* [642, male], has a span of up to 38 mm. Its wings are dark brown with pale yellow marks. The female has light rust-brown wings with variable-shaped patterns fading into one another. The caterpillar is blue-green or green-yellow with three wide white lines down its back. When fully grown it measures about 30 mm. They can completely devastate groups of conifers, generally attacking trees over 25 years old. The moth flies by day, and rests with the wings folded together vertically like those of the butterflies.

The **Slender Treble-bar Moth,** *Anaitis plagiata* [643], measures approximately 40 mm in span. It has greyish-violet wings, the front pair of which is adorned with dark brown and reddish-brown markings. The caterpillars live on the **St. John's Wort,** *Hypericum.* They appear in two generations. The moths fly from May to July and the second generation from the end of July to the beginning of October. It is a European species and extremely common on warm, sunny slopes and clearings.

The **Hawk-moth** family *Sphingidae,* contains over 1,000 species, the majority of which are nocturnal. They have powerful, streamlined bodies with a pointed end to their abdomens and powerful muscles in the thorax. The fore wings are elon-

XXXIII The Oleander Hawk-moth, *Daphnis nerii*, with a span of 8 cm.

XXXIV The Death's-head Hawk-moth, *Acherontia atropos*. Immigrates into Europe and there gives birth to the infertile autumn generation. Span about 10 cm.

gated, narrow and pointed; the rear wings are generally small. The hawk-moths are the best fliers of all the *Lepidoptera*. They shoot along like arrows, but can also hover humming in front of the flowers, with wings vibrating like propellers, while they suck the nectar from the blooms, like the hummingbirds, with their long tubes. The hawk-moths generally measure between 60 and 100 mm in span, though the largest species are up to 20 cm across. Some of the tropical species are dangerous pests to cultivation. Their caterpillars are large and plump; many of them have the habit of rearing up the front end of their body when they are disturbed, so that they somewhat resemble a sphinx. This is the reason that the whole family has received the name *Sphingidae*.

The grey-brown **Pine Hawk-moth,** *Sphinx pinastri* [644, pair], has a span up to 80 mm. Its colour matches exactly that of the bark of a pine tree, on which its caterpillar feeds. It is a familiar and on the whole a harmless species of central Europe.

The caterpillars of the hawk-moths generally pupate underground. The chrysalides of some species have a special sheath for what is destined to become the proboscis. In the case of the **Convolvulus Hawk** this sheath is particularly noticeable, as can be seen from our illustration [645].

644

645

646

647

The **Convolvulus Hawk-moth,** *Herse convolvuli* [646], has a span of up to 11 cm. Its proboscis is considerably longer than the whole of the rest of its body. Its wings are grey with fine black markings, and across the sides of the abdomen there are pink and black stripes. It likes to suck the nectar of tobacco flowers. The caterpillars feed on convolvulus leaves. The moth is found in southern Europe, Africa, Asia and Australia. In some years central Europe is invaded with swarms of these harmless moths; in fact they fly as far north as the Arctic Circle.

In the same way, the well-known **Death's-head Hawk-moth,** *Acherontia atropos* [colour plate XXXIV], from southern to central Europe. In the rest position it folds its wings horizontally over the body. It has striking yellow-and-black rear wings and a black-and-yellow striped body. On its back it bears a pattern reminiscent of a death's head [see p. 360]. On unusually warm July nights, the death's-heads will move northwards and often pass right over the Alps. In central Europe they lay their eggs on potato plants, the tea tree and woody nightshade. The green or or brown caterpillars pupate and the adult moth emerges before the autumn of the same year. Illustration [647] shows a death's-head which has only just emerged, and is still soft. Once the wings have hardened, they are held in the horizontal posi-

648

tion adopted by all hawk-moths. Illustration [648] shows the red-brown chrysalis of the death's-head buried in the earth. It measures about 60 mm. The death's-head has a short proboscis – too short for it to be able to suck the juices from flowers. For this reason, it sometimes invades bee-hives, attracted by the smell of honey, where it is stung to death [649].

649

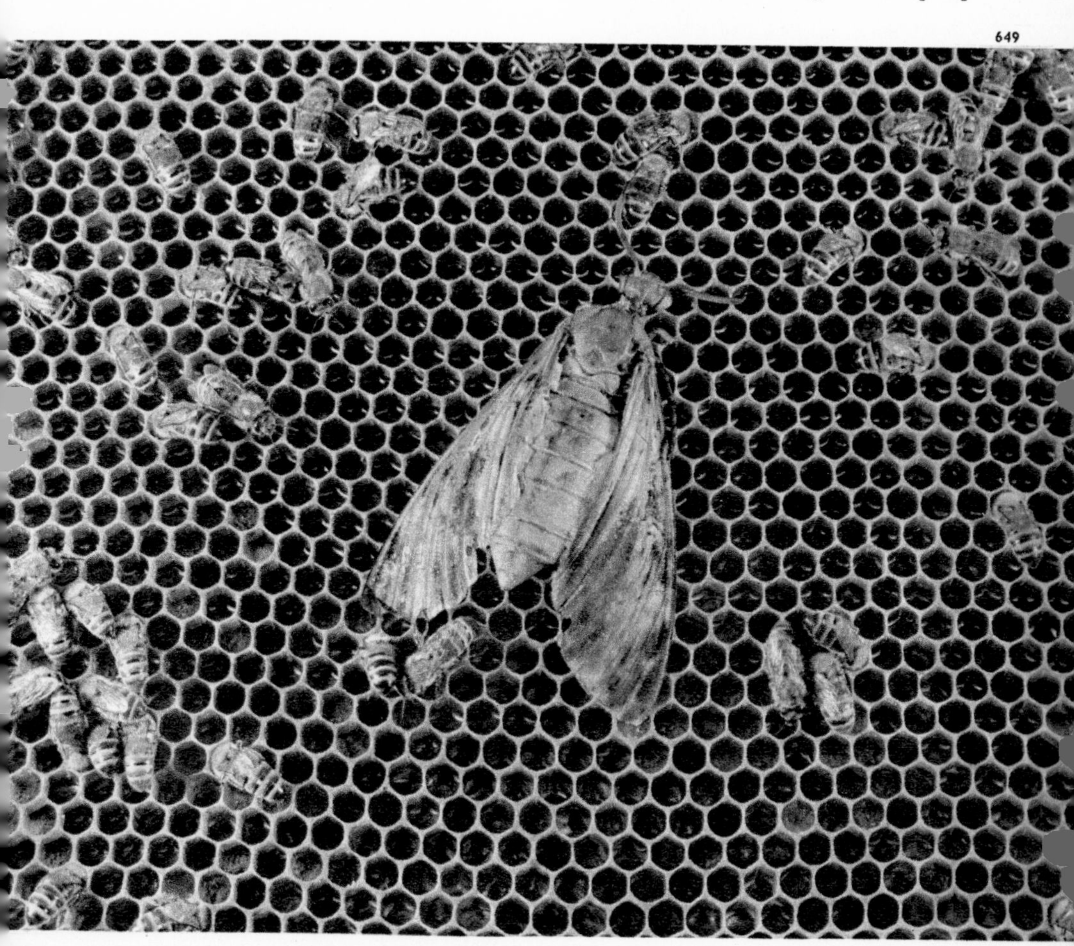

650

651

A much rarer visitor from the south is the beautiful **Oleander Hawk-moth,** *Daphnis nerii*. It has a span up to 10 cm. Its coat is a mixture of shades of green and pink [p. 360, and colour plate XXXIII]. Its habitat is southern Europe and Africa. In illustration [650] there is a moth which has only just emerged from the chrysalis and has not yet lowered its wings and spread them horizontally. It settles on the oleander, which forms the diet of the caterpillar. The latter is a yellowish-green colour with a white stripe running down each side of its body and two blue spots with black edges just behind the head. It pupates inside a strong cocoon [651] on the ground beneath the fallen leaves and twigs of the oleander bush. The chrysalis is rust-brown with a black pattern and measures about 45 mm. Although the caterpillars of this moth are found here and there in eastern central Europe, the chrysalis cannot survive the winter in the open. The oleander hawk-moth is very common in Africa.

652

The central European **Privet Hawk-moth,** *Sphinx ligustri* [652], often has a span of more than 10 cm. The fore wings are brown and black, the rear wings and the body pink and black. The caterpillar [colour plate XXXII, upper picture] lives principally on lilac and privet.

Illustration [653] shows some brightly coloured American species, and one Asiatic hawk-moth: on the left top is **Pholus labruscae,** a migrant

653

hawk-moth which can be found anywhere between Canada and Patagonia. It has a span of 11 cm. The fore wings are green, the rear wings blue, black and red. On the right top is **Compsogene panopus ssp. celebensis** from the island of Celebes in Indonesia. It has a span of 13.5 cm. Below left is **Pholus tersa,** with a span of 85 mm, and on the right **Pholus fasciatus,** spanning 10 cm; both are to be found in Brazil. Illustration [654] shows some

654

migrant hawk-moths from the Old World: on the left above the **Oleander Hawk,** *Daphnis nerii,* (see p. 357). At the top on the right is the **Striped Hawk-moth,** *Celerio lineata var. livornica.* This one has a span of about 8 cm and is coloured brown, brownish-pink and grey-white. The caterpillars feed on vines and other plants, such as **Bedstraw,** *Galium;* **Snapdragon,** *Antirrhinum;* **Spurge,** *Euphorbia* and **Dock,** *Rumex.* This moth populates large areas of the earth and reaches us from the Mediterranean area.

On the left below is **Celerio vespertilio,** which spans 65 mm. Its fore wings are grey, the rear wings pink and black. Its range of distribution extends from Africa to Armenia, and includes the Mediterranean. From here it migrates into central Europe. The caterpillars feed on **Willow-herb,** *Epilobium* and **Bedstraw,** *Galium.*

On the right below is **Deilephila (= Pergessa) alecto,** which measures about 70 mm in span. The fore wings are a light brownish-pink, the hind wings pink and black. The caterpillar feeds on plants of the rose family. It is present in the warmer parts of Europe, in Asia as far east as Turkestan and Malaya.

On the right in the middle is the **Death's-head Hawk-moth,** *Acherontia atropos* [pp. 354-355].

The **Elephant Hawk-moth,** *Deilephila elpenor* [656], spans about 60 mm. It is pink and olive-green and the pink hind wings are black at their roots. Its caterpillars feed on **Balsam,** *Impatiens* and **Willow-herb,** *Epilobium.* It is a Eurasian species.

The **Spurge Hawk-moth,** *Celerio euphorbiae* [colour plate XXXVI, below], has a span of 60 to 70 mm. Its caterpillar [655] is black and red with a row of yellow spots along the sides.

656

655

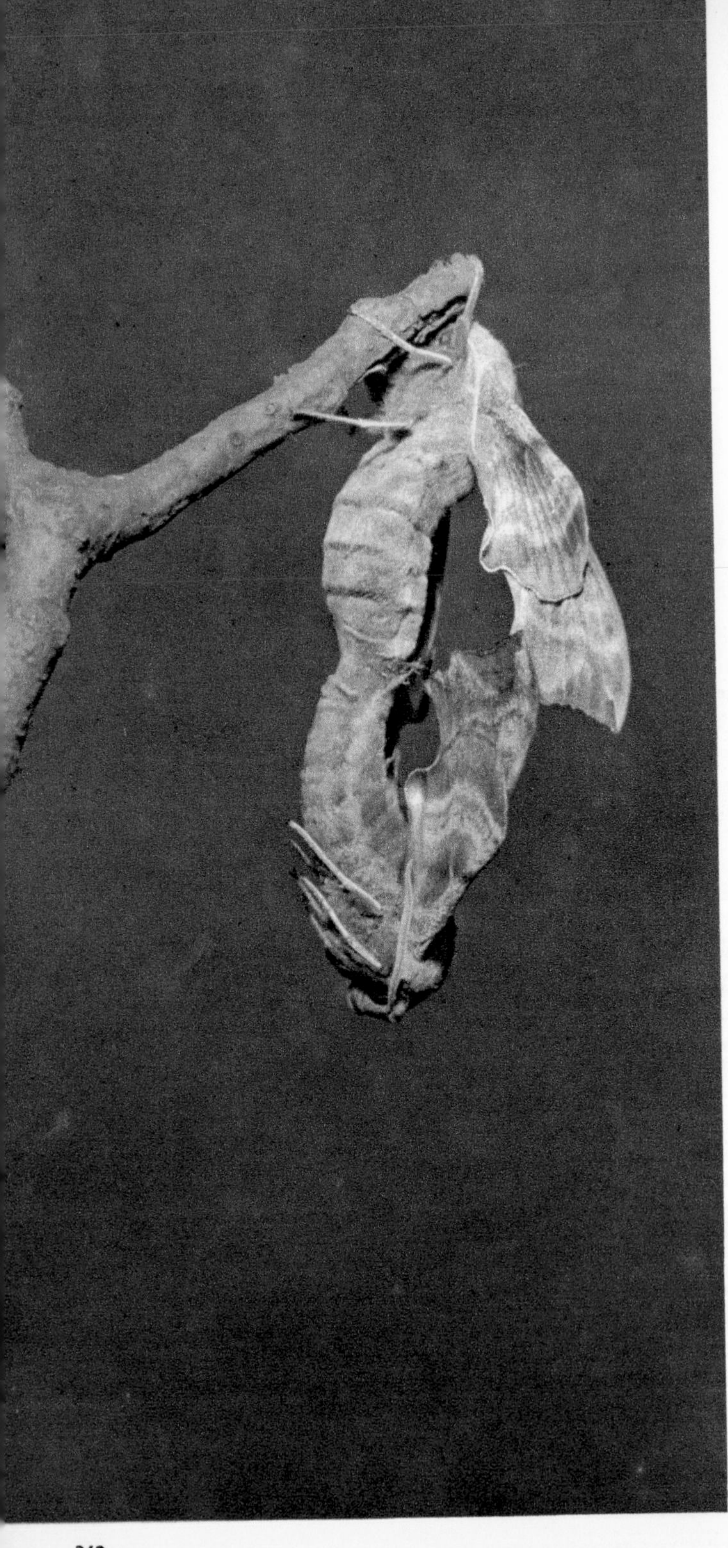

657

Another species belonging to this family is the **Eyed Hawk-moth,** *Smerinthus ocellatus* [657, pair, the male beneath], which has a span of 80 mm. Its fore wings have a pattern in grey and brown, the rear wings a light pink with a blue and black eye-spot. The caterpillars feed on **Willow,** *Salix;* **Poplar,** *Populus;* **Pear,** *Pyrus,* **Plum,** *Prunus* and **Apple,** *Malus.* The eyed hawk is a Euro-Siberian species. A very attractive moth is the **Lime Hawk,** *Dilina (= Mimas) liliae* [658]. It has a span of some 60 mm. Its fore wings are olive-brown and pink, the rear wings light yellow and brown. The caterpillars feed principally on the leaves of lime trees, but also on those of willow, oak and alder. This too is a Euro-Siberian species. The caterpillars are green, with a light back stripe. The body surface is rough and equipped with a tail thorn.

658

659

660

The **Poplar Hawk-moth,** *Laothoë (= Amorpha) populi* [659], has a span of about 75 mm. Its colouring is highly variable, but generally resembles a dry leaf. Both the fore and hind wings are serrated; the fore wings are coloured a variety of brown, grey-brown, olive-green or red shades. On the hind wings there is a large, rust-red patch. The caterpillars feed principally on poplar, aspen, willow and ash. It is distributed throughout the whole of Europe and in Asia as far east as Siberia.

The **Oak Hawk-moth,** *Marumba quercus* [660], has a span of up to 10 cm. It is coloured various light and dark shades of ochre. Its caterpillars live on the **Cork Oak,** *Quercus cerris,* and certain other oriental species of oak. It is principally an inhabitant of the Mediterranean countries, spreading into the warmer parts of central Europe, Asia Minor and the Near East.

661

Illustration [661] shows at bottom left the **Humming-bird Hawk-moth,** *Macroglossum stellatarum.* It has a span measurement of 45 mm. Its fore wings are grey; the hind wings rust-coloured. The caterpillars live on the **Bedstraw,** *Galium.* It migrates into central Europe from the south, and perishes in the autumn; although specimens have been observed making the return journey southwards.

At the top right is the **Hawk-moth,** *Macroglossum croaticum.* It has a wingspan of 40 mm, coloured rusty brown and green and is an inhabitant of southern Europe and Asia. The northern-most of its distribution is the southern slopes of the Alps.

On the left of the picture at the top is the **Broad-bordered Bee Hawk-moth,** *Hemaris fuciformis* = *lorricerae.* It has a wingspan

662

of some 40 mm. Its wings are transparent as glass; there is a dark area on the fore wings. Its caterpillars live on the honeysuckle. It is present in Europe and the nearer parts of Asia. On the right below is the **Narrow-bordered Bee Hawk-moth,** *Hemaris tityus = scabiosae*. In appearance it closely resembles the previous species, except that it has no mark in the middle and the dark border round its wings is narrower. It has a wing measurement of about 40 mm. The caterpillars feed on scabious, devil's-bit scabious and wild teasel.

The **Willow-herb Hawk-moth,** *Prosperpinus proserpina* [662], has a span of about 40 mm. Its fore wings are olive-green, the hind wings yellow with a black border. Its caterpillars feed on willow-herb and the evening primrose.

663

664

XXXVa Eyed Hawk-moth, *Smerinthus ocellatus*. Span up to 8 cm. Europe.

XXXVb Caterpillar of the Tree-of-Heaven Atlas Moth, *Attacus edwardsii*.

XXXVI *Attacus edwardsii* inhabits the valleys of the Himalayan foothills and certain other areas of Northern India.

665

The family of **Emperor Moths** or **Eye-spotted Moths,** *Saturniidae* = *Attacidae*, is distributed in about 1,200 species throughout the world, with the exception of New Zealand and small islands in the Pacific Ocean. The emperor moths are medium to large moths with broad wings, which are usually decorated with brightly coloured eye-spots. In some genera the hind wings terminate in sharp points, while others have transparent, glass-like "windows", which are free of scales, on their wings. The caterpillars are usually equipped with thorny, wart-like growths and pupate inside silk cocoons, which in the case of some species are used in textile-manufacturing. One of the commonest species, distributed throughout the whole of Europe, is the **Lesser Emperor Moth,** *Eudia pavonia*. The female [664], which is larger than the male, has a span of about 65 mm; the male is smaller by some 7 mm. The female is brown and grey, the male darker and rust-coloured on the wings. The moths emerge as early as April after spending the winter as pear-shaped chrysalides. The fully grown caterpillars [663] are green with black stripes and orange coloured warts. They feed on the leaves of sloe, blackberry, hornbeam, dog-rose and other trees and bushes.

The largest member of the *Lepidoptera* found in Europe is the **Greater** or **Viennese Emperor Moth,** *Saturnia pyri* [665, male]. It has a span measurement up to 14 cm. Its colouring consists of a whole kaleidoscope of shades of brown, from the palest yellow-brown to the near-black of the eye-spots. The caterpillars feed on the leaves of various forest and fruit trees. When disturbed they pull the front part of their body together, at the same time emitting a rattling sound. The Viennese emperor lives in the warmer parts of central Europe and is present in large numbers in southern Europe, North Africa and Asia.

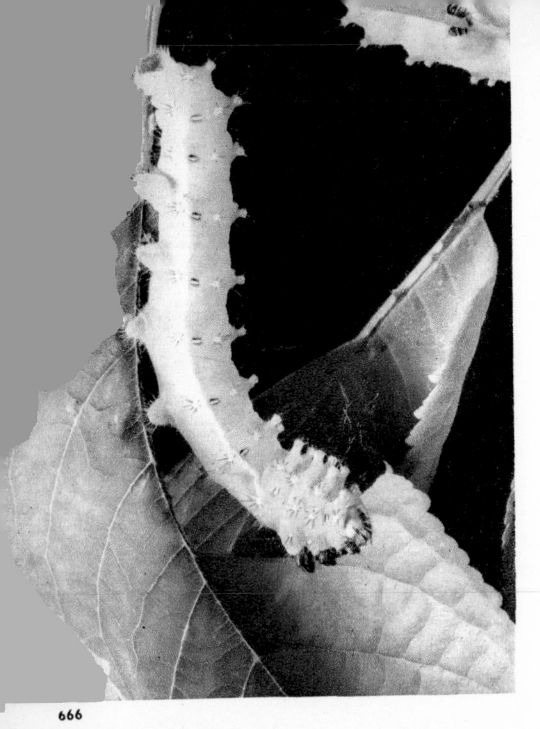

The caterpillar of the **Viennese Emperor Moth** [666] is light green. The body bears light blue warts with long tufts of hair. They pupate in elongated cocoons—about 60 mm in length—which are attached to a tree trunk or a wall. The cylindrical dung of the caterpillar, which feeds on the leaves of the wild cherry, disintegrates in water into individual segments of a graceful regular pattern [667].

Northern India and southern China form the range of an extraordinarily beautiful moth called **Loepa katinka** [668]. It has a span of about 80 mm. It has lively wings, with pink and violet and dark brown markings. The eye-spots on its wings are brown.

666

667

669

670

Illustration [669] shows a moth of the **Chinese Oak Silkworm,** *Antherea pernyi,* on a cocoon which it has just left. It has a span of about 11 cm. Its colouring is extremely variable, but the basic colour is yellow-grey to red-brown. The round eye-spots on the wings are translucent. The green caterpillar [670] feeds on oaks, but also on other trees. In China they are cultured as silkworms, the cocoons being manufactured into a kind of silk shantung. The range of this species is the extensive area between Amur and southern China.

The related Japanese species **Antherea yamamai** [671], closely resembles the Chinese oak silkworm. It can also be cultured in the warmer parts of Europe. This species yields silk thread through its cocoon. The greenish-yellow caterpillar feeds on oak and chestnut leaves.

71

672

673

The head of the **Chinese Oak Silk Moth** [seen from front, 672] has broad, combed antennae, the surface of which looks extremely large. The antennae of the female are simpler and are equipped with two rows of short bristles.

Another species of silk moth is the North American **Robin Moth,** *Hyalophora (Platysamia) cecropia* [673, female]. It has a span of up to 13 cm and is dark brown, white and brick-red. The hind wings are traversed by a brick-red band, and in place of eye-spots it has half-moon-shaped marks. The light green caterpillars live on cecropia and other deciduous woods. In many countries it does damage to fruit trees.

In some silk moths the rear wings are extended into an elongated tail, which is particularly long in the male. Only one of these species is found in Europe, namely **Graellsia isabellae** [674, male; 675, female from underneath]. They span some 80 mm and are blue-green with translucent, brownish-red veined wings. The eye-spots in the middles of the wings are translucent, with black, blue, yellow and red rings. The caterpillars live on conifers. The range of this moth is restricted to a particular area in the mountains of the Sierra Guadarrama situated in central Spain, although it has also been observed in a mountain valley in southern France.

674

675

677

The tails on the hind wings of the **Malagasy Silk Moth,** *Argema mittrei* [colour plate XXXIX], are almost three times as long as the body. The female [676] has shorter tails, although the surface area of the wings is greater. The wingspan is in the order of 20 cm. The wings are bright yellow with red-brown markings and multi-coloured eye-spots. As a result of its bizarre appearance it is one of the best-known exotic moths and one of the most highly prized objects of all collectors.

Another species which has long tails is **Actias isis** [677]. It is about the same size as the previous species and reddish-brown with yellow-washed markings. It lives in South-East Asia and on the neighbouring islands.

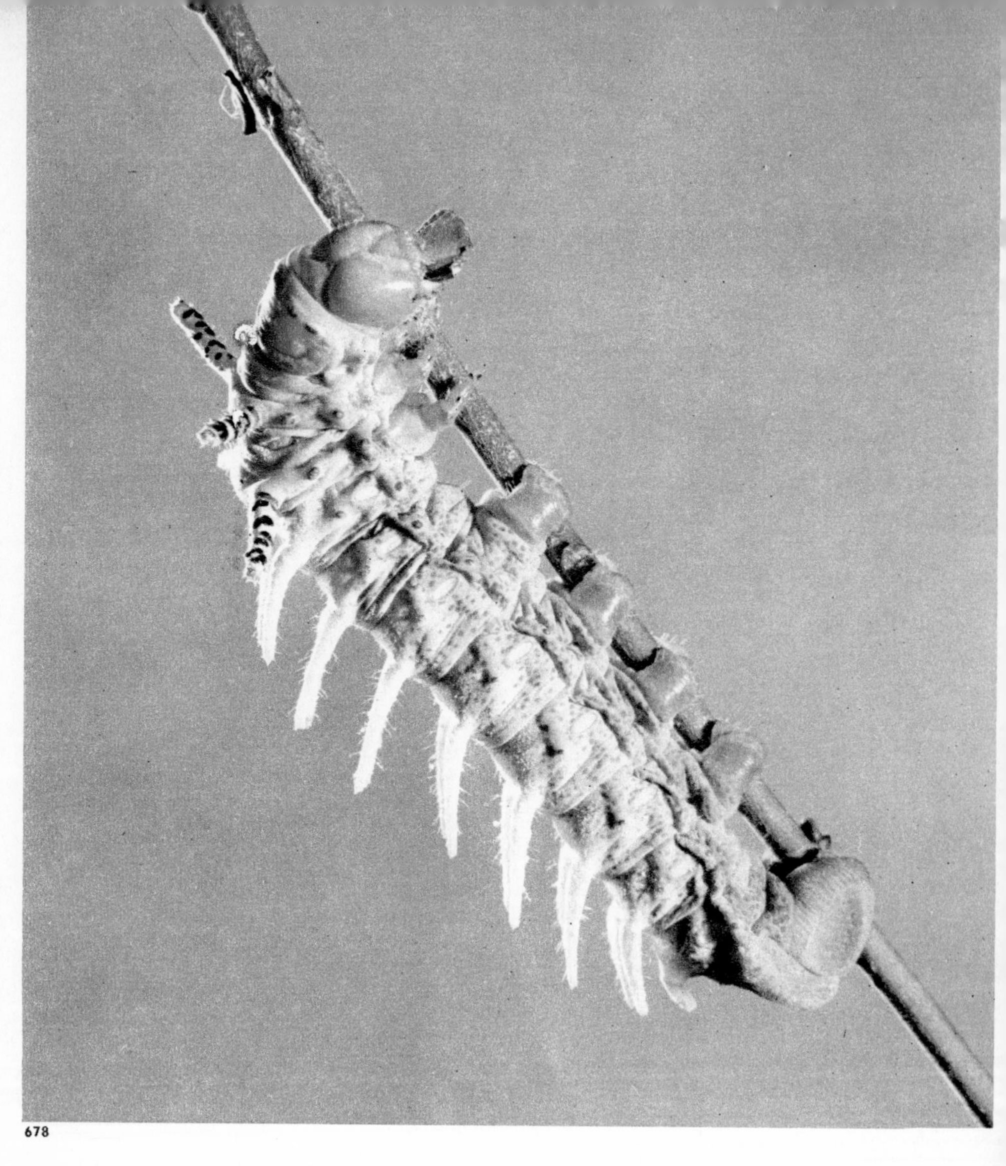

678

One of the largest silk moths, also one of the largest of all the *Lepidoptera*, is the gaily coloured **Atlas Moth,** *Attacus atlas silhetica* [679, male]. The span of its wings is often more than 25 cm. Its colouring is composed of a variety of browns, pinks and reds. The small "windows" on the wings are translucent. The genus *Attacus* is distributed in India, China and on the islands of South-East Asia.

Its caterpillars have a row of long growths on their backs and are covered with a dust-like coating of white wax. They feed on the leaves of various East Asian trees. In Europe they are kept by silkmoth rearers on the **Tree of Heaven,** *Ailanthus*. The caterpillar illustrated [678] belongs to the species **Attacus edwardsii,** which is from northern India. When fully grown it measures about 80 mm. Underneath the coat of wax it is a light bluish-green in colouring, with red marks on the sides of the rear pair of feet [colour plate XXXVb, XXXVI].

In the group of silk moths on illustration [680] there are three of the largest and most beautifully coloured species: on the right **Rothschildia jacobeae** from tropical Argentina and southern Brazil, with a wingspan of about

680

11 cm. The wings are brown and pinkish-violet with translucent "windows."
At the bottom is **Philosamia cynthia:** it measures about 13 cm in span. The moth is olive-brown with a violet band across the wings. It has been brought in from the Far East, and has established itself in the warmer parts of Europe and North America. Its caterpillars live on the **Tree of Heaven,** *Ailanthus glandulosa*.

Above left is the North American species **Telea polyphemus** (male). It has a wingspan of about 12.5 cm and is light brown with translucent frames in the striking blue-yellow eye-spots on the fore and rear wings.
A European species of the subfamily *Hemileucini* is the **Nail-mark,** *Aglia tau* [681, female]. It has a span of about 80 mm and the male about 60 mm. The female is ochre-brown, the colouring of the male somewhat darker, tend-

ing towards rust-brown. The mark on the wings is not translucent. The caterpillars live chiefly on beech, for which reason its range is limited to certain regions. The moth flies by day, very fast and at a great height.

The same subfamily is represented in America by **The Bull's Eye Moth,** *Automeris io* [682, male, and colour plate XXXVIIb]. It has a span of about 60 mm, and the female of about 80 mm. The caterpillars are capable of inflicting great damage on cotton plantations and fruit trees.

681

682

683

Some American species of emperor moths have neither eye-spots nor translucent frames on their wings, and in others these are only hinted at. However, they often have a very interesting wing-shape. In the genus *Arsenura* the fore wings project forwards to a point like those of the Asiatic atlas-moths. Many species of the genus *Dysdaemonia* do not have the "windows" on their wings, nor the ringed patterns; the hind wings extend into a point at the sides, like those of the swallow-tails.

Arsenura romulus [683, top left] has a span of about 15 cm. It is grey-brown with dark shading and lives in Brazil.

Dysdaemonia mayi [683, bottom right] spans 12.5 cm between the tips of the hind wings. The moth is a dark greyish-brown.

The wingspan of the hind wings of the North American species **Moon Moth,** *Tropaea luna* [685, male above, female beneath], is in the region of 90 mm. The moth is light green; the purple-brown border on the fore wings is connected with an eye-spot. The hind wings when spread out describe the shape of a bow along the sides. The body is thickly covered with a white coat of hair. The green caterpillar [684] has pale red warts; it feeds on the leaves of the walnut tree.

On colour plate [XXXVIII] is shown a similar Asiatic species, the **Indian Moth,** *Actias selene.* It has a span of up to 12 cm.

684

685

XXXVIIa The Brazilian saturnid, *Dirphia multicolor*. Male. Span 8.5 cm.

XXXVIIb Male of the Bull's Eye Moth, *Automeris io*, from the U.S.A. The caterpillars live on fruit-trees and other trees.

XXXVIII *Actias selene* is distributed from Japan through China as far west as Ceylon. The caterpillars feed on the foliage of various trees e.g. the walnut.

687

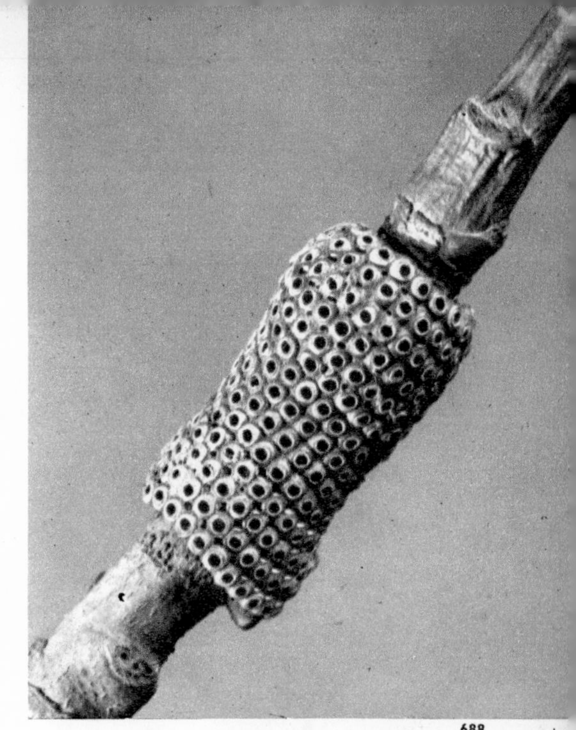

688

689

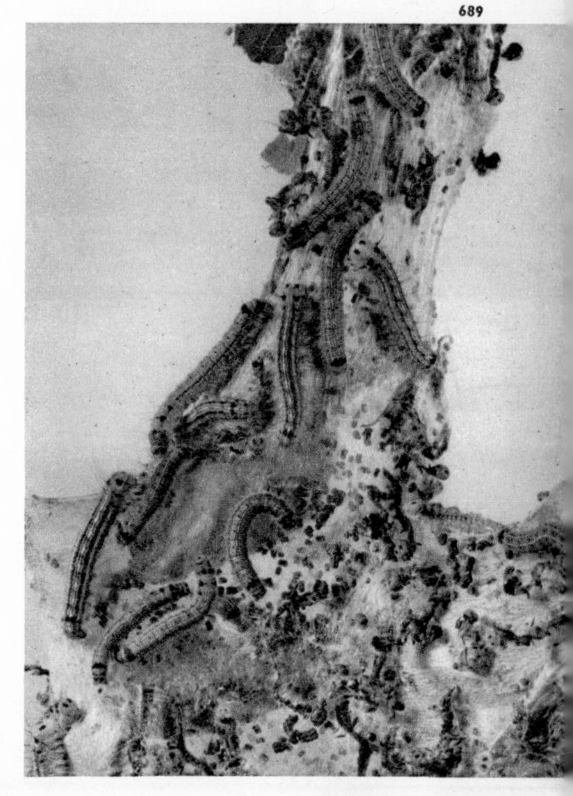

The European **Hook Tip,** *Platypteryx (= Drepana) falcatoria* [686], is a member of the *Drepanidae* family. It measures approximately 30 mm in span, coloured an inconspicuous brown and resembles a dry leaf. It inhabits deciduous forests in two generations and in central Europe flies from April to August. Its caterpillars feed on alder and birch.

The moths of the **Eggar** family *Lasciocampidae,* have a plump body and short wings. The proboscis is lacking in both sexes; the antennae are combed.

The **Lackey,** *Malacosoma neustria* [687, male], spans about 30 mm. Its colouring is composed of various shades of pale ochre to red-brown. The caterpillars feed chiefly on apple, plum and oak trees. They increase in vast numbers and are capable of totally stripping fruit trees of green leaves. The female lays her bark-coloured eggs in a characteristic form, in firmly cemented rings up to 1 cm in diameter, attached to a twig about the thickness of a pencil. Illustration [688] shows empty egg-cases after the caterpillar has emerged. In central Europe this takes place in the early spring. At first they live communally in a large web [689]. Later they scatter along the branches and eat the growing leaves. They are dark brown with a blue head, blue stripes down the side and white lines on the back.

The **Small Eggar,** *Eriogaster lanestris* [690], measures about 35 mm in span. It is chocolate-brown with light spots and patterns. The end of the abdomen is thickly covered with hair. The female of this species lays her eggs in ring clusters and covers them with a grey anal wool. The caterpillars live in clearly recognisable webs on deciduous trees. The moths emerge from barrel-shaped chrysalides sometimes only after the elapse of several years. This species is found in central and northern Europe.

The **Fox Moth,** *Macrothylacia rubi* [692, male], has a span of 50 to 60 mm. The female is grey-brown, the male reddish-brown with two white stripes across the fore wings. The male flies during the day. The female lays her eggs

690

691

692

on low bushes [693]. The caterpillars [691] have a coat of brown hair, while the ring divisions are yellow and black [691]. In late autumn we can frequently find them in the grass, where they look for a suitable place to hibernate. If they are disturbed they curl up into a ball. In the spring these caterpillars feed again for a short time on low plants and then pupate. The fox moth is a very common species on the plains and fields of Europe.

693

694

The **Pine Lappet,** *Dendrolimus pini* [694], spans up to 70 mm. Its colouring is variable, but it generally has ash-grey fore wings with broad brown bands and a white mark. However, specimens are also found coloured grey-red, red-brown, red-yellow, black and brown or silver-grey, and within these there is a wide range of shades of the basic hues. It lives in dry coniferous forests and is capable of doing enormous damage. The caterpillars feed on pine needles, hibernate and pupate only when spring comes round. The pine lappet is a central and northern European species.

The **Drinker Moth,** *Philudoria (= Cosmotriche) potatoria* [695, pair], measures about 60 mm in wingspan. The female is ochre yellow, the male reddish-brown. Both have a dark, obliquely sloping line across the fore wings and two small white marks. It is an extremely rare species, which lives in central Europe on damp, shady meadows or in damp deciduous forests.

The **Lappet,** *Gastropacha quercifolia* [696, male], has a wingspan of up to 75 mm. It is a coppery red-brown and bears three dark, zig-zag lines. The outer edge of the fore wings is serrated. The moth rests with its wings pointing forwards and outwards. Its caterpillars are grey-brown, with two bluish stripes across the body behind the head. When fully grown they measure about 80 mm. Illustration [698] shows a caterpillar on a twig—the head is pointing downwards. It is less perfectly camouflaged when seen from the side [699]. It feeds chiefly on the sloe, mountain ash, apple, plum and willow. The caterpillar hibernates and then pupates in the spring.

696

697

698

699

A powerful species which lives in Italy, the Balkans, Asia Minor and the eastern Mediterranean is **Pachypasa otus** [697, male]. It is grey-brown, with two sharply zigzagging dark lines on the fore wings. The antennae of the male are broadly combed at the base, so that they resemble "owls' ears". The ends of the antennae are narrowed. The caterpillars feed on cypress and oak. This moth has been cultured since classical times by the inhabitants of the island of Kos, who prepare silk from it, which was already known to Aristotle. The famous Greek physician Galen recommended the use of this silk in surgical operations.

700

701

The family *Endromididae* consists of the **Kentish Glory,** *Endromis versicolora.* The male [701] has a span of some 55 cm. The wings and body are rust-brown and the fore wings bear a black and white pattern. The female [702] has a wingspan of about 80 mm; its colouring is a very light brown with white and dark brown markings. The caterpillar [700] is green with white oblique strokes and is completely hairless. It feeds on birch, alder, beech, lime and hazel-nut trees in May and June. The moth sometimes emerges in central Europe as early as February, but usually in the middle of March. It flies in birch groves and mixed forests and its range extends throughout Europe with the exception of the Mediterranean countries, from England to Siberia.

704

The Indian species **Brahmaea wallichii** [703] comes from the *Brahmaeidae* family, which is distributed in Asia, southern Europe and tropical Africa. The moth has a span of 15 cm and ranges in colouring from the lightest to the darkest shades of brown. The pattern on the wings is one of the most complex and unusual of all *Lepidoptera*. The caterpillar is yellow-green with yellow-brown and black markings. In the earlier stages of its development it has extremely long, glossy black, crooked growths of an extraordinarily bizarre appearance, on top of and behind the head. Illustration [704] shows a caterpillar in the penultimate stage. With the next moult the growths will disappear.

705

706

The eastern palaearctic genus *Bombyx* contains the **True Silkworm Moth,** *Bombyx mori*. It was already cultured thousands of years ago, especially in China, and for this reason we do not know what its original appearance was like. However, since the white specimens of cultures here and there produce a grey-brown one, it is possible to assume that originally the species was dark in colour and has only become pale in the course of domestication. It may be that the original form was identical with the present-day species **Theophila mandarina.** The latter is an inhabitant of Asia and strongly resembles in both form and colouring the dark silkworms. The cultivated moths may be completely white, with a cream-coloured pattern, or dark grey-brown; however, within these limits there are also the widest possible variations. The female [705, above] is larger than the male [705, below] and measures about 45 mm in span.

707

708

The caterpillars normally feed on mulberry leaves, although they will accept lettuce as a substitute. They are about 80 cm long, whitish in colour with an ochre pattern [707]. They pupate in egg-shaped cocoons which have a narrow waist in the middle and are of a variety of light shades and colours. The silk thread from which the cocoon is spun forms the raw material of the natural silk industry. When the chrysalides are about ten days old they are killed by hot water or hot air, and the thread is then mechanically unwound, after first being softened in water. Each cocoon yields anything from 300 to 1,500 m, and already a length of 4,000 m has been recorded. The section of a cocoon with the chrysalis and the shed skin (exuvia) inside it [706] measures about 35 mm. The female lays grey, lentil-shaped eggs about 1.5 mm in diameter [708].

709

710

The family *Thaumetopoeidae* contains the well-known **Processionary Moth,** *Thaumetopoea processionea* [709, male], which inhabits the oak forests of the warmer parts of Europe. It has a span of up to 30 mm. The females have a covering of anally spun thread over their eggs. Touching the threads or the cocoon of this web can cause inflammation of the skin or the eyes. The caterpillars live communally in a large web nest and make their way to the feeding grounds in long processions, close one behind the other. The sack-like caterpillar nest [710] is easily recognisable by the deformation on the trunk of the oak. The caterpillars pupate underground as well as in the communal nest [711]. The moths fly between July and September and the caterpillars appear between May and June. They are known to increase in such numbers as to cause great damage, above all to oaks, but also to other trees.

712

713

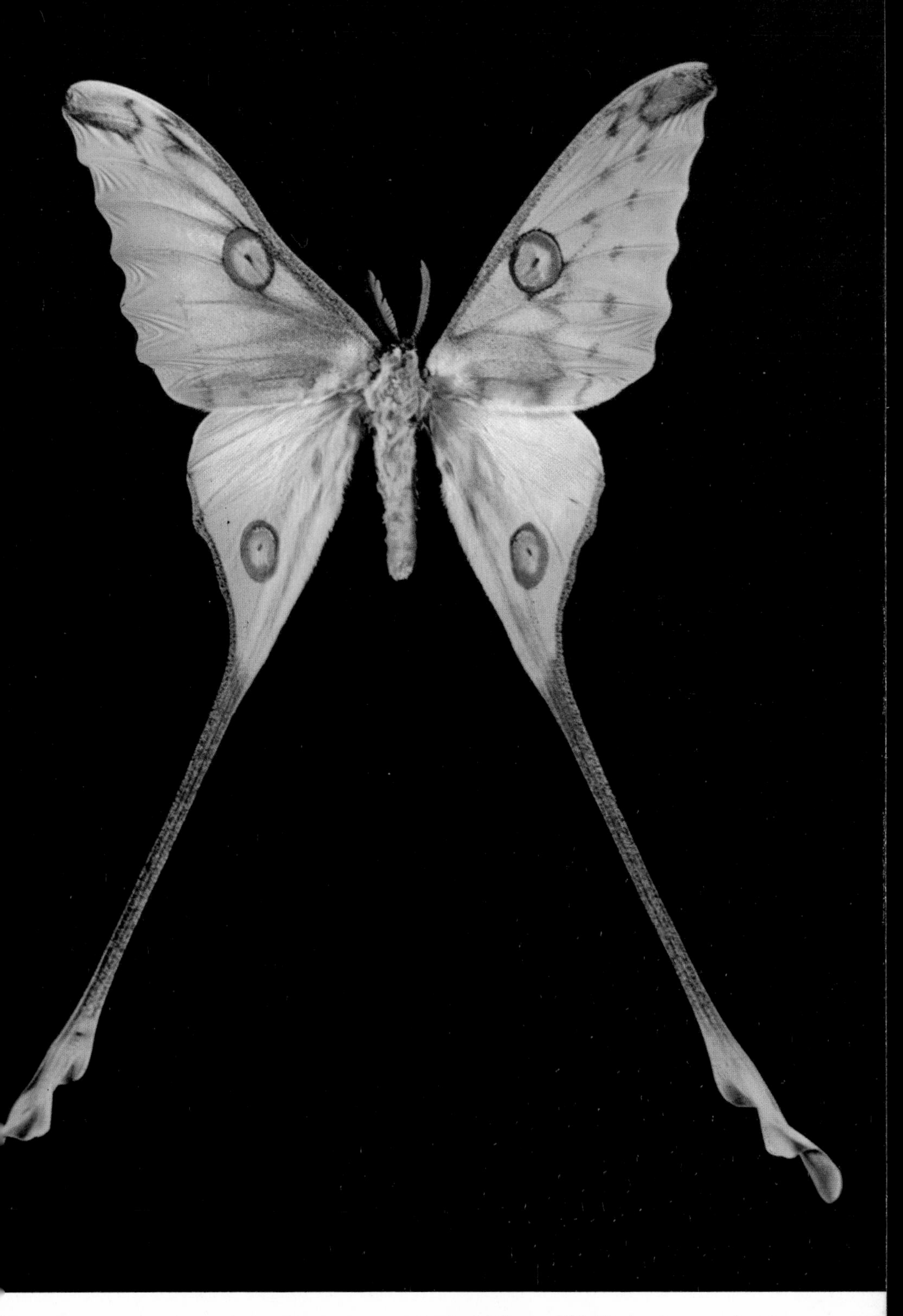

XXXIX One of the largest butterflies in the world is the Malagasy Silk Moth, *Argema mittrei*, from southern Madagascar.

XLa Owlet moth, *Agarista agricola,* from eastern and northern Australia. Span 6.5 mm.

XLb Crimson Underwing, *Catocala elocata.*

714

The moths of the **Tussock-moth** family *Lymantriidae,* are usually light in colour, with white, brown and grey tones. They are small to medium-sized and have an atrophied proboscis. The males possess broad, combed antennae and are smaller than the females. The caterpillars of some species are notorious pests of forest and fruit trees, and those of the genera *Orgyia* and *Dasychira* are distinguished by long, colourful pencils of hair on the back, and flattened bushes of hair. The caterpillar of the **Vapourer,** *Orgyia antiqua* [712], is among the most beautiful of all known caterpillars. It measures about 30 mm and is black, red and yellow in colour, with rich yellow pencils of hair on the back. It feeds on the leaves of wild plum, rose, oak and willow, as well as on pine-needles. The moth is rust-brown with a white patch on the fore wings. The females are plump and wingless.

The **Pale Tussock-moth,** *Dasychira pudibunda* [713, male], has a span of up to 45 mm and is grey-brown. Its very attractive yellow caterpillar [714] has black insertions which are just visible between the ring-segments of the body and a red hair-pencil on the eleventh segment. It feeds chiefly on beech, birch and oak.

715

The **Black Arches** or **Nun,** *Lymantria monacha,* [716, female], has a span of some 45 mm. The female is whitish with a black pattern. The abdomen is general pink.

The males are smaller and have combed antennae. The caterpillar [715] is grey-brown, not particularly conspicuous, and difficult to distinguish from the bark of the pine tree, in the crevices of which it pupates. It feeds principally on fir trees, but while breeding *en masse* it does not despise deciduous trees. It is greatly feared as an infester and pest of forests, especially in pure spruce forest plantations. It is widely distributed throughout Europe and has also been imported into North America.

The **Gipsy Moth,** *Lymantria dispar* [718], has a span of up to 70 mm. The male is brownish, the female cream-coloured; both bear a dark pattern. The dark, hairy caterpillars have red and blue warts and feed chiefly on oaks and fruit trees. The females cover their eggs [717] with rust-coloured downy hair, which gives

716

717

them the appearance of a fiery sponge fungus—hence its German name, the sponge fungus moth. Under favourable conditions it propagates *en masse,* and in these circumstances the gipsy moth becomes a pest of some 150 different species of plants. Its range extends across Europe from central Scandinavia to North Africa and into Asia. It has also been imported into North America, where it has effected catastrophic damage.

718

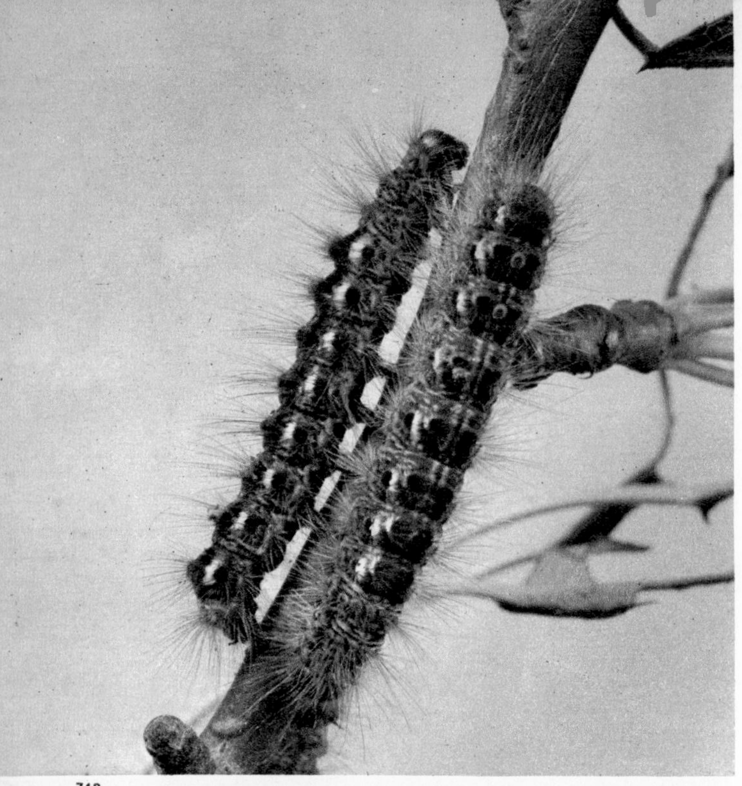

719

720

The **Brown-tail,** *Euproctis chrysorrhoea* = *phaeorrhoea* [720, female, from above], has a span of about 35 mm; the female is completely white and has a tuft of golden-brown, woolly hairs at the end of the abdomen, with which it covers its egg batches. The male is smaller, white, with combed antennae and a rust-coloured abdomen. The caterpillar [719] is dark with a white and red pattern and rust-coloured urticating hairs. Fully grown it measures about 30 mm. At first they live communally in a web-nest, but later they disperse. They feed mainly on the undersides of leaves; later they consume the whole leaf, down to the veins. When they reproduce *en masse* they can strip fruit trees, oaks and other deciduous trees bare of foliage.

Illustration [721] shows the caterpillars of the **White Satin Moth,** *Leucoma (= Stilpnotia) salicis*, a few days after emerging. When fully grown they measure about 45 mm and are very gaily coloured: dark with a red and yellow pattern and a row of white marks on the back. They propagate *en masse* and strip the foliage from poplars. The moth is white and has a span of some 50 mm.

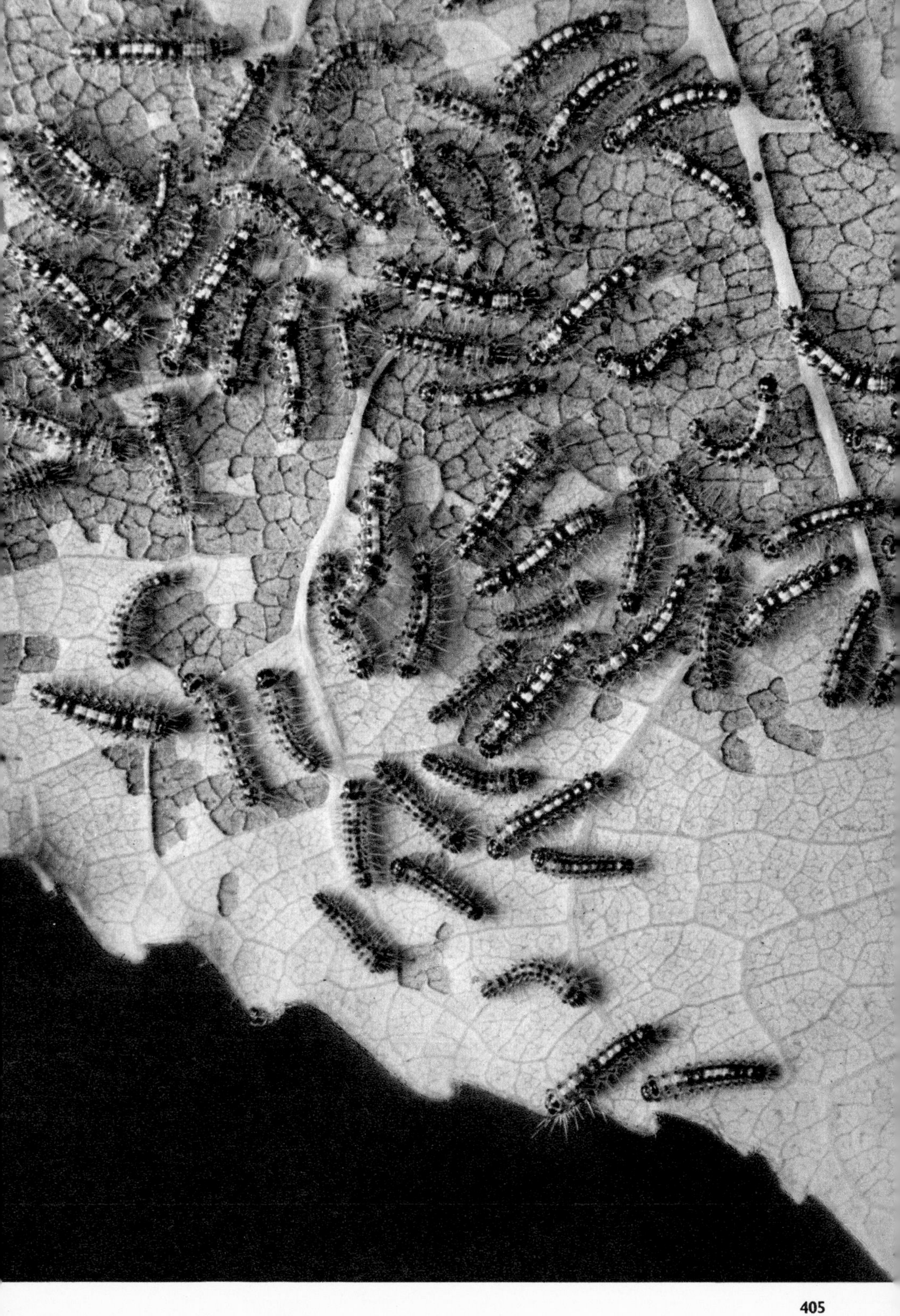

722

The family of **Owlet Moths,** *Phalaenidae* = *Noctuidae,* is very numerous: it contains over 25,000 species, with a world-wide distribution. The noctuids contain extremely small, medium-sized and large species. All have a hearing organ on the last thoracic segment and most of them have a well-developed proboscis. The great majority of noctuids are inconspicuously coloured and are so well adapted to their surroundings that it is impossible to distinguish them as they sleep during the day on the bark of trees or among plants. On the other hand, there are some very brightly coloured moths among the tropical species. Some of them are migratory and almost all of them are remarkably good fliers.

In the tropics there are many species of the genus *Phyllodes*, which is a good example of mimesis.

The fore wings of the 14.5 cm long **Phyllodes** from Celebes [722, left] completely resemble a dried leaf in form and colour. The rear wings have a pink patch.

Episteme lectrix [722, right] is one of the brightly coloured noctuids which are domiciled in the warm zones. It is coloured dark brown with yellow, orange and blue markings and spans some 70 mm. It flies by day over flowers and its range is throughout the warmer parts of China.

Thysania agrippina [723] has the greatest wingspan of all the *Lepidoptera* in the world: 30 cm. It is light greyish-brown with dark markings. Very little is so far known about its life.

724

725

726

The range of the large noctuid **Adris tyrannus** [724] extends over central India, the Malay peninsula, China and Japan. It has a span of about 90 mm. The shape and colour of the fore wings resemble a dry leaf. The orange-coloured hind wings bear a black pattern.

The most beautiful noctuids of the temperate zones are the underwings. Their fore wings are camouflaged with a protective colouring and pattern, while the hind wings are often very colourful.

The **Poplar Crimson Underwing,** *Catocala elocata,* has a span of up to 80 mm. The fore wings are grey-brown, the hind wings red with a black band. Illustration [725] shows the moth shortly after it has alighted and before it has folded its wings; on illustration [726] it is seen in the rest position, in which it sleeps through the day. It is present in central and southern Europe and the neighbouring parts of Asia. In some places the moth occurs towards the end of summer even in large cities, such as Prague, in central Bohemia.

The **Clifden Nonpareil,** *Catocala fraxini* [717], is the largest European species and, indeed, one of the largest members of the genus in the world. It has a span of up to 93 mm. The fore wings are grey; the hind wings black with a broken blue central strip and white edges. The caterpillars live on poplar, aspen, birch, ash, oak, elm and willow. The moth is one of our most beautiful larger noctuids.

The eggs of the Clifden nonpareil [729] are between 1 and 1.3 mm across and 0.7 to 0.9 mm high. Each has 25 to 30 ribs.

The caterpillar of the **Dark Crimson Underwing,** *Catocala sponsa,* [728] feeds on oak leaves. Fully grown it measures about 70 mm. It is grey or red-brown and simulates the bark of the oak tree. The moth is grey-brown with dark crimson hind wings, which have a black band across them. It is found in oak and mixed forests in south-east and southern Europe, Asia Minor and North Africa.

728

729

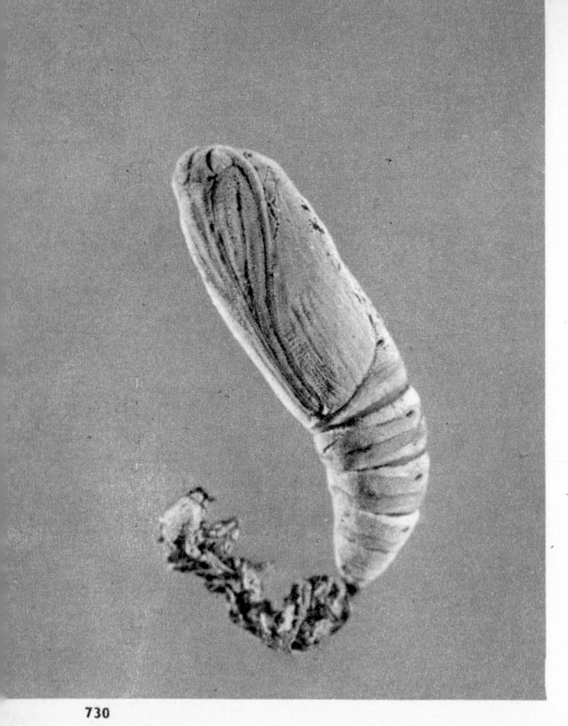

The chrysalides of the underwings are found in a loose web between two leaves which have been spun together. They are frosty white in appearance.

Illustration [730] shows the chrysalis of the **Dark Crimson Underwing.** The tiny eggs are typical both in shape and size of the various species of moths and butterflies. In many species the surface structure of the eggs bears a distinctive pattern of ornamentation. A good example of this diversity of patterns is seen in the eggs of some American species of underwings, all of which belong to the genus *Catocala* and are from the U.S.A.: **C. concumbens** [731], with a spider's-web structure, 1.5 mm in diameter; **C. neogama** [732], 0.95 mm in diameter; **C. recta** [733], 0.15 mm in diameter; and **C. coccinata,** 1.25 mm in diameter, with a loose-weaved structure.

730

731

732

733

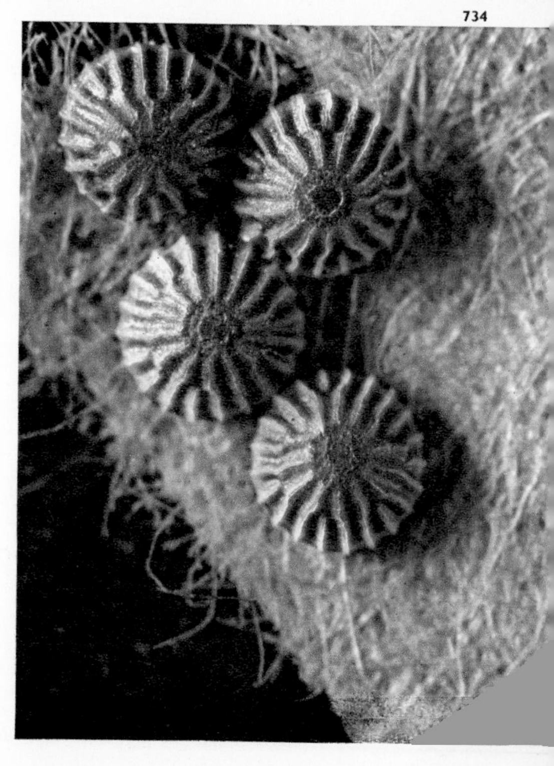
734

735

736

The **Monk** or **Shark Moths,** *Cucullia*, are small and medium-sized noctuid moths, most of which are of an inconspicuous colouring. Some species have silver marks on their fore wings—the so-called "mirrors"—and they all have a tuft of long hairs on the thorax, pointing forwards and somewhat resembling a monk's cowl, which is particularly noticeable in some species. From this they have received their German common name. The caterpillars are often very brightly coloured.

The **Shark Moth,** *Cucullia umbratica* [735], has a span of 40 mm. It is grey-brown; the caterpillars live on low plants, especially on the *Compositae*, such as milkweed and dandelion.

The caterpillars of the **Water Betony Shark Moth,** *Cucullia scrophulariae* [736], measure about 45 mm in length; they are white, brown and yellow and feed on figwort. The moth is brown, with serrated edges to the wings. The water betony shark moth is a common Eurasian species.

The **Grey Monk** or **Field Monk,** *Cucullia campanulae* [737], is comparatively rare in central Europe. It has a span up to 45 mm. Its fore wings are silver-grey with black lines, the hind wings light brown. The caterpillar [738] bears yellow and black spots on a white background. It feeds on the leaves of the **Harebell,** *Campanula rotundifolia*. The grey monk lives in undulating fields and also in higher mountainous districts.

737

738

739

The **Scarce Wormood Shark,** *Cucullia artemisiae* [739], has a span of about 40 mm and is grey-brown. It lives in the plateaux and hill country of central and south-eastern Europe and Asia Minor.

740

Its caterpillar [740] is greenish-brown with red growths, so that it resembles the flowers of the mugwort, on which it feeds. There are three caterpillars in our picture.

741

742

The **Viper's Bugloss Moth,** *Epia (= Dianthoecia = Hadena) irregularis* [741], has a span of some 30 mm. Its fore wings have an olive-green film over them and bear a rich pattern in different shades of brown; the hind wings are light brown. The caterpillars feed principally on **Catchfly,** *Silene otites*, and **Babies' Breath,** *Gypsophila fastigiata*. This species lives in warm, dry places in fields and plains on a chalk soil, but also on the plateaux and high ground of southern Europe and the neighbouring parts of Asia. It is rare in central Europe.

The **Common Wainscot Moth,** *Mythimna (= Leucania = Sideris) pallens* [742], has a span of 35 mm. It is a light ochreous-yellow. The caterpillar feeds on various plants, but chiefly on **Ryegrass,** *Lolium perenne*. It is commonly found on damp meadow land, and has two generations in the year. The moth is fond of the juices of the blossom on lime trees. The common wainscot is a central and northern European species.

The **Green Sandgrass** or **Malachite Moth,** *Calotaenia (= Jaspidea) celsia*

[743], measures some 40 mm in wingspan. The head, thorax and fore wings are light green, the markings on the wings brown, and the hind wings are grey-brown. The caterpillars live on grasses such as **Spiky grass,** *Nardus stricta*, *Calamagrostris epigeios*, **Aromatic Grass,** *Anthoxanthum odoratum* and *Deschampsia caespitosa*. The Malachite moth is found in dry pine forests on sandy soils. It is comparatively rare in central Europe and more common in eastern Europe and the nearer parts of Asia.

The **Bulrush Wainscot Moth,** *Nonagria typhae*, pupates inside the hollow stem of a bulrush [744]. The moth itself is grey-brown and has a span of about 40 mm. The caterpillar, which is brownish and grows to a length of some 50 mm, bores a way into the stem of a rush, where it pupates. The stems which contain chrysalides can be recognised by the round entry holes. The bulrush wainscot is distributed generally throughout central Europe, wherever bulrushes are to be found. It is also found in the temperate zone in Asia, as far east as Siberia.

743

744

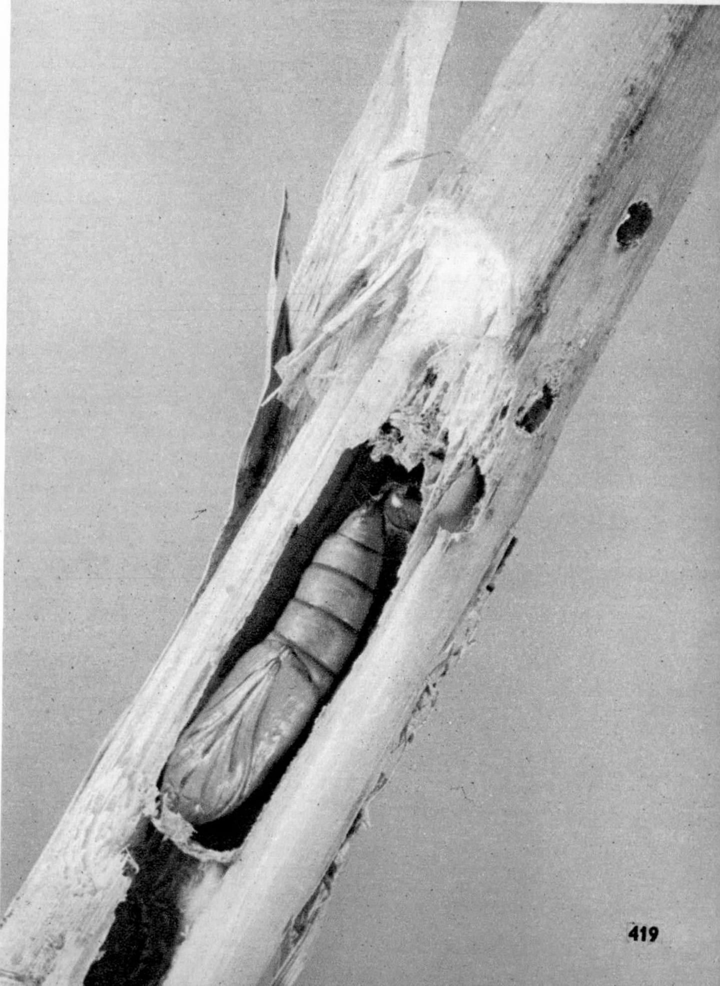

One of the most attractive of the noctuids is the **Angle Shades Moth,** *Phlogophora (= Brotolomia meticulosa* [745]. It has a span of over 40 mm. Its fore wings are green, pinkish-red and brown, with a dark pattern; at the outer edges they are serrated and indented. The brownish-grey caterpillar feeds chiefly on stinging nettles, dead nettles, ferns and other soft-leaved plants. The angle shades live in damp places, near water, on the eaves of forests and in gardens. Sometimes it hibernates and goes through two generations. It is a Eurasian species.

The **Herald Moth,** *Scoliopteryx libatrix* [746], has a span of more than 40 mm. Its wings are brown and grey; the fore wings have a film of violet over them, the middle part being cinnamon-brown with a white patch and a double white band running across the wings. It flies in two generations, the caterpillars feeding on willow and poplar. The herald moth hibernates in cellars and eaves. Illustration [747] shows it in a damp cave, 15 m underground, covered with drops of water. It inhabits central and northern Europe.

746

747

The **Green Silver Lines,** *Hylophila (= Bena) prasinana* [748], has a span of some 35 mm. Its fore wings are a lively green, with silver stripes running across them and red fringes round the edges. The hind wings of the male are yellow, of the female white. The caterpillars are green, yellow and white striped, and feed mainly on beech leaves, but also on those of the oak, birch and other trees. It is a central European species.

The **Dark Spectacle Moth,** *Abrostola triplasia* [749], a Eurasian species, has a wingspan of about 35 mm. Its fore wings are greyish to grey-brown, with black markings. The caterpillars feed on stinging nettles. They are found chiefly in well-shaded places beset with stinging nettles, especially on undulating country and the foothills of mountain ranges.

The **Silver-Y Moth,** *Phytometra (= Autographa) gamma* [750], has a span of about 35 mm. It is grey or reddish-brown, with a silver marking in the middle of each of the fore wings resembling the letter Y, hence its Latin specific name. The caterpillars, which come out in two and even three generations in a year, feed on plants of the mint family *Labiatae.* Sometimes they increase in such numbers that they wreak havoc among clover and other food plants, and also to cultivated garden plants. The moth flies during the daytime, and the caterpillars hibernate. In the early summer one generation migrates into central Europe from North Africa.

749

750

751

752

The **Large Yellow Underwing,** *Triphaena (= Agrotis = Rhyacia) pronuba* [751], has a span measurement of up to 60 mm. The colouring of this species varies somewhat, the fore wings being grey-brown to black, with lighter or darker markings. The rear wings are yellow with a narrow, dark grey line along the edges. The caterpillars are brownish and feed on various soft-leaved herbs, such as dandelions. They hibernate, only pupating with the arrival of spring. The moth sleeps during the day in various different kinds of hiding place: often inside houses and the specimen illustrated on the underneath of a wicker basket. It is an inhabitant of Europe and the neighbouring parts of Asia.

The caterpillar of the **Figure-of-eight Moth** or **Blue-head,** *Episema (= Diloba) caeruleocephala*, is blue-green with black, orange and yellow spots. When fully grown it measures about 40 mm. It attacks hawthorn, sloe, fruit trees and various shrubs. The moth is grey with a light and dark pattern and measures about 40 mm in span. The figure-of-eight is an inhabitant of central and south-eastern Europe and the neighbouring parts of Asia.

The **Grey Dagger Moth,** *Acronycta (= Apatele) psi* [753], has a span of about 37 mm. Its wings are grey with black dagger-like markings, and afford an extremely effective protective colouring against the bark of fruit trees and weathered wood. The caterpillar carries a few scattered hairs and is coloured black, red, yellow and white. On its back, in

754

755

756

the first quarter of its body, it bears a horn about 5 mm long. It lives on rose, plum, lime and other deciduous shrubs and trees on high ground in the whole of Europe and the nearer parts of Asia.

The **Tiger-moth** family *Arctiidae,* contains over 6,000 species with a range extending over all five continents. They are small to medium-sized, generally very brightly coloured.

The **Feathered Footman,** *Coscinia striata* [754], has a span of about 35 mm. Its fore wings are yellow with black stripes; the hind wings ochreous and dark grey. The caterpillars go into hibernation and feed on various grasses and herbs. The feathered footman is found in dry, grassy places throughout Europe and the neighbouring parts of Asia.

The **Crimson Speckled Moth,** *Utethesia (= Deiopeia) pulchella* [755], measures about 35 mm. Its wings are whitish, the fore wings being speckled with crimson and black, the rear wings bordered with a dark grey line. It is a cosmopolitan species, inhabiting especially warmer countries. Its caterpillars feed on rough-leaved herbs. It migrates into central Europe from the south.

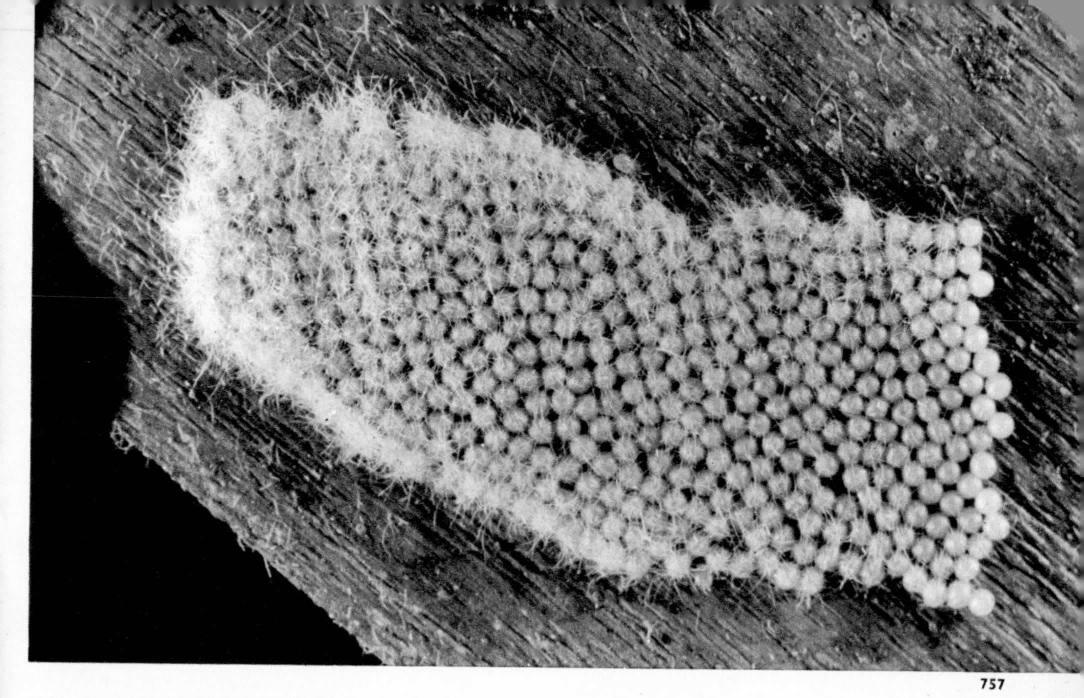

757

The **American Weaver Tiger-moth,** *Hyphantria cunea* [756, pair], has a span of some 35 mm. It is white, sometimes with black spots. The caterpillars are polyphagous—they will feed on almost anything. They live on over 120 species of plants and form an extremely harmful pest of all deciduous trees, not only in America, but also in the warmer countries of Europe, into which they have been introduced. Illustration [757] shows a batch of the green eggs, enlarged, and [758] shows chrysalides behind the bark of a mulberry tree.

758

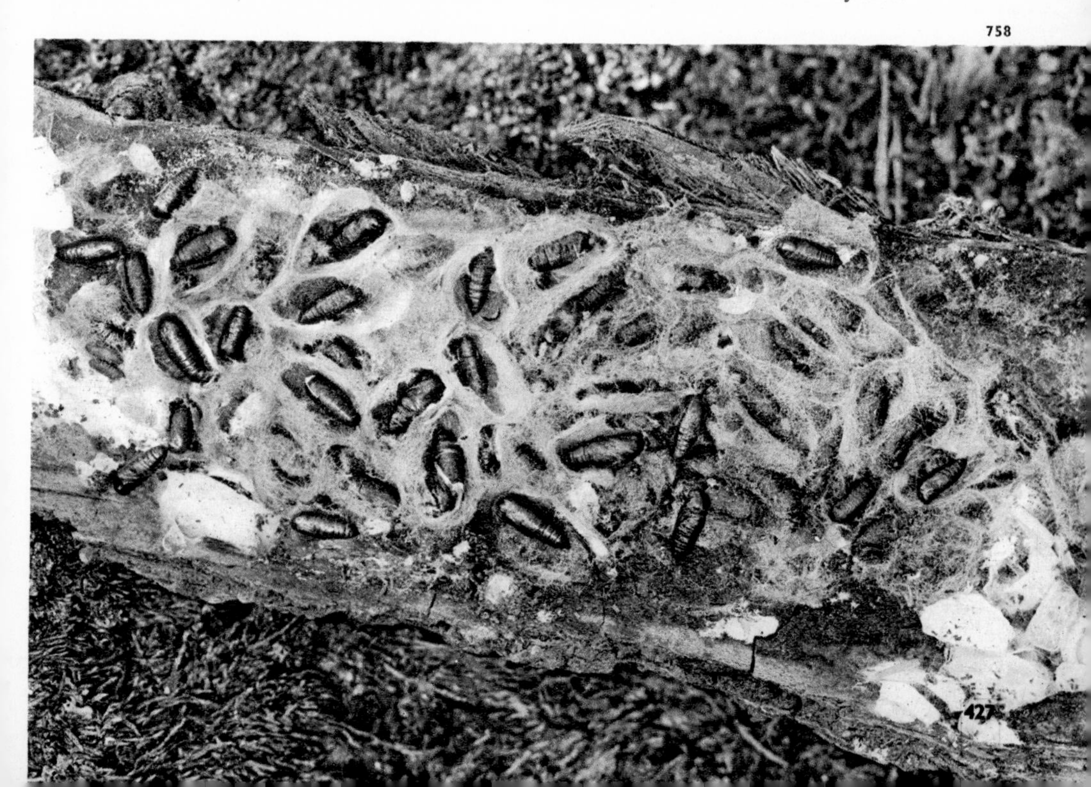

759

760

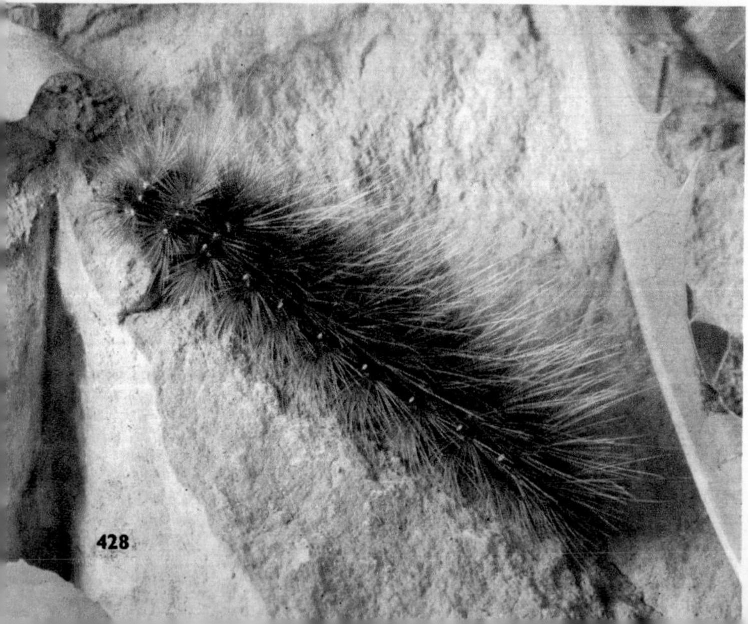

The **Court Lady** or **Brown Tiger-moth,** *Hyphoraia (= Arctia) aulica* [759], measures about 35 mm in span. Its fore wings are reddish-brown with yellow marks, the hind wing ochre-yellow with black spots. The caterpillars hibernate, and feed polyphagously on a variety of grasses and low herbs, being particularly fond of dandelion leaves. They are found on warm, sunny slopes with steppe vegetation in hilly and mountainous country in central and south-east Europe and the nearer parts of Asia.

The **Garden Tiger-moth,** *Arctia caja* [761], has a span measurement of up to 60 mm. The fore wings are coffee-coloured with yellowish-white markings. The abdomen and the hind wings are usually brick-red with glossy blue-black markings. The hairy, dark brown caterpillar [760] is polyphagous, hibernates and pupates in a loosely spun cocoon. It can be seen taken on dandelion leaves. The garden tiger is a Eurasian species and is very widely disseminated in central Europe.

761

762

The **Scarlet Tiger,** *Panaxia (= Callimorpha) dominula* [762], has a span of 53 mm. The fore wings are black with a greenish gloss, and yellow markings. The hind wings are scarlet with black markings. It lives in Europe and Asia in damp forests. In central Europe the caterpillar hibernates. It feeds on stinging nettles, dead nettles, forget-me-nots, blackberry leaves and other plants.

764

Our illustration [763] shows a selection of the most beautiful European tiger-moths. Top left is the **Purple Tiger-moth,** *Rhyparia purpurata.* Its fore wings are yellow with dark grey markings, the hind wings purple with black marks. It is an inhabitant of Europe and western Asia.
Immediately below it is **Arctia fasciata,** which has yellow-white fore wings bearing a black pattern. The hind wings are red and yellow with black spots. It is an inhabitant of south-west Europe.
At bottom left is the **Bright Tiger-moth,** *Arctia hebe.* The fore wings of this species are pale yellow, with black bands and patches, the hind wings red with black marks. It is a Eurasian species.
Top right is the **Jersey Tiger-moth,** *Euplagia (= Callimorpha) quadripunctaria (= hera).* Its fore wings are glossy dark brown with yellow bands, the hind wings being ginger-coloured with black marks. It is a mountain species of central and southern Europe, which is also found in the nearer parts of Asia.
On the right below is the **Engadine** or **Yellow Tiger-moth,** *Arctia flavia.* The fore wings are almost black, with a yellowish-white pattern, the hind wings yellow with dark grey markings; the abdomen is red at the sides.
The large moth in the middle is the **Augsburg Tiger-moth,** *Pericallia matronula.* It has a span of up to 85 mm and is the largest European tiger-moth. Its fore wings are cinnamon-yellow with a yellow pattern along the leading edge; the hind wings are yellow with black patches. The abdomen is red. Today it is only very rarely encountered, but years ago it was widely distributed throughout central Europe. The caterpillars hibernate twice before they pupate. They feed on shrubs, but also on other plants, such as the dandelion.
A tropical species of tiger-moth is shown in illustration [764]—**Ephestris nulaxantha.** It has a wingspan of 70 mm, is black and yellow and comes from Brazil and Columbia.
The family *Uraniidae* contains about 100 species of tropical moths. Some of them fly by day, others at night. Some of the diurnal

765

species are so brightly coloured that they are nowhere equalled in the whole animal kingdom.

Urania leilus [765, top], with a span of about 70 mm, has an iridescent colouring of black, blue-green, blue, green and white. This is a migratory, day-flying moth, which flies very swiftly at a great height and rests in the tops of trees. It is a native of Central and South America, although some exceptionally attractively coloured specimens have been taken also in Cuba and Jamaica.

Sematura lunus [765, below] has a span of some 70 mm. It is dark brown with whitish-yellow stripes and banded wings. On the tip of each hind wing it bears two eye-spots. It flies at night and lives in Panama and the northern half of South America.

The marvellously beautiful **Chrysiridia madagascariensis** *(= Urania ripheus)* is

XLIa Garden Tiger-moth, *Arctia caja;* when suddenly disturbed it shows the hind wings.

XLIb *Chrysiridia madagascariensis = Urania ripheus.* The caterpillars live on euphorbiaceous plants.

XLIIa Heath-bush or Brown Hairstreak, *Thecla betulae*. Female. A Euro-Siberian species. Span 3.5 cm.

XLIIb Autumn generation of the Map Butterfly, *Araschnia levana* = *A. prorsa*. Central Europe. Span 3.5 cm.

illustrated on colour plate [XL 1b]. It has a span of 85 mm, flies during the day and may be counted among the most beautiful of all butterflies and moths on account of its particularly fine iridescent colouring. It lives in Madagascar. Some very similar related species are found in East Africa.

Nyctalemon menoeticus [766] belongs to the small family *Sematuridae*. The wingspan amounts to some 13 cm. The moth is dark brown with a white stripe running down each wing and white markings. It is a nocturnal insect and lives in southern China and India, on the Malayan Peninsula and the Philippines.

Alcides metaurus [767] has a span measurement of about 10 cm. Its black wings have bands of blue-green on them and are covered with a copper and brassy gloss. It is an inhabitant of northern Australia and flies in daylight.

766

767

768

The **Skipper** family, *Hesperiidae,* consists of butterflies of small size, with short wings, a compact body and a large head with club-shaped antennae. They fly by day over the flowers. They present a group of very heterogeneous species, of which there are about 3,000 distributed throughout the Old and New Worlds.

Sarbia damippe [769], one of the Brazilian skippers, has a wingspan measuring 45 mm. It is dark brown with yellow bands on the wings and fire-red on the head and also on the tip of

769

the hind quarters.

The family of **Blues, Coppers** and **Hairstreaks,** *Lycaenidae,* consists of small butterflies with a cosmopolitan distribution. The family may be divided into three subfamilies: **Blues,** *Plebeiinae,* **Coppers,** *Lycaenini,* and **Hairstreaks,** *Theclinae.*

The **White-letter Hairstreak,** *Strymon* (= *Thecla*) *w-album* [768], has a span of about 30 mm. This butterfly is dark brown and bears on the underside of its hind wings a small, orange, semicircular spot and a narrow, white, W-shaped stripe. Its caterpillars are brown and live on elms [770].

770

771

772

One of the most attractive hairstreaks is the tropical species **Thecla coronata,** with a wingspan of about 50 mm, which makes it also one of the largest. The wings are glossy blue-green on the upper surface, with a black border running round the wings and the tails on the hind wings and scarlet markings at the roots. Illustration [771] shows a male, glossy green with violet bands on the hind wings and red bands on the fore wings. It lives in the tree tops in primeval forests, only rarely coming down to ground level. Its range of distribution consists of Guatemala, Columbia and Ecuador.

Similar forms are found in the family *Erycinidae*. Illustration [772] shows a male **Helicopis cupido.** It measures about 35 mm in span and is light yellow, light brown and black in colouring. This butterfly is an inhabitant of equatorial South America.

The **Chalk Hill Blue,** *Lysandra (= Lycaena) coridon* [773], has a span of 34 mm. The male is a glossy blue on the upper surface of the wings, the female dark brown. The caterpillars feed on the leaves of papilionaceous flowers, and are unusual in that they exude a sweet secretion from dorsal glands, which is eagerly sought after by ants. The caterpillars of some other blues exude a similar-smelling substance, so that they not only enjoy the protection of the ants, but even live inside their colonies and feed on the ant larvae. Other species feed on plant-lice.

773

774

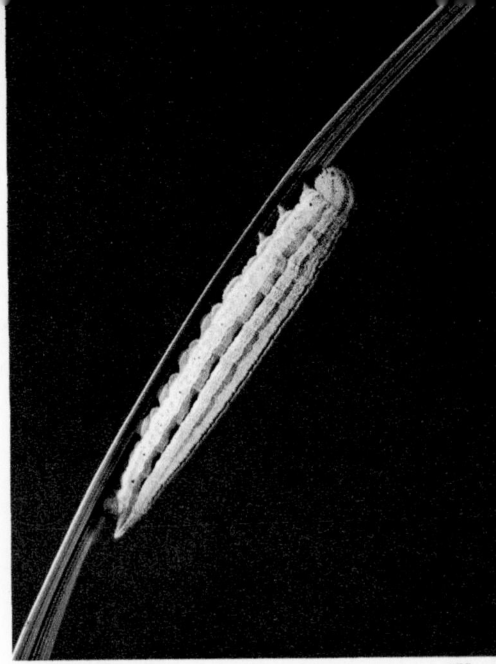

775

776

The family of **heaths, marbled whites** and **graylings,** *Satyridae,* contains approximately 2,000 species of medium-sized butterflies of inconspicuous colouring from all over the world. They are mostly a lighter or darker shade of brown. As a rule they have eye-spots on their wings of varying sizes. As with all the other day-flying butterflies, the ends of their antennae are flat and broad at the sides of the thorax. At the base of the wings are situated hearing organs (tympanums).

The **Marbled White,** *Melanargia galathea* [776], measures in span about 45 mm. It is an exception to the rest of the family in that the eye-spots on the upper side of the wings are not clearly developed. It is yellowish-white and brown, mottled in a pattern reminiscent of a draughts board. On the underside of both the fore and hind wings there are a large number of small, circular eye-spots. It is an inhabitant of sunny plains, slopes, fields and forest clearings, and flies only on sunny days.

Illustration [774] shows the chrysalis of the **marbled white** shortly before the emergence of the moth. The caterpillars are mostly hairless, coloured brownish or greenish, generally with dark stripes down their backs. The rear end of the body is commonly pointed, often ending in a short fork.

On illustration [775] is a caterpillar of the **Small Grayling,** *Hipparchia aelia = Satyrus alcyone.* It feeds on coarse grasses, and also hibernates. This butterfly is a Eurasian species.

In the tropics of South America live *Satyridae* of the genus *Callitaera*, which have transparent wings. **Callitaera pireta** [777] has a span of about 55 mm. The membranous wings have a brownish venation; on each of the hind wings, which have a strong pinkish tint, there is a black eye-spot with a brown border. This species is an inhabitant of Ecuador and the upper Amazon.

The Indo-australoid family *Amathusiidae* contains the species **Taenacris selene** [778], which has a span of some 90 mm. The fore wings are light grey on the upper surface, with a dark brown border; the hind wings whitish, and similarly equipped with a dark border. On the top surface of the hind wings are situated a pair of large eye-spots, and on the under side two pairs. The butterfly lives in the southern Moluccas. Like all butterflies of this family it avoids strong sunlight and prefers to fly at twilight.

777

778

779

780

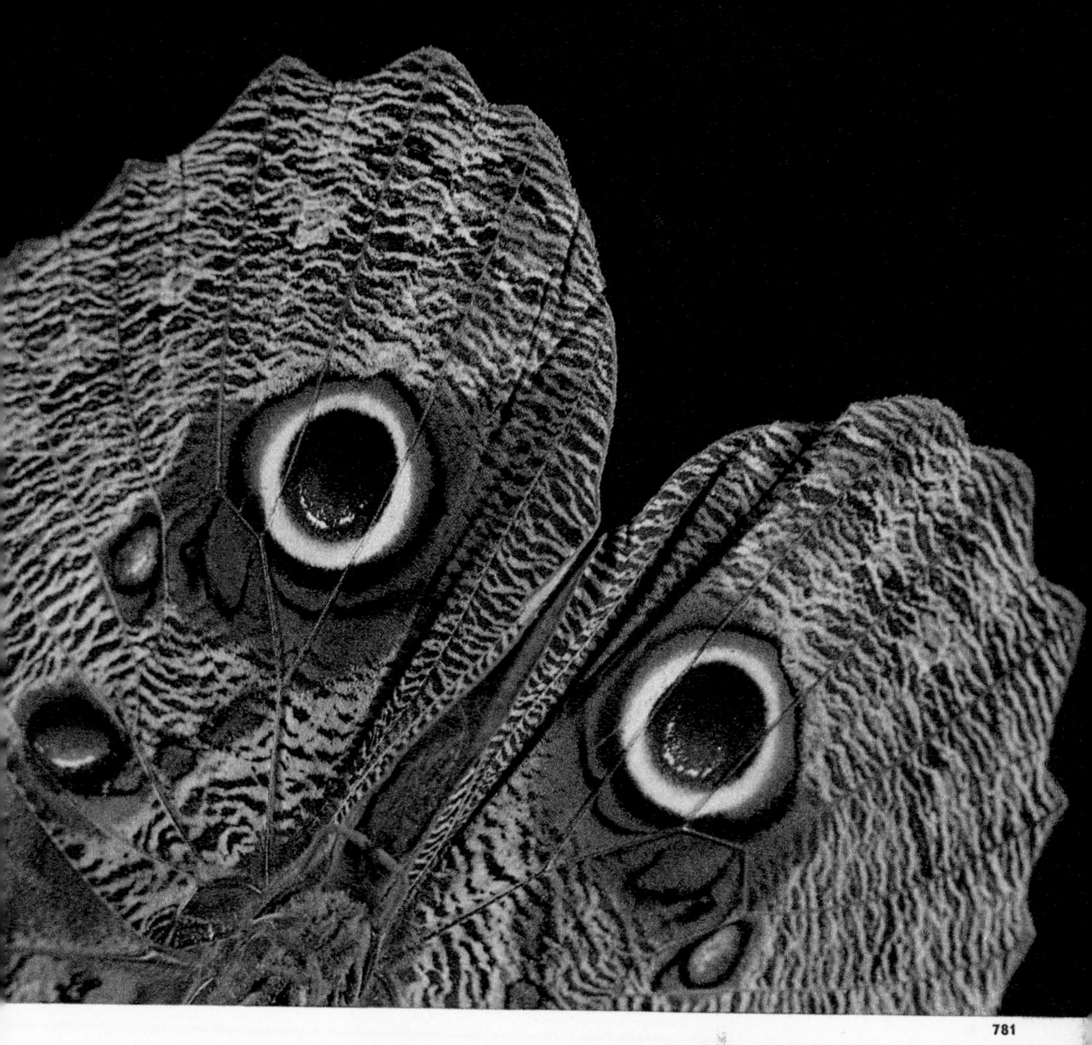

781

Among the largest species of butterflies from subtropical China may be numbered **Sticho-phthalma formosana** [779], with a wing-span of 10 cm. Its wings are light brown on the upper surface with a prominent dark brown design around the borders of the wings; the underneath is covered with a pale rust colour and has a design of narrow oblique stripes and an irregular line of little eye-spots. This butterfly lives in Taiwan.
The Indo-Australasian *Amathusiidae* are closely related to the family of *Brassolidae* from the tropical regions of the New World. The members of this family are mostly dark-coloured and range in size from larger than average to giant. They live in the impenetrable tropical forests, where they spend the day hiding in the shadows, and only fly out into the open at twilight. The caterpillars feed on monocotyledonous plants.
Caligo eurilochus [780] has a wing-span of 14 cm. The upper side has a bluish-violet sheen and a dark brown border to the wings. The caterpillars feed on plants of the banana family *Musaceae.* It is an inhabitant of Honduras and Columbia.
The butterflies of this genus are well known as extremely adroit and quick-reacting fliers.
The under side of the wings of *Caligo* butterflies is decorated with a very delicate linear pattern. In the case of the species *C. eurilochus* this pattern is brown to black, while the ground colouring of the wings is grey. The hind wings each have a large, yellow-ringed eye-spot [781] in the middle of them, reminiscent of those of the noctuids.

782

783

The **morpho** family, *Morphidae,* consists of large, tropical butterflies from South America which have become famous for the iridescent colouring and bright hues of their broad wings. Some 50 species are distinguished, most of which pass their lives among the tree tops in primeval jungles. In some parts of South America perfect specimens are cultured from the caterpillars.

One of the most attractive specimens is **Morpho rhetenor** [782]. It has a wingspan of about 13 cm and is extremely beautifully coloured. Seen in the wild it leaves an unforgettable impression, resembling a flying streak of blue lightning. Since the colour of the wings is constantly changing according to the angle of incidence of the sunlight, it is at one moment blue-green, then again a shimmering dark blue, then blue violet or violet—a constantly changing pattern of wonderful colours. The underside of the wings [783], by way of contrast, is a brown tone, and as in the case of most morpho butterflies, resembles a dry leaf; this is the reason that the butterfly appears to suddenly disappear when it settles and folds its wings. The brightness is suddenly extinguished and the protective colouring of the underside of the wings makes it virtually invisible.

Morpho mineiro has on the dark grey underside of its wings [784] a clearly visible row of little eye-spots. The upper surface is an iridescent blue-green.

784

785

Morpho hecuba [787] is the largest of all the members of this genus. It has a wing-span of about 17 cm. In contrast to the iridescent species it has a simple colour pattern. Its fore wings are red-brown with black markings along the edges; the hind wings almost black with a yellow-white root. The underside [785] is dark brown with red-brown eye-spots having black rings round them, and areas of red and brown in complex patterns which are shot through with silvery stripes and markings.

Morpho catenarius [786] measures about 10 cm in span. The wings are translucent; on the upper side they are milk-white, with a tinge of blue-green, and glossy dark-brown ornamentations; the patterns of the underside show through. Several members of the genus *Morpho* are white-winged in this way. Some of them have a sheen like mother-of-pearl. The broader form of their wings characterises them as slow fliers with little endurance. They live in areas of restricted range.

786

787

788

789

The family *Heliconiidae* is distributed principally in tropical America, though a small number of species live in North America. Members of this family are medium-sized butterflies with long, narrow wings, which are always very brightly coloured and are sometimes glossy. Their most striking common characteristic is the all-pervasive and bug-like smell, which is noticeable even at a distance of ten paces. For this reason these butterflies are avoided by birds, lizards and apes. In some species this evil-smelling body secretion is poisonous, which causes all insectivores to give them a wide berth. A number of species of butterflies which are not related to the *Heliconiidae* have taken on their appearance and colouring, a resemblance which has clearly arisen as a form of protection from insectivorous beasts for these otherwise defenceless species. Their outstandingly good camouflage enables the *Heliconiidae* to fly about ir completely open places and to rest without having to hide. Generally they pass the night in large groups on trees and bushes. They also have a marvellous sense of direction, so that, for example, they return each day to the same place to sleep. Their caterpillars bear two elongated horns just behind the head. They live singly in the tree tops of primeval jungle, where they feed on the leaves of lianas and other creepers. The chrysalides have a number of hook-like thorns as a protective covering on their upper side.

Heliconius leopardus [788] has a span of 90 mm and dark brown wings with broad black borders and patches. It lives in Bolivia.

Heliconius vicina [789] has a span of about 70 mm. It is dark brown and orange; the light markings on the fore wings are yellow. It lives in the Amazon basin.

Another member of the *Heliconiidae* is **Metamorpha dido wernickei** [790]. It spans some 90 mm, is coloured turquoise-green and black, with translucent wings. The undersides of the wings have grey-brown borders and bands with a silver gloss. The caterpillars feed on the passion-flower, *Passiflorae.* It lives in southern Brazil and Paraguay.

The same family also contains **Dione vanillae,** which measures about 65 mm in span. On the upper side it is rust-brown with black markings and patterns. The underside [791] is likewise rust-brown and heavily ornamented with long silvery, shimmering patches. Its colouring is reminiscent of that of the mother-of-pearl butterfly, which is a member of the *Nymphalidae*.

791

In the family of **fritillaries** and **vanessids,** etc., *Nymphalidae,* we group some 5,000 species, distributed throughout the world. The European members of the family may be numbered among the most beautiful and best-known of butterflies.

The **Large Tortoiseshell,** *Nymphalis polychloros* [793], has a span of up to 60 mm. It is red-brown and black, with blue tinges along the edges. The caterpillars live together in large companies under a protective web on elms, willows and poplars. The butterfly hibernates and survives the winter. This species inhabits the whole of Europe, and as far east as Siberia.

The chrysalides of the nymphalids are firmly attached to the underside of branches and twigs. They always hang head downwards. Often they have little horns, and gold or silver markings. On illustration [792] we see the chrysalis of the **large tortoiseshell.**

The **Peacock Butterfly,** *Nymphalis io,* is a very widely known, marvellously beautiful European butterfly. Its caterpillar [794, enlarged] feeds on stinging nettles and hops, and generally lives communally in large groups.

792

793

794

XLIIIa The Fritillary, *Prepona demophon muson*. Male. Southern Colombia. Span 9.5 mm.

XLIIIb The most common Eurasian fritillary is the small Tortoiseshell, *Aglais urticae*. The butterfly hibernates in winter.

XLIV The Birthwort Butterfly, *Zerynthia polyxena,* with a span of about 55 mm. It lives in the warmer parts of central Europe, southern Europe, and Asia Minor.

795

The butterfly hibernates in the winter.
On illustration [795] are shown:
Top left, the **Comma,** *Polygonia C-album,* which measures in span about 45 mm. It is rust-brown with darker patches. The caterpillar feeds on stinging nettles and sallows. The imago of this Eurasian species hibernates.
Bottom left, the **White L,** *Polygonia L-album,* which lives in south-eastern Europe and the neighbouring part of Asia, and occasionally is found in central Europe. It has a span of some 70 mm, and is coloured rust-brown and black, with white flecks. The caterpillars feed on sallow, willow and birch.

Below right, **Polygonia egea,** a yellowy rust-colour, which resembles the comma, except that the mark on the underside of each of its rear wings resembles a J rather than a comma. The caterpillars live on elm, hazel, etc. The imago goes into hibernation. It is a southern European species.
Top right, the **Painted Lady,** *Vanessa cardui,* with a span measurement of 55 mm. It is coloured a light red-brown, with black and white patches. On the underside of the wings it has blue eye-spots. The caterpillars feed principally on thistles and stinging nettles. It is a migrant, driving northwards over the Alps in

796

797

798

May and June, often in millions; in the autumn the second generation journeys back southwards. Its flight is extremely fast.
The chrysalis of the **comma** is grey-brown; on its back it bears silver spots, and a long triangle near the tail [796 and 797].
Closely related to the *Nymphalidae* is the family *Libytheidae*, which is also distributed throughout the world, although only in a few species. **Libythea celtis** [798, top right] has a span of approximately 45 mm. It has dark brown wings with an ochre pattern on them, and two white marks in the corners. The hind wings are silver-grey underneath. It is a Mediterranean species, which also occurs very infrequently in southern central Europe. Its caterpillars eat the leaves of the **Nettle-tree,** *Celtis australis*. Besides southern Europe, the habitat of this butterfly range from North Africa across Asia Minor to East Asia.
Above left in illustration [798] is the **White L** seen from underneath. The white markings on each hind wing form the letter L.
Left below in [798] is the **Comma,** likewise seen from below and showing the mark which gives it its name.

799

800

Around the shores of the Mediterranean lives **Charaxes jasius** [799, above], one of the most attractive known butterflies. It measures up to 75 mm in span. The upper surface of the wings is dark brown, with an ochre-yellow pattern, and a line of blue spots on the hind wings. The underside [799, below] is unusually brightly coloured, with a chocolate-brown, black, silver, yellow-red and violet-blue mosaic. The caterpillars, which are green and equipped with horns, feed on the **Strawberry-tree,** *Arbutus unedo.* In captivity they may be fed on rose leaves.

Eriboea eudamippus formosanus [800] has a span of about 70 mm, and is light yellow and dark brown in colour. It is a remarkably good flier; it sucks the liquid from animal corpses, dung and similar refuse. If it is disturbed, it eventually returns to its original resting place. It lives in Taiwan.

Junonia almana asteriae [801] measures up to 50 mm in span. It is red-brown with black spots on the fore wings. On each wing it bears a large eye-spot. The caterpillars feed on plants of the acanthus family, *Acanthaceae*. This butterfly has a range extending from this side of India through the Malay Peninsula to China and Japan.

Megalura chiron [802] measures approximately 45 mm in span. The upper side of its wings is yellow-brown with dark brown bands; the underside dark brownish-grey or light grey, with a silvery white band down the middle and a blue-violet gloss. On the fore tip of each front wing are situated five or six white spots. The caterpillars feed on the fig and mulberry trees. It lives in Mexico, the whole of Central and South America and the large islands of the Caribbean.

801

802

803

The genus *Protogonius,* whose range extends from Mexico to South America, contains the nymphalid **Protogonius tithoreides** [803]. It has a span of about 80 mm. The upper side of its wings is red-brown, yellow and dark brown, the underside grey-brown with a fine, carefully arranged pattern which makes it look like a dry leaf when it is in the rest position. The tails on the folded hind wings resemble the stalk of a leaf. In this the butterfly resembles the poisonous *Heliconidae,* and is excellently protected. The caterpillar is dark brown, and has a dark-coloured rim along its back and two horns on the black head. It hides from the strong sunlight in a leaf which has been spun together at the sides. At night it emerges, and attacks the leaves of its food plants—principally the pepper genus, *Piper,* and the **American medlar,** *Mespilus americanus.*

The Malagasy species **Salamis dupreyi** [804] measures about 80 mm in span. It is distinguished by broad wings, which have elongations at both the front and the back. They are whitish, with a mother-of-pearl gloss; the tips of the fore wings are black, and there are black spots on all the wings. At the edge of the hind wings there are small eye-spots with a violet centre and red, yellow and black rings. The underside [804] has a remarkably effective protective colouring, superbly adapted to the place where the butterfly rests among the plants. It is a light yellowy-brown, with a dark stripe running down the middle of each wing, which resembles the central vein of a dry leaf. The whole surface of the wings is covered with a variety of smaller and larger markings of an almost fungal appearance.

A magnificent example of mimesis is the **Leaf Butterfly,** *Kallimacha inachus formosana,* the underside of which resembles a dry leaf, even down to the mould and arid rust marks. In

804

805

806

some cases it is coloured a browning-red, sometimes light grey-brown, and in other instances it verges on greenish. On illustration [806], there is a true leaf at top left; the other two "leaves" are butterflies—the one at the bottom brownish-red, and the one at the top grey-brown and olive-coloured. At the bottom of the picture is a specimen with wings spread. It measures about 70 mm in span. The upper surface is brownish-red, with a greeny-blue gloss. The band across the fore wings is orange-coloured; the spots are white. In the middle of each fore wing there is a small transparent window. These butterflies are particularly fond of the juices of over-ripe bananas. **Coenophlebia archidona** [807] has a wingspan of about 95 mm. The underside of the wings are tobacco-brown, and yellow-brown at the edges, with a dark band running diagonally across the wings, resembling the principal vein of a leaf. Shining silver specks complete the impression of a leaf damp with drops of dew. The upper side of this rare American leaf butterfly is rust-brown, with a dark venation and a pattern round the edges of the wings. It is an inhabitant of Columbia and Peru.

Cyrestis thyodamas formosana [808] lives on the island of Taiwan. It has the appearance of a cracked stone. It has a span of 45 mm and translucent, whitish wings with a mother-of-pearl gloss, with ochre-brown bands at the edges. The fine pattern of lines on the wings is also ochreous brown to black. The caterpillars feed on various species of fig.

807

808

809

The American nymphalid **Historis orion** [809] has a very wide range, and is present in extremely large numbers. It is distributed from Mexico southwards to Argentina right across the width of the continent, including the large offshore islands. It has a span of up to 11 cm and is dark brown with two white dots on the fore wings, with a large rust-yellow area in the middle. Its underside has the appearance of a dry, dark-brown leaf. The butterfly is in the habit of alighting on garbage in the neighbourhood of human habitations, can also be found drinking from the puddles on roads and paths after rain, and is particularly fond of sucking the juice of various fruits on the tree. Among the most highly prized of all butterflies are the nymphalids of the South American genus *Agrias*. The upper surface of their wings make a magnificent show with large fields, expressionately patterned in colours of red, blue, yellow and green. On the underneath of the hind wings are complex, extraordinarily colourful patterns. The males have a striking little tuft of hairs on the upper side of each of their hind wings; these are connected with the organs of smell on the sides of the abdomen. **Agrias claudia** [810] has a span of about 80 mm. The large patch on its fore wings is fiery-red, with a blue, brown and black field around the outside. The hind wings are bordered with radiant blue and black. The underneath [811] is gaily coloured in different shades of blue, red, brown, ash-grey and black. The butterfly is a jewel even among its like.

810

811

812

813

814

The small American nymphalids of the genus *Callicore* have wings which are generally velvety-black on the upper side, with golden-green or blue bands. The underneath of the silver-coloured hind wing is marked with black shapes, which often delineate shapes which can be read as Arabic numerals. Most common is eight, but nine, six and nought can also be distinguished. The underneath of the fore wings is generally coloured shining red, with yellow lines at the corners between black bands. The genus *Callicore* has its range chiefly in the mountainous areas of South America between Mexico and Argentina.

The species **Callicore clymena** [812] has a span of up to 40 mm. It has a large red field on its fore wings with a glossy blue band running across them. The underside of the wings [813] shows, clearly legible, the figure 88.

The nymphalids of the genus *Cethosia* are medium-sized and coloured rust-red on top. They have deeply serrated wings. This is particularly striking on the underneath, where the indentations are emphasised by an even deeper, light-coloured line on a dark background, running zig-zag round the edges. The colouring of the underneath of the wings is black and ochre, finely marked with narrow red and blue stripes. The males are as a rule rust-coloured, the females dark brown or sometimes bluish. The caterpillars normally live communally on the **Passion Flower,** *Passiflora.* They are brown or black, with a yellowish or brownish ring-shaped marking. They have two little horns on their heads.

The species **Cethosia gabinia** [814; female] is an inhabitant of Nias Island, near Sumatra. It has a span of 75 mm. The upper surface of its wings is brown with white, or on the hind wings bluish, markings. The roots of the wings are a more intense blue, decorated with black and ochre-yellow. The male is rust-coloured. These butterflies fly gliding over the highest tree tops, alighting in clearings in the jungle to suck the juices of tropical flowers. They are related to the **admirals,** *Limenitidae.*

815

816

The **Admirals,** *Limenitidae,* and **Emperors,** *Apatura*, form a group within the *Nymphalidae* of comparatively large and powerful butterflies, generally of darkish colouring. The admirals have one or several rows of white, shining dots, which are sometimes joined to form bands. The wings of the emperors are bright with all the colours of the rainbow. Their undersides are generally brightly coloured, but nevertheless the colouring is normally so arranged that it resembles a colourful flat surface, and the butterflies are hard to spot in the rest position with folded wings. They are remarkably strong and inexhaustible fliers, and often rise up above the tops of the trees. In the early morning they can often be found drinking the drops of dew from the forest floor. They gather around any place where the sap of a tree is flowing out. The caterpillars are as a rule of an inconspicuous colouring, and have various different growths on their bodies. The majority of species are found in the tropics of the Old and New Worlds, although some very attractive species live also in countries further to the north.

Adolias dirtea javana [815] has a span measurement of 90 mm and is dark brown with yellow spots. These spots also extend on to the body of the butterfly.

Parthenos sylvia sulana [816] has a span of almost 10 cm. Its colouring is brown or brown-green with white markings; the roots of the fore wings are yellow, blue and violet. It is a good flier, which glides along high above the forests without moving its wings. Its range of distribution consists of the island of Sula between Celebes and New Guinea. The species **Hypolimnas missippus** is a striking example of sexual dimorphism. The females resemble certain of the *Danaidae*—a family of butterflies whose members have the advantage of being unpalatable to insectivores. We have here another instance of a defenceless creature being protected by its resemblance to a better protected species. The caterpillars live on plants of the purslane family, *Portulaceae*, and on stinging nettles, *Urticaceae.* They are common throughout the whole Indo-Australasian region, as also in Africa and North and South America.

The male **Hypolimnas missippus** [817] has a measurement in span of about 60 mm. It is dark brown; the blue mirrors on its wings have a violet-blue sheen. It flies very swiftly and with considerable endurance. The female [818] is rust-brown; the point and the leading edge of each of its fore wings are dark brown and decorated with a white pattern. The outer edges of both pairs of wings are dark brown with a doubled row of white markings. This, however, corresponds exactly to the appearance of the species **Danaus chrysippus,** which can only be distinguished from the female of Hypolimnas by its somewhat heavier flight.

817

818

A most attractive nymphalid related to the emperors is the Japanese **Sasakia charonda,** a national emblem in Japan. The male [819] is smaller than the female, measuring approximately 90 mm in span. It is dark brown with yellow and white spots. A spot in the lower corner of the hind wings is crimson-red, and also shows through on the underside of the wings. On top the wings have a wonderful blue iridescence. The female measures some 11 cm, but her colours are not iridescent. The male rests as a rule in one chosen place on a high tree, and drives away all intruders in the form of insects, and even birds. The butterfly is fond of the sap of oaks and chestnuts and can be found sucking at any wound in the bark where the sap is exuded. It drives all other insects away from its source of nourishment, even those which possess a large sting. In July this butterfly is extremely common throughout Japan.

819

820

821

822

The European **Purple Emperor,** *Apatura iris* [820], measures about 60 mm in span; the females are slightly larger. The wings are red-brown with white markings and small red-brown eye-spots. The males have a bright violet-blue iridescence like that of some tropical butterflies. The undersides also of their wings [822] are reminiscent, in their gay colouring and multiplicity of patterns, of those of tropical butterflies. It is grey-brown underneath, with rust-red bands. On the fore wings there is a large eye-spot with an iridescent blue-red "pupil", with black and rust-coloured borders. On both wings there are white flecks and bands, with a mother-of-pearl sheen. The small eye-spots on the hind wings show through on both sides. The green, horned caterpillar feeds on sallow and other species of willow, and occasionally also on aspen. It hibernates on the buds and among the twigs of the sallow, and pupates in the spring. The chrysalis [821] is green and greatly flattened at the sides. The butterfly inhabits damp places on the eaves of forests and woods, on meadows, and in mountain valleys. In the summer they can be found in forest rides on horse dung and a variety of other rotting animal matter. They are comparatively rare in Europe and Asia.

823

The caterpillar of the emperor has a very unusual appearance. Its head appears to have a face. This impression is strengthened by the extremely lively movement from side to side, as if looking around. The two black spots look like eyes, and the horns give a grotesque effect. The caterpillar of the **Small Emperor,** *Apatura ilia* [823], is green, decorated with yellow points and red-yellow stripes.
The caterpillars of the emperors have still other remarkable features. They possess, for example, a quite extraordinary memory. They creep out from their resting places on chosen shoots which have been spun round with a web to form a nest; from there they make their way to their nearby feeding grounds on sallow, aspen, or more rarely poplar. When they are replete they make their way back to the nest from which they started out, to take their rest. In the rest position they resemble certain slugs in shape [824]. The butterfly is somewhat smaller than the purple emperor, and coloured a deep shade of rust-brown. The male has a red-violet iridescence. This butterfly, like the purple emperor, is a Eurasian species.
In central Europe only a few species of the *Limenitidae* are found. They love to bask in the sun. One of them is the **White Admiral,** *Limenitis camilla*. It is dark brown in colour with a white strip across both wings [826]. Its chrysalis [825] is brown-green and ornamented with golden-yellow. On its head it bears two horn-like growths. Its caterpillars feed on honeysuckle, *Lonicera*.

824

825

826

827

828

829

The related species **Neptis rivularis** has dark brown wings with white markings and stripes. The underneath is coloured rust-brown with opalescent marks and bands. Illustration [827] an adult butterfly which has just emerged from the cocoon, which is usually golden-yellow in colour. The caterpillar lives on *Spiraea salicifolia*, at the edge of streams. The butterfly is found in the warmer parts of south-eastern Europe and the adjoining parts of Asia. The most beautiful and largest European species is the **Large Admiral,** *Limenitis populi*. The caterpillar [829] has a series of pinecone-like growths behind its head. It feeds on aspen leaves, and hibernates in a little nest, about 10 mm across, called a hibernaculum, which it spins on twigs of aspen. It pupates in the spring on the upper side of an aspen leaf. The chrysalis [828] is a glossy pale ochre in colour, ornamented with black markings. The pouch on the back of the chrysalis contains a brown-yellow fluid which helps the butterfly to emerge.

83

The female **Large Admiral** [830] measures 75 mm in span, the male 65 mm. The wings are dark brown with a blue-green gloss, the wing-markings white; on the outer edges are found orange half-moons. The undersides have a rust-brown colouring with glossy blue-grey markings. The upper surface of the wings of the male is a less expressive white in colour.

The group of nymphalids which we call the **fritillaries** consists of butterflies with a usually rust-coloured upper surface to the wings and silver markings or bands on the underside. The **Queen of Spain,** *Argynnis (= Issoria) lathonia* [831], measures approximately 45 mm in span. Its upper surface is rust-coloured with black flecks, while the underside of the hind wings is ornamented with large, shining silver marks. The caterpillar, which hibernates in the winter, feeds on plants of the violet family, *Violaceae,* certain plants of the borage family, *Boraginaceae,* and blackberry, *Rubus.* This is a migratory butterfly, which is present in dry places in west, central and southern Europe, as well as the adjoining areas of Asia.

The **Japanese fritillary,** *Argyreus hyperbius* [832], has a span of about 70 mm. It is an ochreous brown-yellow; the fore wings are ornamented with white corners and blue markings. The undersides have a rich mother-of-pearl

832

831

833

gloss and are gay with a multitude of patterns and colours.

Another, smaller group of fritillaries lacks the silver gloss on the underside of the wings.

The **Heath Fritillary,** *Melitaea athalia* [833], has a span of some 35 mm. On the upper surface it is red-brown in colour with a dark pattern; on the underside it is somewhat lighter. It alights in forest clearings and rides. Its caterpillar, which hibernates, feeds on the leaves of cow-wheat, foxglove and woodsage.

The family of *Danaidae* comprises principally

834

butterflies which are residents of the tropics and subtropics. They are protected against insectivorous birds and reptiles by an unsavoury, stinging excretion, which prevents their pursuers from satiating their appetite on them. We have already mentioned more than once that their appearance is imitated by numerous other, otherwise defenceless species. The caterpillars of the *Danaidae* feed mainly on plants of the swallow-wort family.

The migratory American **Monarch** or **Milkweed Butterfly,** *Danaus plexipus* [834], has a span of 85 mm. It is honey-yellow with dark brown venation and borders, which are punctuated with white dots. In the autumn it travels southwards to the subtropics and tropics; in spring it returns in whole swarms, flying as far north as Canada. It is a strong flier of great endurance, which has often been observed far out at sea. Its occasional presence in the Canary Islands, in England, Greece, and even Australia, may perhaps be attributed to the fact that individual specimens have made their own way there; on the other hand, it is also maintained that it has in fact been imported artificially.

835

836

Among the *Danaidae* there are some butterflies which themselves imitate other, likewise unpalatable species. Thus, **Lycorea halia** [835], which comes from southern Brazil, copies in form and colouring the similarly protected *Heliconidae* (page 446). The colouring alone of these poisonous or unpalatable species serves as a danger signal sufficient to frighten off any potential pursuer which has once made their acquaintance. For this reason the unpleasant taste of a butterfly coloured orange with black bands and yellow spots is enough in itself to protect a species having similar colouring from becoming a tasty meal for some insectivore. Some of the *Danaidae* have a much simpler colouring. **Danaus limniace** [836, female] measures 80 mm in span and is dark brown and yellow-green in colour. This species is common in Kashmir, southern China and on Taiwan. An even simpler colouring is that of **Hestia leuconoe clara** [837], likewise from Taiwan. This butterfly is yellow-white in colour with black markings. It measures about 12 cm in span.

The family of **Swallowtails,** *Papilionidae,* is distributed throughout the world. Among the tropical species of swallowtails are numbered the largest and most gaily coloured of all butterflies. The genus of **Apollo Butterflies,** *Parnassius,* inhabits in a large number of species and subspecies primarily the warmer areas of Europe, northern and central Asia, and also western North America. They are medium-sized butterflies with translucent, generally whitish wings, which are ornamented with black or grey flecks and red or yellow eye-spots. Their caterpillars feed on various crassulaceous plants, especially orpine, *Sedum.* The most beautiful and rarest species live high in the inaccessible mountains of central Asia.

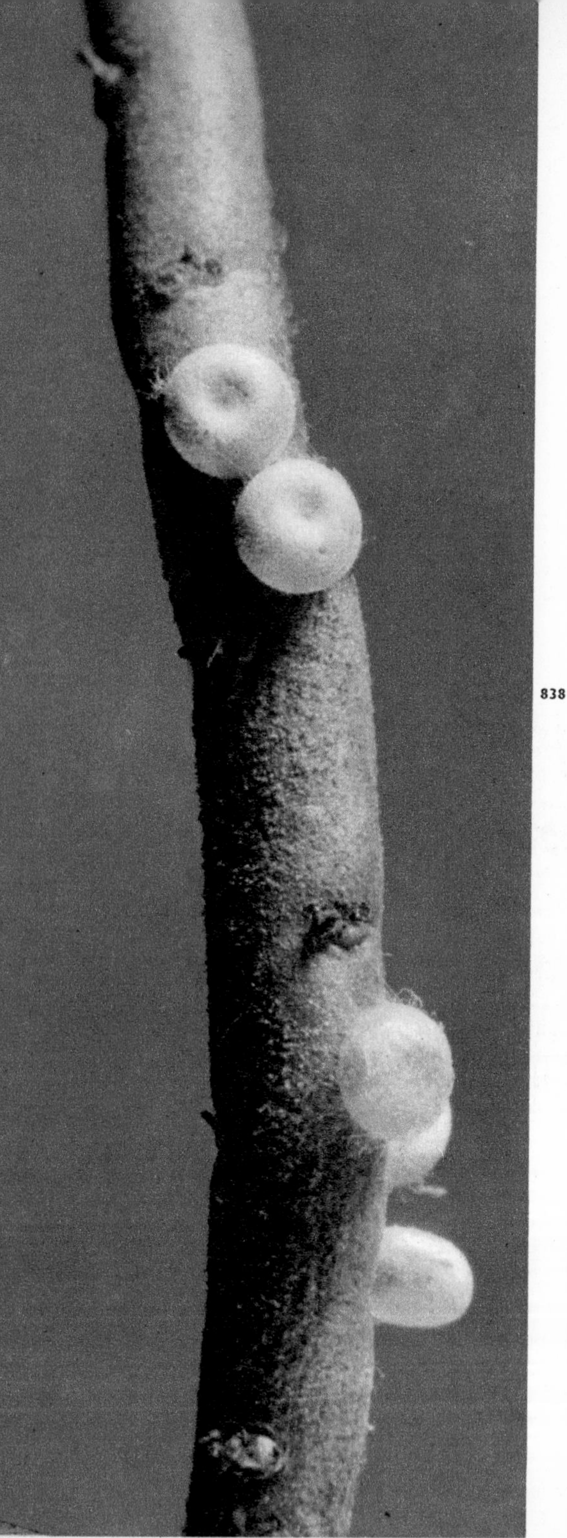

838

839

The most familiar European species is the **Apollo,** *Parnassius apollo* [840, and colour plate XLVb]. It has a span of about 80 mm and is present mainly on chalky hillsides and the higher parts of Europe and Asia. The female lays her whitish eggs, which are about 1.5 mm in diameter, on orpine [838]. The caterpillar develops but generally remains within them during the winter, and only starts to feed in the spring. The fully grown caterpillar [839] is black, with red spots and a fine coat of hairs. It pupates underground. The dark brown chrysalis [841] lies in a tight, hard cocoon, which is thickly covered with a whitish wax dust. The Apollo is protected by nature conservation laws in central Europe.

840

841

The genus *Zerynthia* is known in three species and several varieties ranging throughout the Mediterranean countries to Iran. The caterpillar feeds exclusively on the **Birthwort,** *Aristolochia;* the butterfly is therefore restricted to those few places where this plant grows. Thus the **Birthwort Butterfly,** *Zerynthia polyxena (= hypsipyle)* [844], is found only in the warmer parts of south-east Europe. It has a span of about 55 mm. Its wings are a light yellow colour with a black pattern; along the outer edge of the hind wings there is a row of red marks [colour plate XLIV]. The caterpillar is red-yellow in colour, and has thorny growths. Illustration [842] shows the caterpillar attached to the food plant shortly before pupation. The chrysalis [843] is an inconspicuous light grey and brown, with dark markings. Sometimes the caterpillar draws a neighbouring leaf over the spot with the aid of a silk thread before they pupate, in order to hide themselves under it when they are inside

843

842

844

their cocoons. In the river basins of central Europe this species is threatened with extinction due to the use of insecticides.
The most attractive of the butterflies related to Zerynthia is **Bhutanitis (= Armandia) lidderdalei** [colour plate XLVa]. It has a span of about 10 cm. The caterpillars are restricted to the climbing varieties of birthwort. The butterflies live in the eastern Himalayas, in India, Bhutan, Burma and China.

845

From south-east China comes a small swallowtail with extremely elongated tips to the rear wings: **Leptocircus curius walkeri** [845]. It has a span of only 30 mm. Its black wings have large, completely transparent windows in them.

The marvellously beautiful **Teinopalpus imperialis** [846, female] measures over 80 mm in span. The female is more beautiful than the male—coloured dark green, grey violet and yellow. The butterfly lives in the mountains of south-east China.

846

XLVa *Armandia (Bhutanitis) lidderdalei* lives in the forests of the mountains of Bhutan, of the Himalayas, and Burma.

XLVb The Apollo Butterfly, *Parnassius apollo ssp. carpathicus,* is distributed throughout central Europe as far as the Altai Mountains. Span 8 cm.

XLVIa The caterpillar of the Swallowtail. Central Europe. Length about 3 cm.

XLVIb Bird Butterflies, *Troides (Ornithoptera) croesus,* lives on certain islands e.g. Batjan between Celebes and New Guinea.

847

Some of the most beautiful butterflies in the world belong to the giant **Bird Butterflies** of the genus *Troides* (= *Ornithoptera*). The gaily coloured **Troides paradisea** [847, male] has a span of more than 12 cm. It has velvet-black wings with golden-green, glossy panels and is found in New Guinea.
On Celebes lives the velvet-black and golden-yellow **Troides helena hephaestus** [848, male].

848

849

A butterfly from the Solomon Islands is **Troides victoriae regis** [849]. It measures 15 cm in span. Its wings are velvet-black and relieved by areas of green with a golden gloss. The female bird butterflies are in general larger, but not so brightly coloured. The female of this species [850] reaches a span width of upwards of 22 cm. It is dark brown with light brown or, at the base of the wings, yellow markings.

Troides brookiana [851, male] has a span measurement of 13 cm. It is velvet-black with a scarlet collar; the lighter areas on the wings are a beautiful golden green. On Sumatra, Borneo and the Malay Peninsula this butterfly can frequently be found on damp paths as well

850

851

852

as on rubbish in the neighbourhood of human habitation.

The hind wings of the swallowtails generally extend into decorative tail-like appendages. The Japanese species **Papilio alcinous** [853, male] is dark brown with red marks on the undersides of the hind wings. The female [854] is grey-brown, and so are the marks on the edges of the wings. The caterpillar has a glossy grey-black colouring with a white saddle-like strip; this colouring makes it resemble bird droppings. It feeds on birthwort. In Europe it can be reared successfully on the upright birthwort, *Aristolochia clematis*. The cater-

853

854

pillars flourish on this and will pupate. The chrysalis has a complicated ornamentation on its back [852].

Papilio hesperus [855] measures 12 cm in span. It is black, and the band and the markings on the wings are light green. Its range is from the Congo northwards throughout West Africa.

855

856

857

The Malayan species **Papilio coon** [856] is distinguished by the remarkable form of its wings. It has a span of 11.5 cm. The grey-brown fore wings are narrow and extremely long; the hind wings have tails with large, flattened appendages. They are black, with light grey-brown markings, and possess two yellow marks on the rear end. This swallowtail flies erratically over forest trees on plains, and sips at their flowers.

The South American **Papilio phosphorus gratianus** [857, female] has a span of 80 mm. Its wings are black with small white semi-circles on the outer edges. The fore wings are ornamented with a largish green patch, the hind wings with a group of three beautiful crimson marks with an eye-catching blue iridescence. Such bright colouring is only found otherwise in the Colibris. This butterfly lives in Colombia and eastern Peru. Many tropical American swallowtails do not have noticeably developed tails on their wings. On the inside edge of the hind wings of the males are situated highly developed organs of smell.

Papilio leucaspis [858]

has a peculiar colour scheme. The inner field of both pairs of wings is deep yellow and becomes united when the butterfly spreads its wings into a large pair of triangular shapes. The fore wings have a brown border with yellow spots on the leading edge, the outer edges are hemmed with black and light brown patterns. On the hind wings there is a black area at the back with semicircular markings of bluish scales and a red stripe across it. The black tails have a yellow tip. This butterfly is extremely common on the eastern slopes of the Andes, in Ecuador, Peru and Bolivia.

A marvellously beautiful, large and noble butterfly is **Papilio philoxenus termesus** [859, female]. It has a span of approximately 12 cm. The head and underside of the body are red; the fore wings are grey-black, the hind wings black with a large whitish mark and a series of red spots on the short tails. The underside has the same colouring as the upper—a rare occurrence in the swallowtails.

The hind wings are unusually long, with deep, blunt indentations and, when extended, wider than the fore wings.

858

859

860

In the primeval forests of tropical Africa lives the largest of the African papilionids, which resembles the Indo-australasian Ornithoptera in form and colouring, except that it does not have their gloss. It is **Drurya (= Papilio) antimachus** [860, male], with a span of 23 cm. Its wings are red-brown, brown-black and black in colour; the three spots on the fore

861

wings are whitish.
It resembles the swallows in the way it sails over the treetops of the primeval forests on its long, narrow wings. Its habitat lies in Sierra Leone, the Cameroons and the Congo.

Papilio androcles [861] has a span of over 80 mm. It is dark brown and white, with greenish-yellow roots to the wings.

862

863

Papilio (= Graphium) sarpedon has a span of exactly 60 mm. On its dark brown wings it has a row of light green spots stretching from the tips downwards; both the last ones are greenish-blue. On the hind wings, the first patch is colourless, the others are blue-green, as are the semi-circles on the edge of the hind wings. The underneath is lighter in colouring, but otherwise resembles the top, except that the greenish spots have a mother-of-pearl iridescence, and there are red marks on the hind wings which are particularly striking in the female. Illustration [862] shows a recently emerged female. The yellow-brown chrysalis [863] has a pointed thorny growth on its back. On the top chrysalis the pattern of the wings can already be seen showing through the outer case, which means that the adult butterfly is on the point of emerging. This butterfly is common in the forests and gardens of Japan. The male can often be found sucking plants and moisture along paths and riverbanks.

864

865

One of the best-known European butterflies is the **Swallowtail,** *Papilio machaon* [864]. It measures some 70 mm in span. The wings are yellow with black bordering and markings. On the darker bands of the fore wings are yellow scales which look like dusts; on those of the hind wings the scales are blue. Just above the tails of the hind wings is situated a large red, blue, and black-bordered eye-spot. The underside is similarly coloured, but lighter. The caterpillars feed on umbelliferous plants, especially carrots and the **Burnet Saxifrage,** *Pimpinella saxifraga*. The butterfly pollinates clover flowers and settles frequently on warm stones in summer along the sides of fields. The caterpillars are green-yellow with black and orange markings. When they are disturbed they poke out an orange-coloured, two-pointed fork with a penetrating smell from behind the head. On illustration [865] are caterpillars on the **Wild Carrot,** *Daucus carota*. The butterfly is distributed throughout almost the whole

866

867

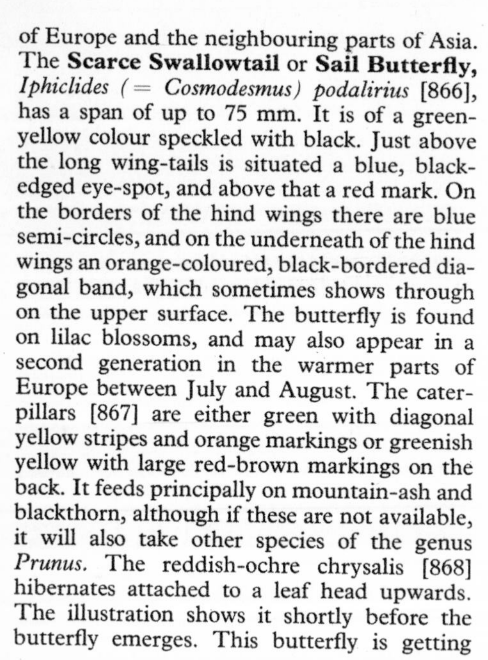

of Europe and the neighbouring parts of Asia. The **Scarce Swallowtail** or **Sail Butterfly,** *Iphiclides (= Cosmodesmus) podalirius* [866], has a span of up to 75 mm. It is of a green-yellow colour speckled with black. Just above the long wing-tails is situated a blue, black-edged eye-spot, and above that a red mark. On the borders of the hind wings there are blue semi-circles, and on the underneath of the hind wings an orange-coloured, black-bordered diagonal band, which sometimes shows through on the upper surface. The butterfly is found on lilac blossoms, and may also appear in a second generation in the warmer parts of Europe between July and August. The caterpillars [867] are either green with diagonal yellow stripes and orange markings or greenish yellow with large red-brown markings on the back. It feeds principally on mountain-ash and blackthorn, although if these are not available, it will also take other species of the genus *Prunus.* The reddish-ochre chrysalis [868] hibernates attached to a leaf head upwards. The illustration shows it shortly before the butterfly emerges. This butterfly is getting rarer and rarer in central Europe, because the senseless destruction of the blackthorn has destroyed its natural biotope. In civilised countries it is protected by law. It is a central and southern European species, also present in the nearer parts of Asia. The southern European **Swallowtail,** *Papilio alexanor* [869], has a span of 70 mm. It resembles *P. machaon,* except that instead of the spots it has diagonal stripes. Its basic colour is deep yellow, the venation of the wings also being yellow. Its caterpillars live on umbrelliferous plants. The butterfly is present in certain areas on the nothern and eastern shores of the Mediterranean, and in Asia.

Illustrations [870] and [871] show a typical caterpillar and chrysalis of the **whites** family, *Pieridae.* The caterpillar [870] is of **Colias australis** from central Europe. It is light green with yellow bands and black spots. It feeds on **Crown Vetch,** *Coronilla varia,* and **Horse-shoe Vetch,** *Hippocrepis.* The butterfly is shown on colour plate XLVIIIa. The chrysalis [871] is that of the **Brimstone Butterfly,** *Gonepteryx rhamni.* It is green and flattened.

868

869

870

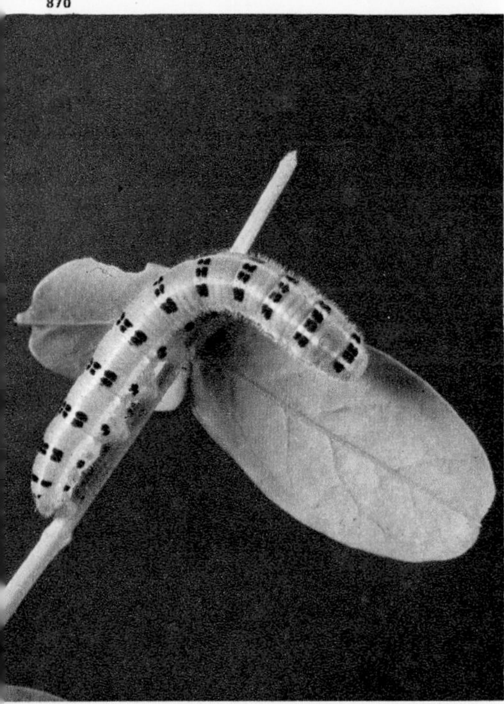

871

872

873

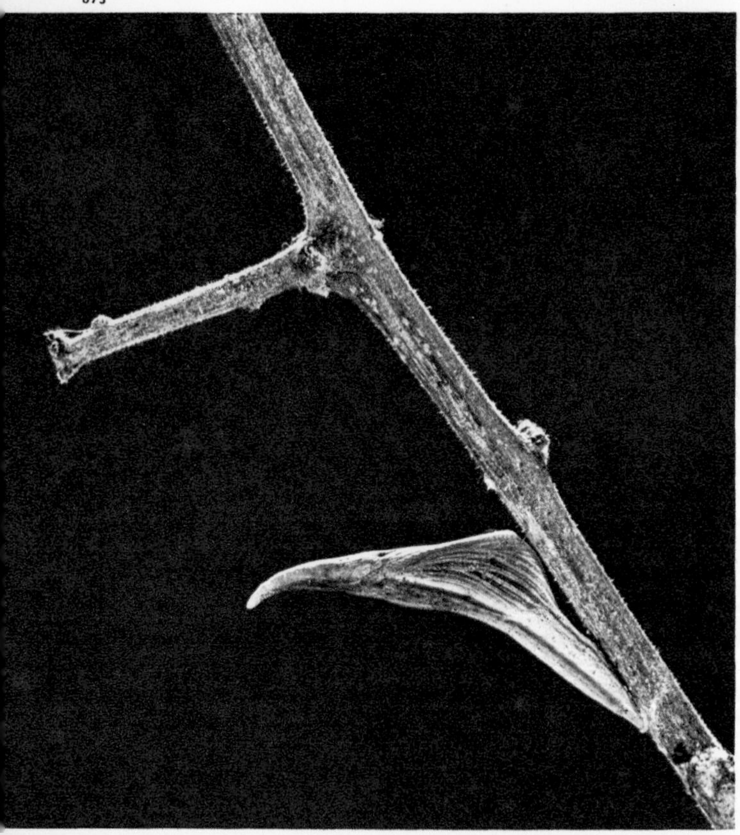

The **Sulphur** or **Brimstone** is one of the best-known butterflies of central Europe. The male is sulphur-yellow, the female a yellowish-white. The caterpillars feed on the black alder, *Rhamnus*.

The **whites** family, *Pieridae*, comprises approximately 1,500 species of butterflies distributed throughout the world. Among them are numbered the most familiar and very commonest of all butterflies. They are nearly all light in colour, exceptions being found chiefly among the tropical species. Some species are mimetic, imitating other butterflies which are exceptionally well protected in their colour, so closely that it is sometimes difficult to determine which family a particular specimen actually belongs to. However, they have one characteristic which is unique among butterflies: the pigments of the scales on the body and wings of these creatures are derivatives of uric acid, so that they can be easily and reliably identified by chemical tests, by which means membership of the pierid family can be simply proved or disproved. The butterflies are mostly small or medium-sized, with the exception of a small number of the tropical *Coliadinae*. Almost all are distinguished by sexual dimorphism in colour and size.

The subfamily **Coliadinae** contains some comparatively large, extremely beautiful tropical species.

From Taiwan comes **Hebomoia glaucippe formosana** [872, male]. It has a span of about 85 mm and has yellow-white wings with black spots. The triangular black-bordered areas on the fore wings are bright orange-red. On the underneath the butterfly has the colouring and markings of a yellowed, dry leaf.

A similarly brightly coloured butterfly, of the subfamily of **true whites,** is the **Orange Tip,** *Anthocharis cardamines,* with a span of about 40 mm. The outer third of the white fore-wings is a bright orange colour in the males, whitish in the females with a dark grey pattern. Both are marbled on the underside with green [874, female]. The caterpillar is green and lives principally on the **Cuckoo Flower Hedge Garlic** and other wild *Cruciferae.* It pupates in a remarkable, horned chrysalis [873]. The butterfly may be encountered during the spring on willows. It is a Eurasian species, as also is the **Black-veined White,** *Aporia crataegi* [875]. It has a span of some 60 mm, is white with a black venation in the wings, and is found in low-lying areas. The caterpillars live communally and hibernate in a spun nest or web. They feed on fruit trees, blackthorn, dog-rose and other plants. This species does multiply in vast numbers and can threaten fruit trees when it does. It is a migratory butterfly with remarkable wing endurance.

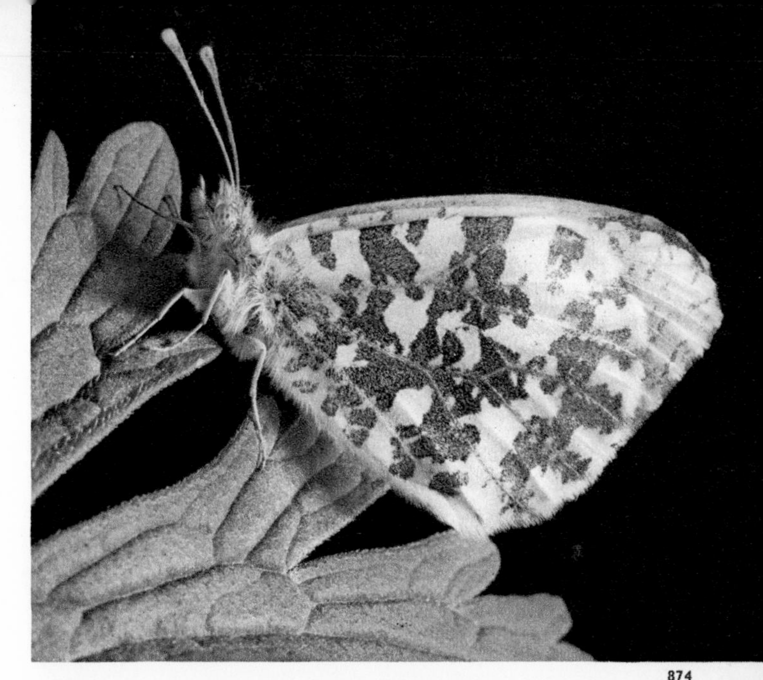

874

875

876

877

The **Cabbage White,** *Pieris brassicae,* has a span of approximately 60 mm; the males are slightly smaller. The upper surface of their wings is white with black points on the fore wings and a similar mark on the rear wings. The female [877] has moreover two spots on the upper surface of the fore wings, together with a wedge-shaped mark. Illustration [876] shows a brood of eggs attached to the underside of a cabbage leaf (enlarged). The yellow-black caterpillars are familiar pests of vegetable plants of the genus *Brassica.* On illustration [878] are shown (also enlarged) the chrysalides of the cabbage white. The range extends throughout Europe, the whole palaearctic region, and North Africa.
Colour plate [XLVIIIb] shows one of the very gaily coloured pierids, the Australian species **Delias aganippe.** It has a span of some 70 mm. The illustration shows it from underneath.

XLVIIa Underside of a male Swallowtail, *Papilio horishanus*. Span 9.5 cm. Flies in July in eastern China and on Taiwan, in areas more than 2,000 metres above sea level.

XLVIIb *Troides (Ornithoptera) priamus* (male) from New Guinea and the neighbouring islands north of Australia. Span about 15 cm.

XLVIIIa *Colias australis* lives on dry slopes in central Europe. Span 4 cm.

XLVIIIb The White, *Delias aganippe,* Female, from Southern Australia. Underside. Span 6 cm.

879

880

The seventh order belonging to the superorder of **nerve-winged insects,** *Neuropteroidea,* consists of the **two-winged flies,** *Diptera.* As far as numbers of species is concerned this order stands fourth in the insect kingdom: it has been estimated that there are upwards of 60,000 species. They are small to medium-sized, and their characteristic feature is that only one pair of wings has developed. The other is transformed into balancers (halteres), which serve as gyroscopic organs, controlling the insect in flight. They are by and large diurnal creatures; only the gnats and craneflies also fly at night. The *Diptera* feed mainly on plant juices and other sweet liquids. Many species, however, are parasitic, and live on the blood of mammals. Among these there are several well-known and feared carriers of various diseases. The larvae are legless and feed either on plant and animal remains or are also parasites. Many of them pupate in what is called a puparium—a barrel-shaped structure, from which the adult insect eventually emerges. Some gnat pupae live in water and possess the power of movement, while other groups bear a certain resemblance to butterfly chrysalides.

The first suborder, the **Gnats,** *Nematocera,* consists of two-winged insects with threadlike antennae which are as long as or larger than the head. They are particularly well developed

881

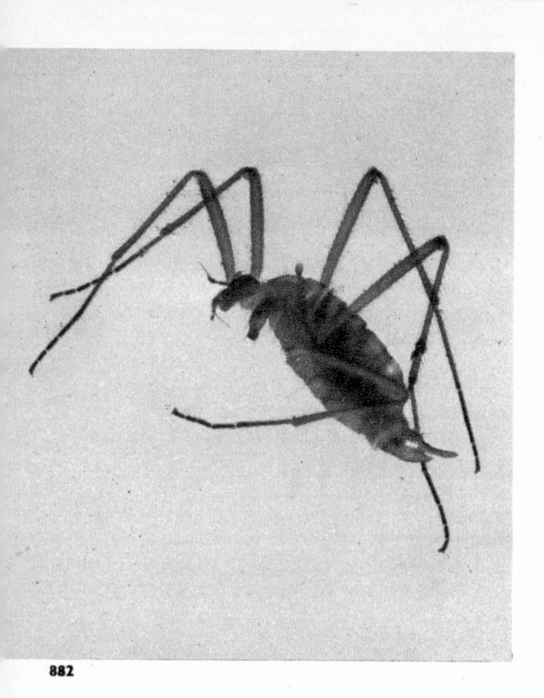

882

883

in the males. The pupae are free-living—that is, they are not enclosed in barrel-like cases.

The family of **Craneflies** or **Daddy-long-legs,** *Tipulidae,* contains first of all the **Great Cranefly,** *Tipula oleracea.* The female [879] is up to 23 mm long and coloured yellow-grey. The larvae feed on the roots of plants, and in times of mass reproduction they form a serious threat to the farmer, attacking fields, meadows and gardens [880, right a larva, left a pupa]. The larval stage spends the winter in hibernation underground. The cranefly is a resident of all of the temperate parts of Europe and the nearer parts of Asia.

A member of the subfamily of **Crested Craneflies,** *Flabelliferinae,* is **Tanyptera atrata** [881, male]. The first segments of the abdomen are red in the female; the males have crested or combed antennae. The larvae feed in compost, especially in rotting oak stumps. It is a harmless forest species, found in Europe and western Asia.

Chionea hrabei [882] is a member of the family *Limnobiidae.* It is about 5 mm long. It is pinkish-grey, wingless, and can only crawl along very slowly. Its habitat lies in the snow-fields of the Carpathians and the High Tatra in Slovakia.

The family of **Gall-midges,** *Cecidomyidae/Itomididae,* consists of over 4,000 species of very small insects, some of which are so tiny

884

that one can hardly believe in their existence. Their larvae create plant galls.

The **Blackberry Gall-midge,** *Lasioptera rubi,* brings up spherical galls on blackberry and raspberry bushes [883].

The familiar **Gall-midge,** *Bayeria capitigena,* lives on cypress spurge, on the tips of which it causes round, red galls [884].

Illustration [885] shows the trunk of a willow tree which has recently been attacked by larvae of the **Willow Gall-midge,** *Helicomyia saliciperda.*

A member of the **hairfly** family, *Bibionidae,* is the **St. Mark's Hairfly,** *Bibio marci* [886, female], a black midge about 13 mm long. It may be found on garden fences in March. It develops out of dark-coloured larvae about 2 cm long, which live underground for two years. They cause considerable damage by feeding on the tender roots of cultivated plants and germinating seeds. The St. Mark's hairfly is present in the whole of Europe and western Asia.

885

886

887

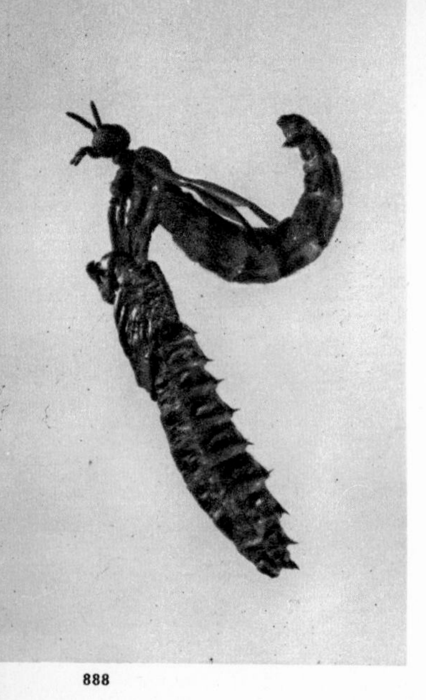

888

889

A species often found in flowering hedgerows of central Europe is the **Garden Hairfly,** *Bibio hortulanus*, [887]. The female has an orange body and brownish, transparent wings. The males are completely black, with a covering of whitish hairs. It measures about 13 mm. The larvae feed on plant roots in the same way as the last-mentioned species. The garden hairfly is present in North Africa as well as throughout Europe and western Asia.

A completely harmless member of the hairfly family is **Penthetria funebris.** It measures roughly 11 mm, and is coloured black with brown wings. The males often have reduced, very short wings. Illustration [888] shows an imago emerging from its pupa; illustration [889] shows a male with reduced wings. This hairfly lives on damp plant-growths. Its larvae develop in compost, in damp straw, and elsewhere. The wingless males run about on the ground.

The family of **Biting Midges,** *Ceratopogonidae*, consists of small two-winged flies, most of which bite and suck blood like true gnats. Their larvae are predatory, and develop in damp surroundings, as a rule in water. The larvae of the species **Probessia bicolor** [891] has at the end of the abdomen, near the breathing organs an array of lashes, with which it clings to the surface of the water.

The **True Midges** or **Dancing Midges,** *Chironomidae,* are mostly harmless. Their larvae develop in fresh water and form a major part of the diet of various fish. They

890

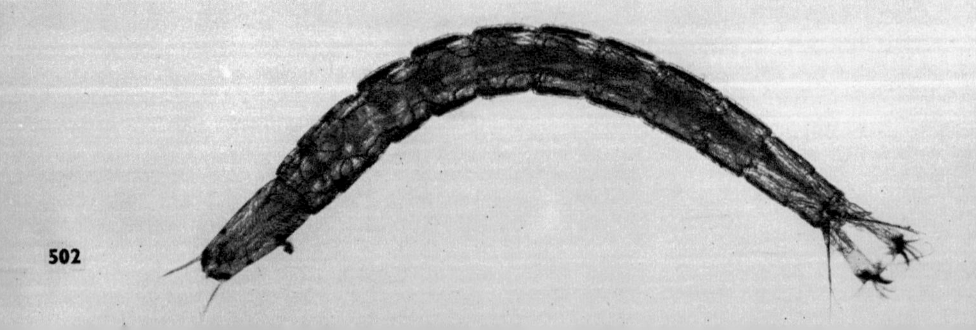

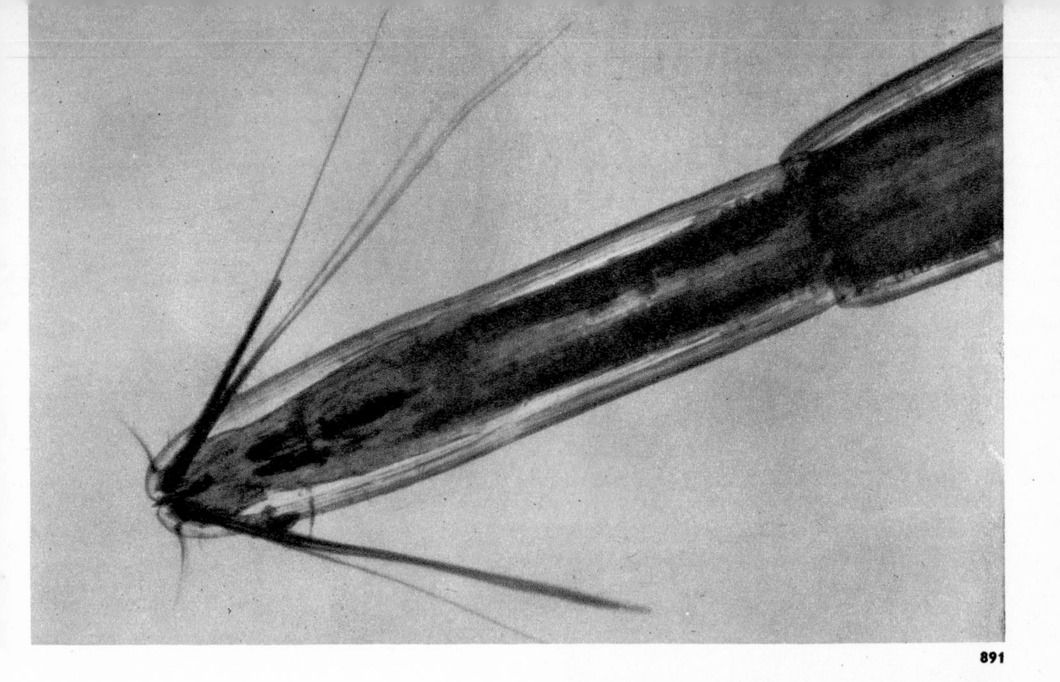

891

893

themselves feed on algae and other small parts of plant matter. The majority live on the bottom in little spun nests.
The larvae of the genus **Corynoneura** [890] measure only 3 to 5 mm, and live in ponds

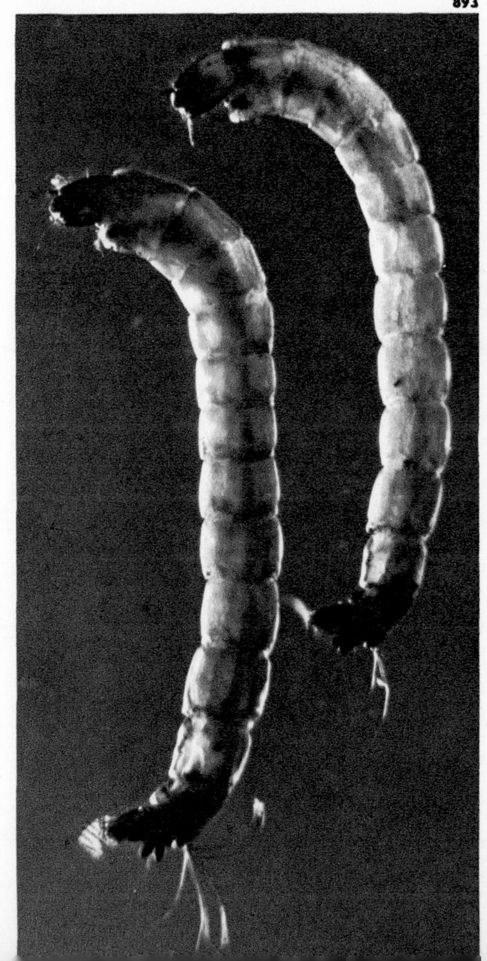

892

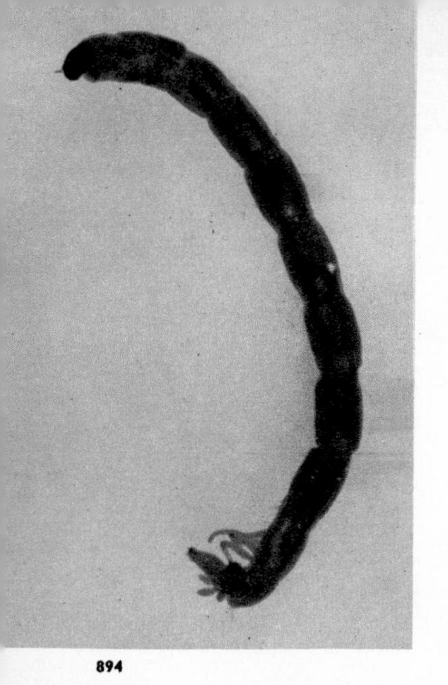
894

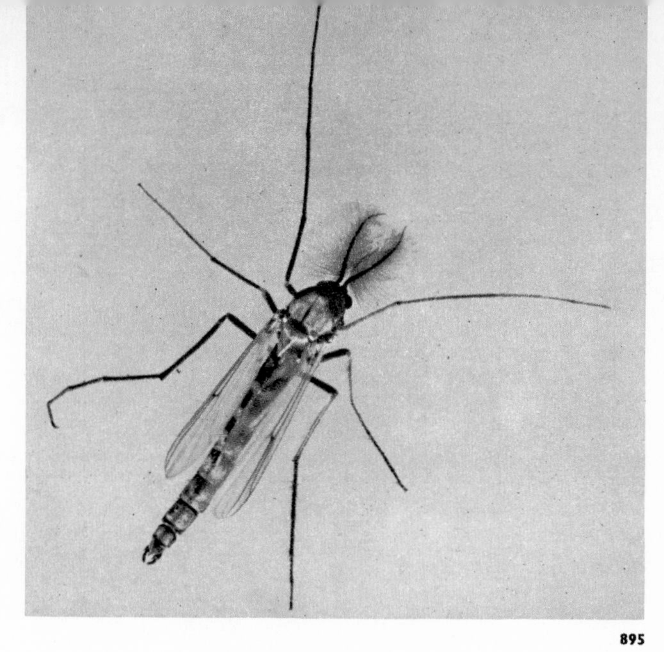
895

among the water plants. At the end of the abdomen there is a breathing tube and hooks with which it attaches itself to its mud-tube [892].

The yellow-green larvae of the genus **Cricotopus** [893] are about 5 mm long.

The best-known midge is the **Feather Midge,** *Chironomus plumosus,* the larvae of which [894] are red, about 15 mm long, and widely used as food for aquarium fishes. They live on the bottom of ponds in a little mud tunnel.

The male of **Chironomus plumosus** [895] measures about 10 mm, and has combed antennae. It is found almost everywhere in

896

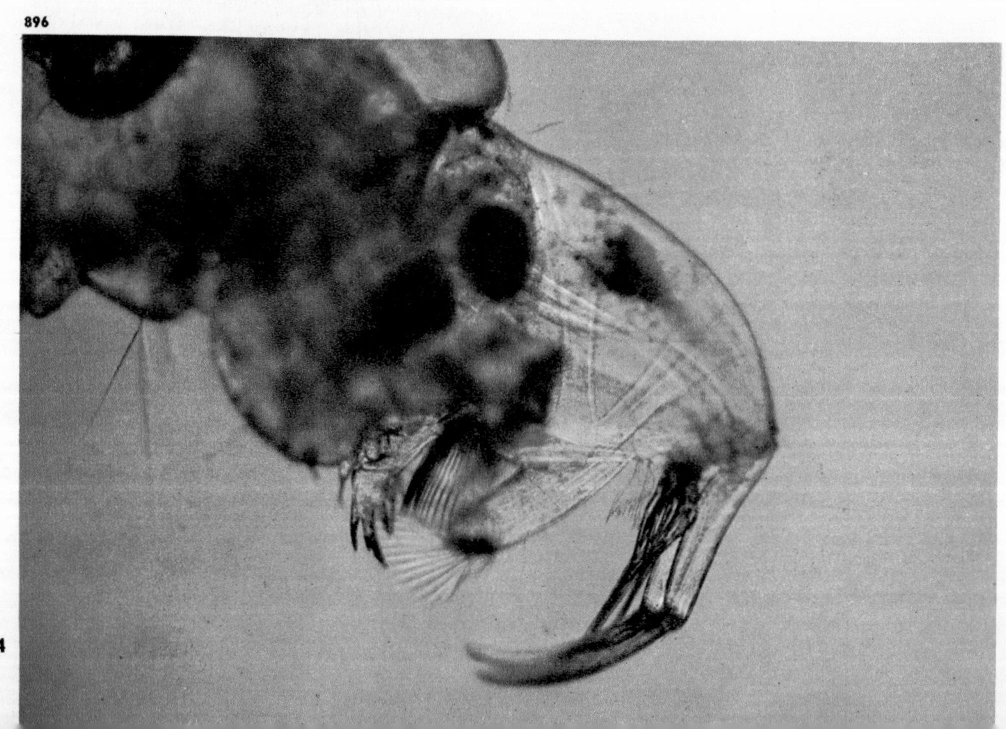

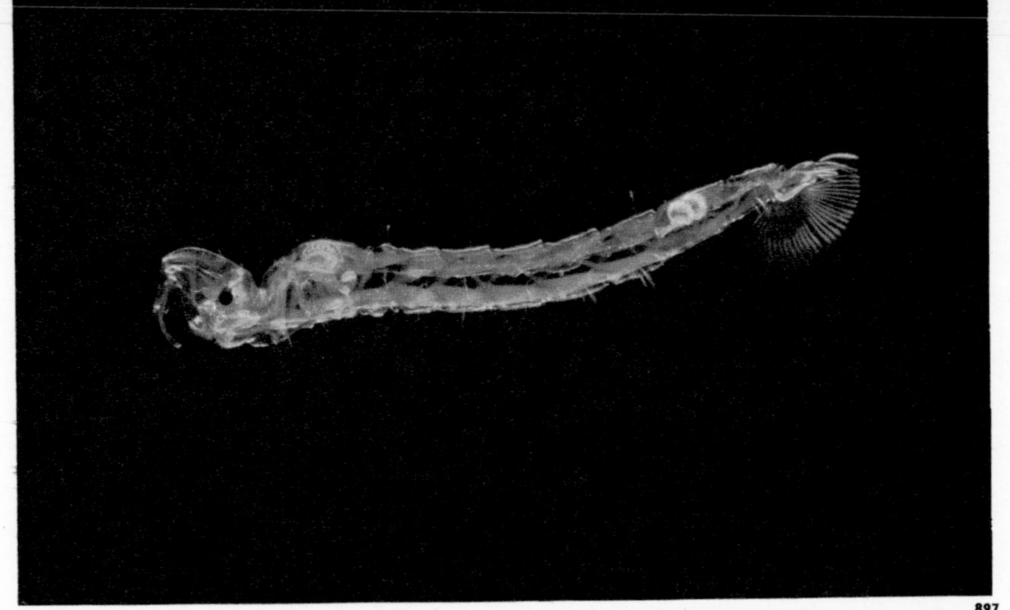

897

fresh water. Another favourite element in the diet of aquarium fish is the transparent larva of **Chaoborus plumicornis,** *Corethra*. These are known as **Phantom Larvae.**
However, they make high demands on the cleanliness of the water, and can still best be found in mountain lakes. They are predators, swimming freely about with whiplike movements and attacking prey as large as they are themselves. Illustration [896] shows the head, [897] the whole larva, about 11 mm long.
The larvae of the **Common Mosquito** [898] measure about 8 mm. They swim head downwards.

898

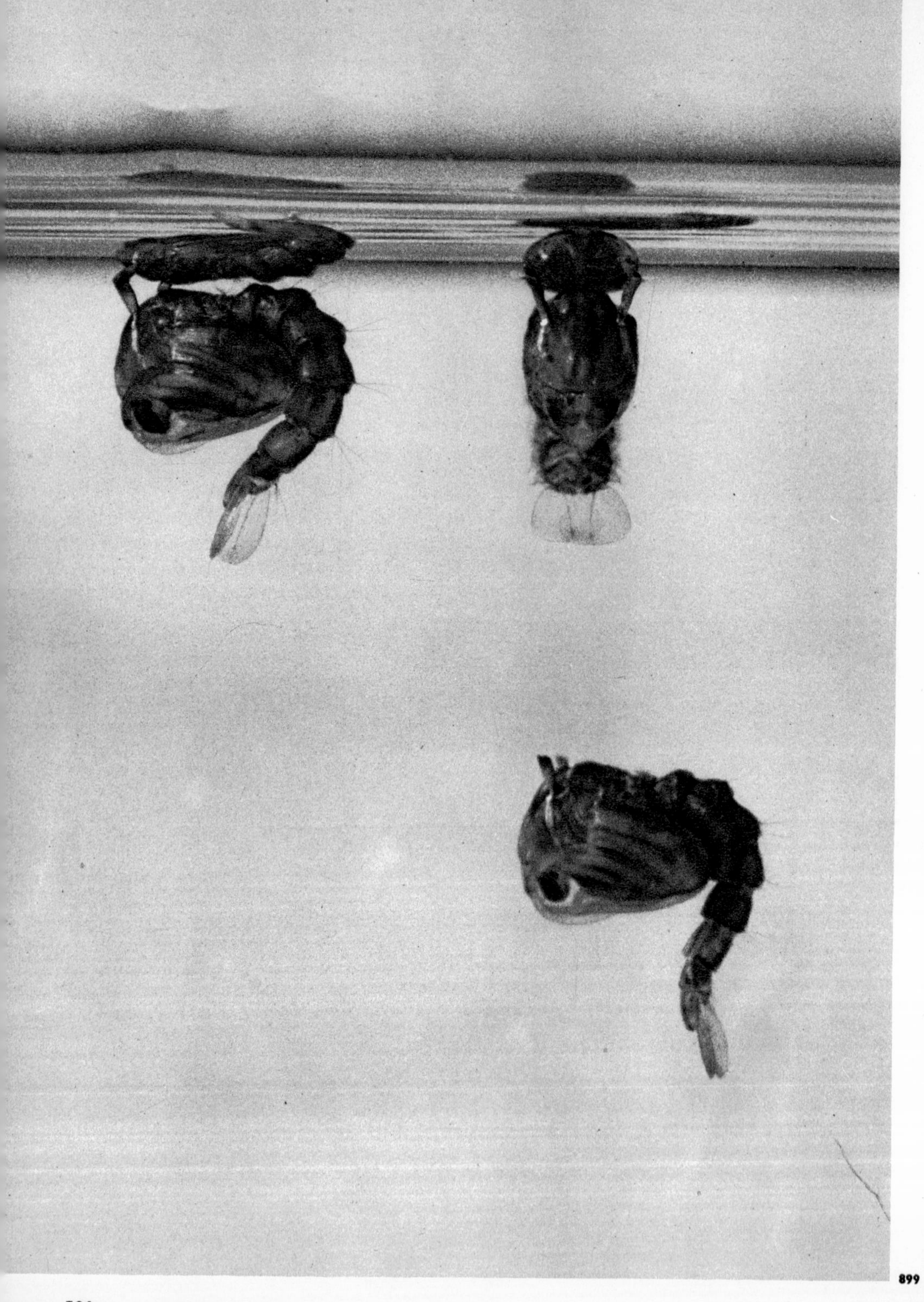

899

In west and central Europe the gnat larvae appear in early spring in still fresh waters. The females, who hibernate in caves, cellars and hollow trees, lay their eggs on the surface of the water. The larvae feed on plant and animal matter, and after a period of growth, metamorphose themselves into free-swimming pupae, which in contrast to the larvae swim with their heads upwards. They take in air through two breathing tubes, and swim with the aid of the raft-like tail [899].

The family of **Mosquitoes,** *Culicidae,* is distributed in more than 2,000 species throughout the whole world, even in the Arctic. The imagines are irritant, and in some cases even extremely dangerous bloodsuckers; in some species only the females bite, in others the males also. Besides the unpleasant bite inflicted, they also present a threat to health as carriers of diseases, of which malaria and yellow fever are the most dangerous, being spread by species of *Anopheles* and *Aedes* respectively. In Europe the most common and best-known mosquito is the **common Gnat** or **House Mosquito,** *Culex pipiens.* Illustration [900] shows a female, illustration [901] a male. This mosquito is found in a number of biological variations, which are sometimes regarded as species. This problem of taxonomy has not yet been resolved. One of them, for example, sucks only the blood of birds, and attacks no other creature.

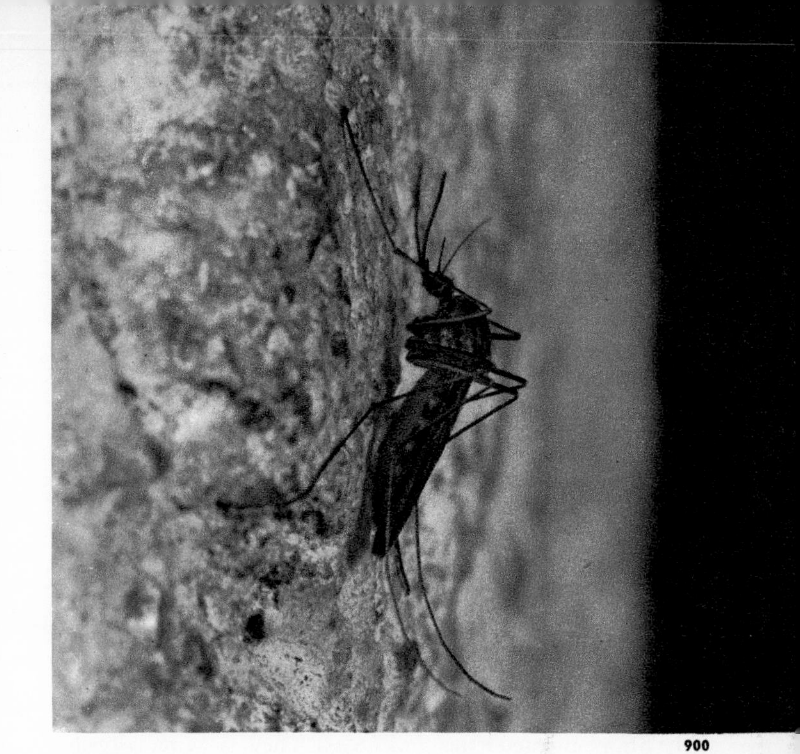

900

901

902

The second suborder of the *Diptera* consists of the **Short-horned Flies,** *Brachycera.* They have short, tripartite antennae; the pupae are usually encased in a barrel-shaped case. The larvae (maggots) are entirely legless.

The family of **Soldier Flies,** *Stratiomyiidae,* contains about 1,500 species of flies. One of the most attractive and, in central Europe, commonest is the **Chameleon Fly,** *Stratiomys chamaeleon* [902]. It measures some 15 mm, is yellow and black in colouring, and completely harmless. In summer it can be found on umbelliferous plants. On the shield and on the sides of the thorax it bears small, spur-like thorns.

The grey larvae of the **Soldier Fly** [903] are about 35 mm long. They develop in muddy and polluted water, but also outside the water

903

in moss, underground, or behind the bark of trees. It gets its nourishment from animal and vegetable refuse; they pupate in the last larval moult.

The family of **Horseflies** and **Clegs,** *Tabanidae,* has world-wide distribution in almost 2,500 species. The males feed on the juices of flowers and honeydew, whereas the females have a well-developed proboscis, with which they suck the blood of warm-blooded creatures, not excluding human beings.

The **Blind** or **Thunder Horsefly,** *Chrysops caecutiens* [904], has eyes with a golden-green glaze. It lives near water, and inflicts a very painful bite on human beings, especially active on warm days. It grows to a length of about 10 mm.

A similar species, also from central Europe, is **Chrysops relictus** [905]. It is about 11 mm long, and lives in the same manner, in the same areas, as the blind horsefly.

904

905

906

The largest European horsefly is the **Common Gadfly,** *Tabanus bovinus* [906]. It measures about 25 mm, and is coloured black and yellowish brown. The female sucks the blood of horses and cattle, but, unlike other members of this family, seldom attacks man. Its larvae lead a predatory existence underground.

Some horseflies are carriers (vectors) of infectious diseases of animals. Thus, for example, in Egypt a horsefly of the genus *Tabanus* carries a fatal disease of horses known as surra; in central America another species is responsible for the transmission of another such disease, derrangadera. Horseflies also carry splenetic fever and tularaemia.

A very unpleasant pest

907

908

even for humans is the **Cleg,** *Haematopota pluvialis* [907]. It is about 13 mm long, grey, and has speckled wings. It flies especially frequently on sultry days or in light drizzle. Its bite is remarkably painful.

The family of **Robber Flies,** *Asilidae,* comprises over 400 species of predatory flies which are distributed in the hot and temperate zones of every continent. They hunt their prey—other insects—on the wing, and catch the victims between their fore legs, sucking out their juices with their sharp, hard proboscis.

The **Robber Fly,** *Laphria flava* [908, male], measures almost 25 mm. It is black with a yellow-brown abdomen, and a coat of hair. It hunts in forest clearings in west and central Europe.

The **Robber Fly,** *Laphria gibbosa* [909, pair], grows to a length of almost 30 mm. It has a thick coat of black and grey-yellow hairs, and is likewise an inhabitant of central Europe. It too hunts in forest clearings.

909

910

911

912

Another predatory fly of central Europe is **Choarades gilva** [910, male]. It measures 20 mm, is black, and coloured golden-yellow on the upper surface of the abdomen. The larvae [911] are a peculiar shape, are about 25 mm long, and live in the burrows of wood-eating (xylophagous) insects in coniferous bark. After a period of roughly three years they develop in loam or compost into pupae, which have hard horns on their heads [912].

The striking hornet-like **Assassin Fly** or **Robber Fly,** *Asilus crabroniformis* [913], is about 25 mm long and black and yellow in colouring with speckled wings. It sweeps low over fields of stubble or meadows, hunting its prey—which consists mainly of small dung beetles of the genus *Aphodius*, which it seizes, carries off to a hiding place, and sucks out their juices.

The **Bee-fly** family, *Bombyliidae,* contains over 2,000 species of flies, most of which are brightly coloured and have a metallic gloss. The most common and most striking is the west and central European **Large Bee-fly,** *Bombylius major* [914]. It measures some 12 to 16 mm; its proboscis is straight and almost as long as the rest of its body, which is covered with a thick coat of brownish-yellow hairs. The wings are dark-brown for the front half, transparent at the back. Humming like the hawk-moths, it hovers in front of the blooms sucking the nectar out of them. It only settles in order to rest. The larvae live parasitically in the nests of wild bees and other *Hymenoptera*.

914

913

915

916

The family of **Hoverflies,** *Syrphidae,* contains about 4,000 species with cosmopolitan distribution. They are small to medium-sized, gaily coloured, and perform an important function as pollinators of flowers due to the way in which they visit the blooms. Some of them may be observed on any summer's day "hanging" in the air—they are enabled to remain poised in the same place by the extremely rapid vibration of their wings. The larvae of the hoverflies develop in a number of different environments—in faeces, in humus, behind the bark of trees, but also parasitically in the nests of honeybees and bumblebees. They are omnivorous and feed on various plant and animal remains. Many species are predators, and are welcomed because they destroy aphids.

Lasiopticus pyrastri [915] measures upwards of 13 mm. This species whitish yellow and black in colour, is very useful in that its larvae feed on plant lice. The larvae are up to 15 mm long, grey or greenish in colour, with a dark pattern [916].

Helophilus trivittatus [917] grows to a length of almost 20 mm, is black and yellow in colouring and in summer feeds on umbelliferous plants. The larvae develop in damp humus.

917

919

918

The familiar **Syrphus ribesii** [918] measures about 12 mm, and is black and yellow. It pollinates in the spring the blossoms of fruit trees. The larvae free redcurrants, drapaceous fruit trees and various vegetables from aphids. It has been observed that one hoverfly destroys daily up to 100 aphids. It alights in the middle of a colony and sucks the juices from one after another. Illustration [919] shows a larva of **Syrphys balteatus** among some **Plant Lice,** *Aphis fabae,* on a **Nasturtium,** *Tropaeolum maius.*

Some hoverflies of the genus *Volucella* strongly resemble bumblebees, as for example the **Bumblebee Hoverfly,** *Volucella bombylans* [920]. It is almost 15 mm long, and coloured black with a rust-coloured tip to the end of the abdomen. In summer it is commonly observed on flowers. The larvae live parasitically in the nests of bumblebees.

The genus *Eristalis* contains, among other species, the familiar, widely distributed **Dronefly,** *Eristalis tenax.* The breathing tubes of its larvae are encased in a long caudal appendage. It is found in latrines. Somewhat smaller —about 13 mm long—is the **Dronefly,**

Eristalis arbustorum [921]. It is black and yellow-brown in colouring. As an imago it flies round the blooms in summer like the other members of this family, although its larvae too have their development in latrines.

920

921

922

The family of **Thick-headed Flies,** *Conopidae,* comprises about 500 species of small to medium-sized insects which resemble certain wasps or bees in the shape of their bodies. They live on flowers, and lay their eggs in bees, wasps, or grasshoppers. The larvae develop parasitically within the abdomen of their victims. The thick-headed flies of the genus **Physocephala** [922] are about 14 mm long. They are present in west and central Europe. The **Fruit-fly** family, *Trypetidae,* contains some 1,500 species of small, often very gaily and beautifully coloured flies. They generally feed on blossom and blooms. The larvae develop either on the blossoms of various plants and trees—for example, the *Compositae*—where they hibernate as pupae, or in some other part of the plant, often in the flesh of the fruit. The **Cherry fly,** *Rhagoletis cerasi* [924], grows to a length of some 4 mm, and is black and yellow with speckled wings. Because of the fact that they lay one egg on each of the stalks of young cherries in May-June, it is a very

923

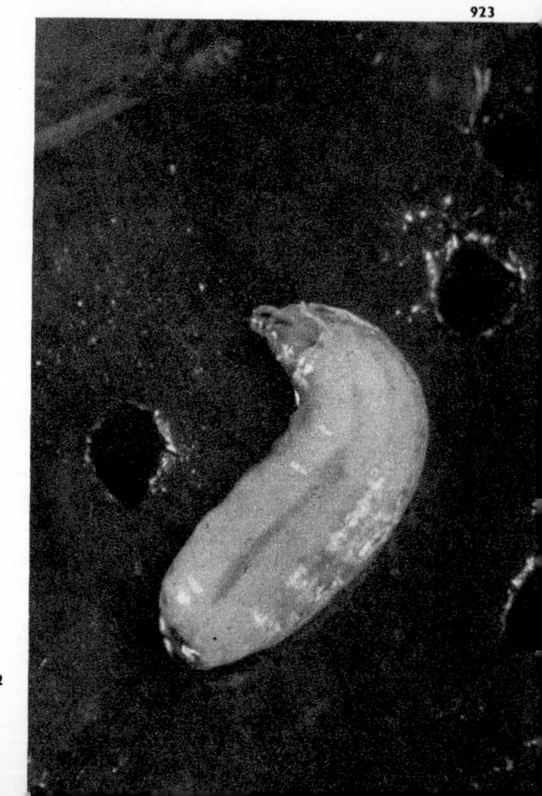

924

familiar pest, which in some years can cause almost 100 per cent ruin of the crop. Illustration [923] is an enlarged photograph of one of the larvae on a ripe cherry. The pupae are about 4 mm long, light yellow [925], and they hibernate just beneath the surface of the earth in the vicinity of cherry trees.

The larvae of the family of **Leaf-miner flies,** *Agromyzidae,* eat tunnels between the upper and lower surfaces of leaves, leaving a characteristic pattern visible on the leaf. In this way the larva of the **Leaf-miner fly,** *Phytomyza lappina,* tunnels a way through the leaves of the burdock [926].

925

926

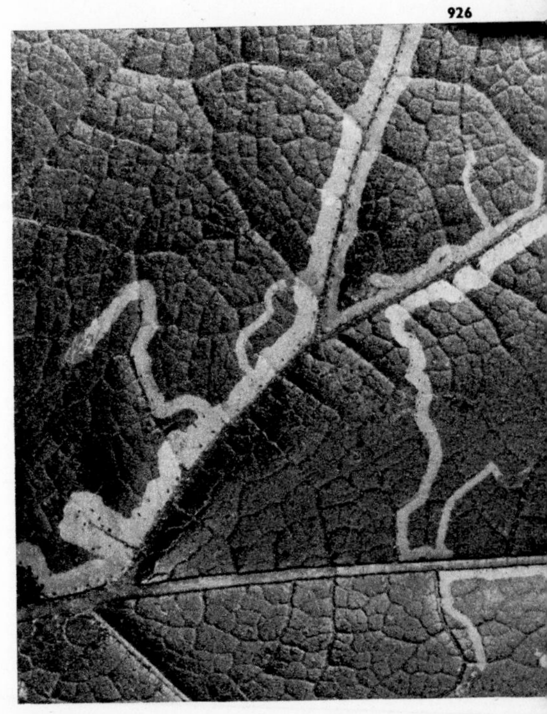

927

928

The best-known member of the family *Braulidae* is the cosmopolitan **Bee-louse,** *Braula caeca* [927, 928]. It measures about 1.5 mm, is rust-brown, and wingless. With its powerful, hooked legs it clings fast to the skin of the bee, which carries the parasite with it—even on its flights away from the hive. It feeds on the same nourishment as the bee, and lays its eggs within the hive. The larvae likewise feed there on honey. The bee-louse is only really a noticeable pest when it starts to multiply *en masse.* Two well-known species of this family (one of them from Africa) are placed by some authorities, with certain justification, in the related family of **Hunch-backed Flies,** *Phoridae,* of which some species infest the nests of termites and ants in a similar manner.

The group of flies known as **Keds** or **Louse-flies,** *Pupipara,* consist of parasites which in the adult stage feed on the blood of various different mammals and birds. The **Forest Fly,** *Hippobosca equina* [929], which belongs to the family *Hippoboscidae,* is parasitic on horses. It measures up to 8 mm, is flecked with brown and yellow, always has wings, and sucks the blood principally of horses, but on occasion also of cattle, and sometimes by some aberration, of man and dogs. It can be found in large numbers around horses, especially on the thin skin under the tail. It can be found in the warmer regions of Europe, but is more common in the warmer parts of Asia.

The **Deer Ked,** *Lipoptera cervi* [931], measures about 6 mm, is grey-brown in colouring, and bears wings during the summer. It attacks deer, roes, and other ruminants, attaching itself to them, shedding its wings, and getting inextricably entangled in the pelt of the host. There it can be found throughout the year, even in winter, both in the adult form and as pupae, from which colourful adult insects will emerge in the following spring. The winged specimens [930] frequently attack man in summer and autumn, aiming mainly at the hair of the head.

A similar species is the **Sheep Ked,** *Melophagus ovinus,* which infests the hides of sheep, and is wingless throughout its life. Originally a Eurasian species, it is now distributed throughout the world.

929

930

931

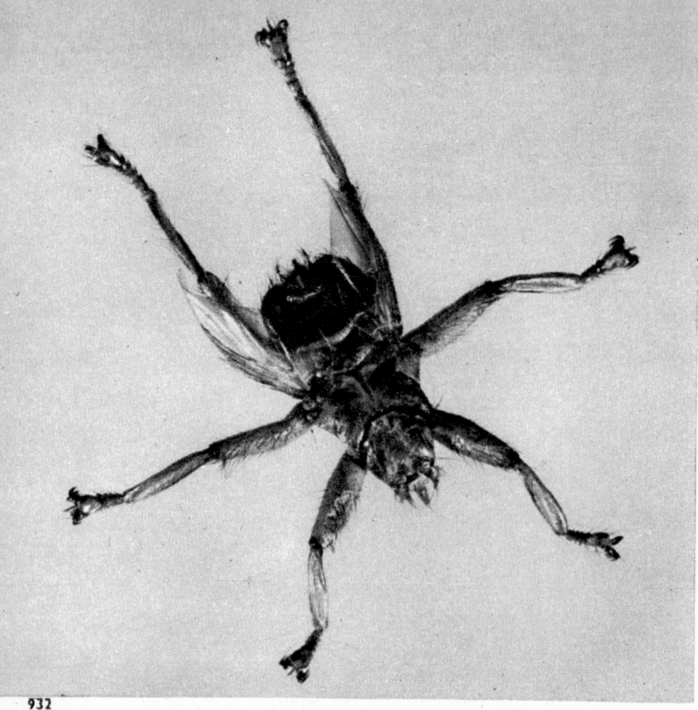

932

933

Many species of keds infest the feathers of birds. They are very swift fliers, and move equally quickly over the bird's body, sucking the blood. If threatened they fly away, to return when the danger is past. The **Ked,** *Crataerhina pallida* [932], is parasitic on swallows. Its wings are only short, that is to say atrophied; it attacks its victims mainly in the nest, and so weakens them through sucking their blood that the birds are sometimes unable to continue flying and fall to the ground. As the swallows leave their nests they scatter a number of keds, which sometimes attack man. We show a female shortly after she has laid her eggs. On illustration [934] we see a female with the fully grown larva in her body. The latter develops in the body of the mother until it is ready to pupate immediately after being laid.

A ked which inhabits the nests of house martins under the eaves of houses is *Ornithomyia biloba* [933]. It measures about 6 mm and attacks particularly the young fledgelings in the nest, sometimes in large numbers. The specimens shown are females with larvae in their bodies.

The **Bat-fly** family, *Nycteribiidae,* contains about 50 species of wingless, pupae-bearing (pupiparous) insects, which infest the pelt of bats. These they move about with great agility, rather like spiders, so that it is extremely difficult to catch them. They are coloured rust-brown. *Penicillidia dufouri* [935] is about 5 mm long. It is found in winter on sleeping bats, especially on those of the genus *Myotis*.

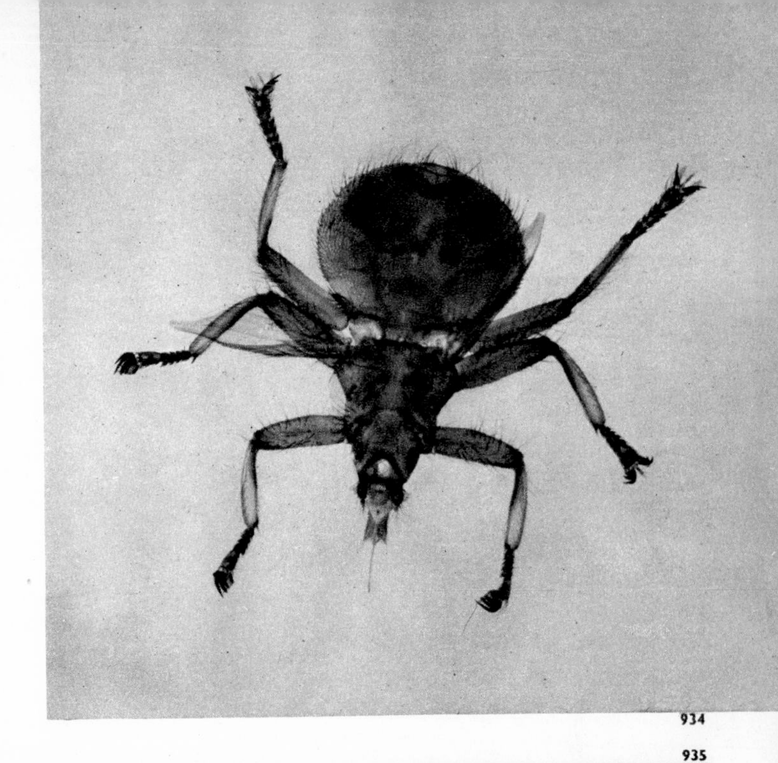

934

935

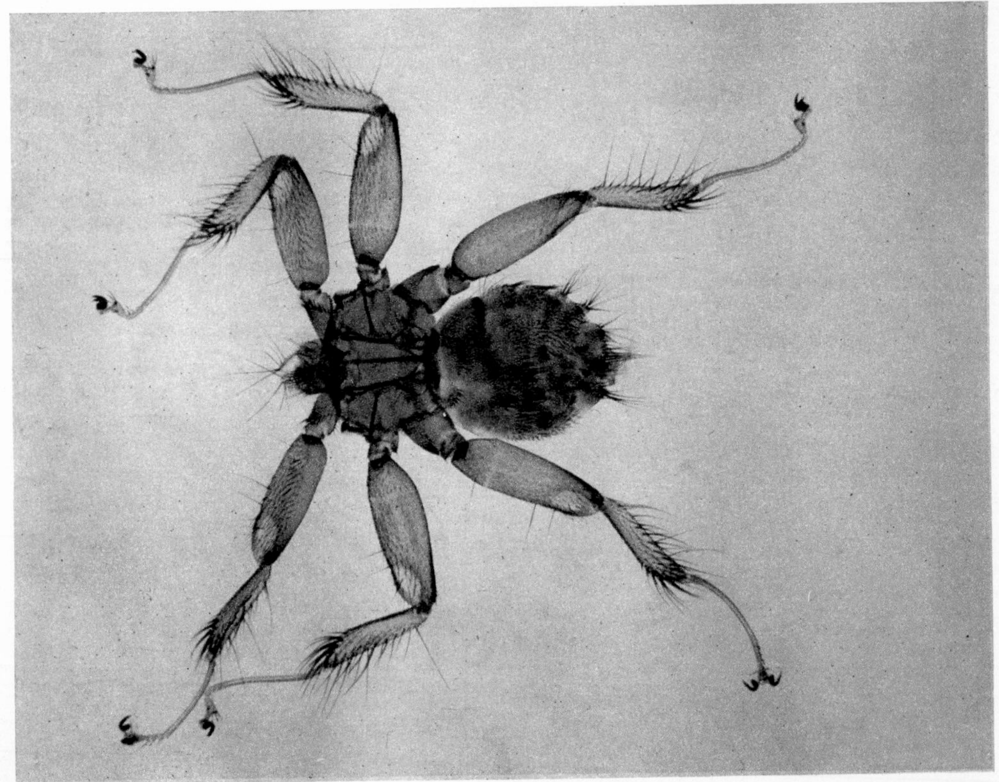

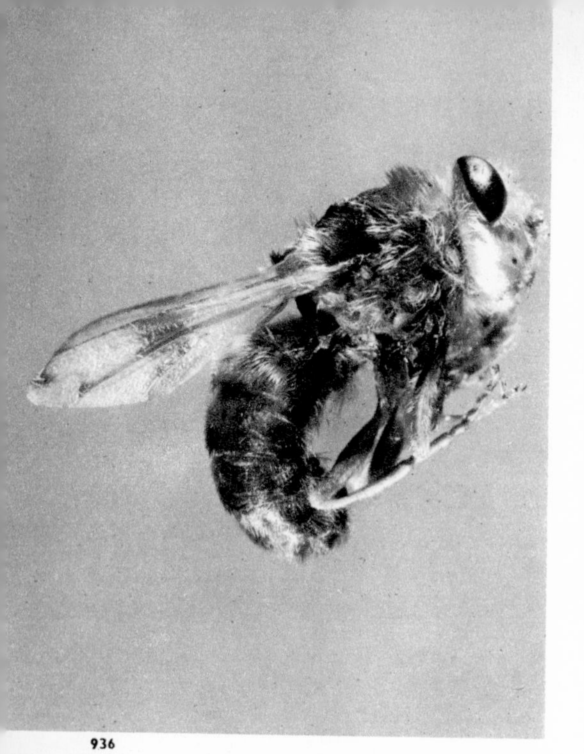

936

937

938

939

940

941

One of the most common members of the **Bot Fly** family, *Gastrophilidae,* is the **Horse Bot Fly,** *Gastrophilus intestinalis* [936, 937]. It measures up to 16 mm, has a brownish coat of hair, red eyes, and speckled wings. The rear segment of the thorax is dark; the end of the abdomen pointed. The mouthparts are atrophied, for the adult insect takes no nourishment. The females attach their eggs to the fore legs and breast of horses, donkeys, and mules. The larvae get into the intestines through the mouth of the animal, which licks them off its legs. Then they establish themselves there in the mucous membrane of the stomach and the duodenum, where they cause suppurating abscesses, in which they live parasitically for 10 months. Then they are ready for the next metamorphosis, leave the host with its faeces, pupate underground, and after about 40 days emerge as flies. The horse bot fly has a cosmopolitan distribution.

The most dangerous **biting flies** of the family *Muscidae* are the African members of the genus *Glossina.* The **Tsetse Fly,** *Glossina morsitans* [938], measures about 12 mm. It carries the disease known as nagana, which affects nearly all domestic animals and many related wild animals. The males as well as the females suck blood. A similar species, the **Sleeping Sickness fly,** *Glossina palpalis,* transmits the dreaded sleeping sickness among humans.

In the temperate zones, a common species is the **stable fly,** otherwise known as the **Biting Horse Fly,** *Stomoxys calcitrans* [939]. It is about 7 mm long, and speckled grey and black. It is distinguished from the housefly principally by the forward-pointing direction of its piercing proboscis. It attacks horses, cattle and mankind, not only by biting and blood-sucking, but also by transmitting various diseases. Its larvae have their development in dung. The distribution is cosmopolitan.

A regular guest in human habitations during the summer is the **Lesser Housefly,** *Fannia canicularis* [940]. It flies tirelessly, circling round the light-fixtures on the ceiling. It is somewhat smaller and slimmer than the common housefly, with a similar colouring, but is not so persistent. Its larvae, which are set with bristles and spines, live on decaying matter and in lavatories. The barrel-like pupae resemble the larvae, except that they are slightly larger and have no power of independent movement. In illustration [941] the largest creature shown is a pupa; the others are larvae.

942

944

943

The **Common Housefly,** *Musca domestica* [942], is up to 10 mm long, and coloured grey-black. It does not bite or suck blood, but manages nevertheless to make a sufficient nuisance of itself through its extreme persistence, and is dangerous to man in that it settles not only on human foodstuffs but also on carrion, dung, and so on. It can transmit a number of diseases which are caused by bacteria, viruses, or parasites. They reproduce in vast quantities: a single pair of flies could bring forth many billions of offspring, were it not for their enemies. The larvae feed chiefly on vegetable, animal, and decaying matter. They are up to 15 mm long [943], while the barrel-like puparia [944] are about 8 mm. They are found in rubbish heaps, and every kind of rotting substance, even underground. The common housefly is a cosmopolitan species, and can be found everywhere where there are human habitations.

The **Blow Fly** or **Bluebottle family,** *Calliphoridae,* contains among others the **Greenbottle Flies** of the genus *Lucilia*. They are coloured golden-green or golden-blue, and may be found on carrion, dung, or rotting meat. They prefer to lay their eggs on living creatures, or in festering wounds on animals or humans. The larvae burrow their way into the living flesh and feed on it (myiasis).

The **Forest Blow Fly** or **Toad Greenbottle,** *Lucilia silvarum* [946], is up to 9 mm long, and has a coppery and green gloss. The female lays her eggs in the eye-rims and nostrils of toads and frogs. The larvae consume within a few days the entire eyes and eye-parts of the living host, blinding it, although not until the larvae have eaten their way up to the brain does the tormented creature actually die. The fully grown larvae bury themselves in the ground beneath the dead toad, and pupate.

Some similar flies of related genera cause myiasis when their eggs are, by chance, swallowed with the food of various hosts. The tropics boast a number of species which cause dangerous diseases in this way in human beings.

The larvae of the **Caterpillar Flies,** *Tachinidae,* are likewise parasites. They are distributed in several thousand species throughout the world, but especially in the temperate zones. Generally, they lead the life of internal parasites in various insects: butterflies, beetles, bees, wasps and ants, and grasshoppers. The flies live on flowers; their maggots develop in the larvae of these insects, gradually hollowing out the interior of their still-living host. Since some species have come to specialise only in some injurious pests as hosts, not only are they highly regarded by foresters and agricultural economists, but they have even been deliberately employed in the fight against insect pests, with successful results. However, most of them have polyphagous larvae, which usually choose to live as parasites in the caterpillars of a wide variety of butterflies. It is certainly true that, together with the ichneumons and gall wasps, the caterpillar flies are responsible for the gradual decline and even disappearance of many of our species of harmless, gaily coloured butterflies, which not so long ago were quite common. So the culture of caterpillar flies brings not only benefits, but can also do damage.

The **Hedgehog Fly,** *Echinomyia fera* [945], measures some 16 mm, and has a striking hedgehog-like appearance. It also preys upon lepidopterous larvae.

945

946

947

948

One member of the **Warble Fly** family, *Oestridae,* is the **Deer Bot,** *Cephenomyia stimulator* [947]. It measures approximately 15 mm, lives in forests, and its larvae develop in the gullet of wild deer. It has a thick coat, and resembles a yellow-and-brown bumblebee. It has a black stripe running diagonally across its back. The female introduces its tiny larvae into the nostrils of the host, where they consume the tissue and the mucous membrane. Later, they move further up the nostrils and down the gullet of the host. The creature, which is tormented day and night by the larvae in its nose and throat, snorts and coughs, and not infrequently suffocates [948]. The fully grown larvae leave their host, and pupate in thorny puparia [949], which measure about 20 mm. In late summer the adult insect emerges, and may be observed soaring high over the forest trees.

The **Sheep's-nostril fly,** *Oestrus ovis* [950], measures some 12 mm. It is yellow-grey, with black spots; the range extends throughout the Old and the New Worlds. It can commonly be found on the fences around grazing land. The females deliver their larvae in the same manner as described above in the nostrils of domestic and wild sheep. From these they make their way into various nearby cavities in the head, especially into the brainpan, where they live on the mucus of the

949

950

951

brain. The host animal suffers considerably as a result of the constant irritation, and becomes greatly weakened. The grown larvae make their way through to the outside again, are blown out by the sheep, and pupate in the earth.

The **Cattle Warble Fly,** *Hypoderma bovis,* is about 15 mm long and resembles a bumblebee. It has two black stripes on its yellow-and-brown body. The larvae [951] develop to their final stage in swellings known as "warbles" beneath the skin of cattle. Fully grown they measure up to 50 mm. At the approach of these flies the cattle show considerable agitation and seek panic-stricken to escape. This species occasionally also attacks man.

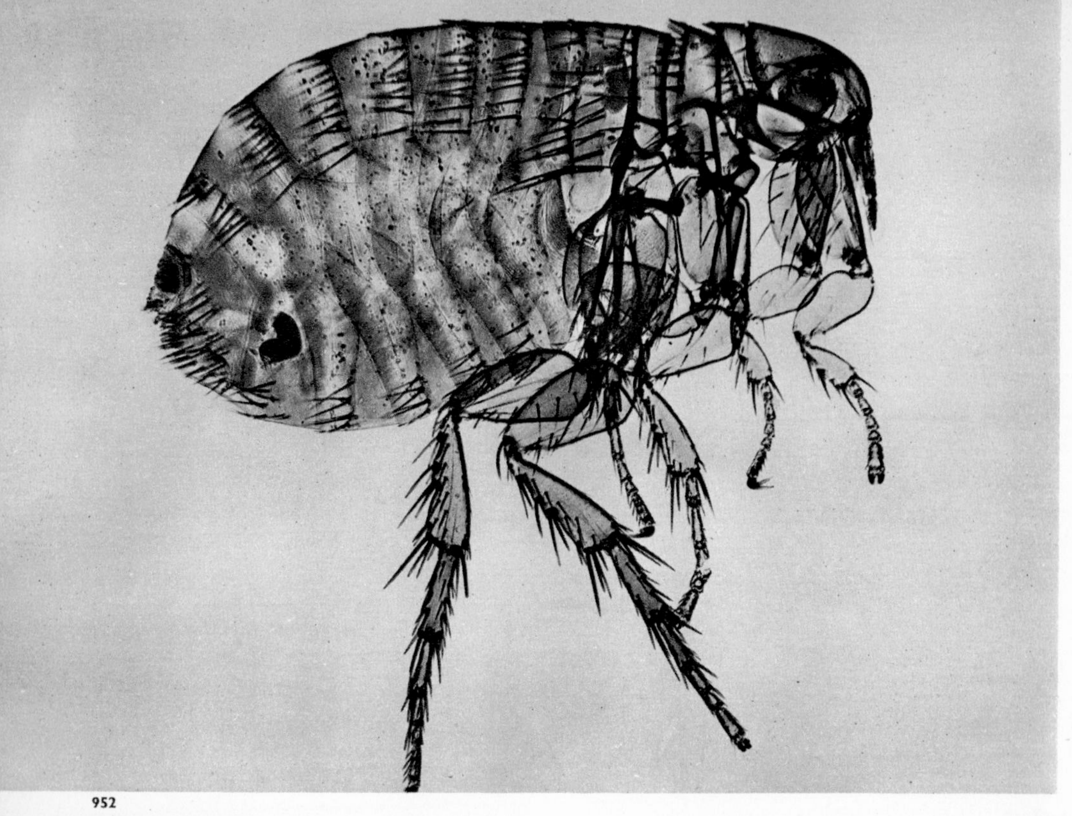

952

953

954

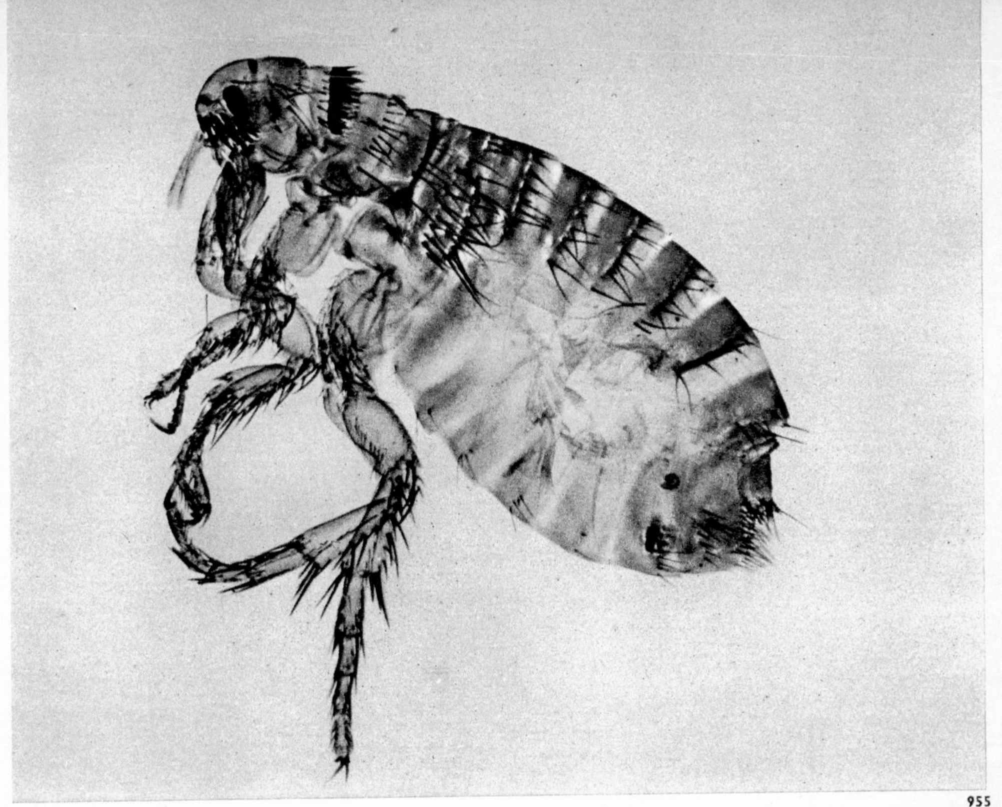

955

956

The eighth order the **Fleas,** *Aphaniptera*, comprises about 1,400 described species, found in all parts of the world. They are wingless insects, and all of them are parasites.

Among the most dangerous is the **Indian Plague Flea,** *Xenopsylla cheopis,* [952, female]. It measures about 2 mm and sucks the blood principally of all species of rats. However, since it changes hosts when it is possible to do so, and sometimes transfers to man, it is a carrier of plague, and in favourable countries is the chief agent of spreading an epidemic. In the Middle Ages it caused the death of some 25 million people in Europe. During the last great epidemic in India, which broke out in 1896, over 10 million people died of plague inside 20 years.

A slightly larger species is the **Human Flea,** *Pulex irritans* [953 and 954]. It measures 3 mm and lives on man, but also on pigs, dogs, and cats. It too is a carrier of infectious diseases.

The **Dog Flea,** *Ctenocephalides canis* [955, female], measures approximately 3.5 mm. From dogs and cats it occasionally springs on to human beings. All these three species of flea belong to the family *Pulicidae*.

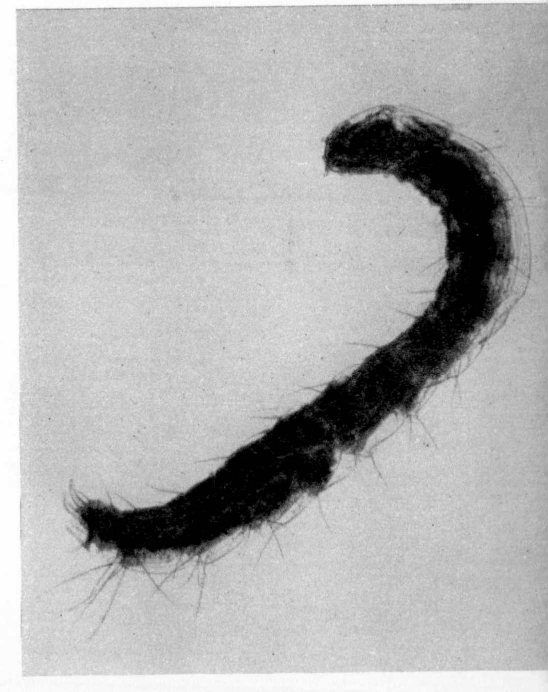

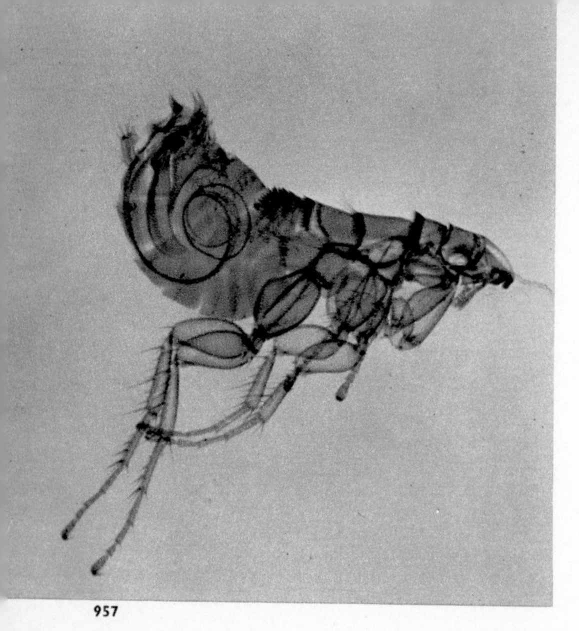

957

958

959

The legless flea larvae [956] develop in places where dirt collects. They feed chiefly on the excreta of fully grown fleas, which contain the remnants of blood.

The yellow-and-red flea *Nycteridopsylla pentactena* [957, male] belongs to the family *Ischnopsyllidae*. It measures about 2.5 mm and lives on bats. On its back it has five little combs; it has a small, hook-shaped head, a constricted thorax and large abdomen [958, female]. This species can be found in Europe during the winter on hibernating bats.

The largest flea in the family *Hystrichopsyllidae* is **Hystrichopsylla talpae.** The male [959] measures up to 4 mm; the female, which is larger [960], up to 6 mm. It is a species of so-called nesting flea, which dwells underground, in the nesting burrows of moles, shrews and voles, and on various species of mice. On its back it has four hard chitinous combs, of which the last two are interrupted by gaps.

In view of the ease with which they change host, the fleas may be regarded as very dangerous. They are very mobile, passing with great rapidity from host to host, and many species pass from animals to humans. The best protection against them is cleanliness in one's surroundings, and personal hygiene. The sanitary standard of present-day living conditions in civilised countries has threatened their viability. The taxonomic position of the flea is still much disputed. It has from time to time been classified in almost every insect order, even the butterflies. At the turn of the 19th and 20th centuries there were many hefty polemics about the "flea problem", more than ten thousand articles and books being written about them. A decisive solution has still not been reached. It seems likely that they may properly be linked with the *Diptera* and *Neuroptera*. As far as their origin is concerned, they are clearly a very ancient order of insects.

960

SYSTEMATIC CLASSIFICATION OF INSECTS

Class	Insecta	Insects	
Subclass 1	**Collembola**	**springtails**	**Primitive Insects**
Subclass 2	**Protura**	**half-insects**	
Subclass 3	**Diplura**	**double-tails**	
Subclass 4	**Thysanura**	**bristletails**	
Subclass 5	**Pterygota**	**winged insects**	

Superorder 1	**Ephemeroidea**	**mayflies**
Order	Ephemerida	mayflies
Superorder 2	**Perloidea**	**stoneflies**
Order	Plecoptera	stoneflies
Superorder 3	**Libelluloidea**	**dragonflies and demoiselle-flies**
Order	Odonata	dragonflies
Superorder 4	**Embioidea**	**foot-spinners**
Order	Embiodia	foot-spinners
Superorder 5	**Orthopteroidea**	**straight-winged insects**
Order 1	Saltatoria	grasshoppers
Order 2	Phasmida	stick-insects
Order 3	Dermaptera	earwigs
Order 4	Diploglossata	double-tongues
Superorder 6	**Blattoidea**	**cockroaches**
Order 1	Mantodea	mantids
Order 2	Blattaria	cockroaches
Order 3	Isoptera	termites
Order 4	Zoraptera	zorapterans